Lecture Notes in Computer Science 12519

More information about this subseries at http://www.springer.com/series/7407

Belgacem Ben Hedia · Yu-Fang Chen ·
Gaiyun Liu · Zhenhua Yu (Eds.)

Verification and Evaluation of Computer and Communication Systems

14th International Conference, VECoS 2020
Xi'an, China, October 26–27, 2020
Proceedings

Springer

Editors
Belgacem Ben Hedia ⓘ
CEA List, DSCIN, LECA
Gif-sur-Yvette, France

Yu-Fang Chen ⓘ
Academia Sinica
Taipei, Taiwan

Gaiyun Liu
Xidian University
Xi'an, China

Zhenhua Yu
Xi'an University of Science and Technology
Xi'an, China

ISSN 0302-9743   ISSN 1611-3349 (electronic)
Lecture Notes in Computer Science
ISBN 978-3-030-65954-7   ISBN 978-3-030-65955-4 (eBook)
https://doi.org/10.1007/978-3-030-65955-4

LNCS Sublibrary: SL1 – Theoretical Computer Science and General Issues

This Springer imprint is published by the registered company Springer Nature Switzerland AG
The registered company address is: Gewerbestrasse 11, 6330 Cham, Switzerland

Preface

These proceedings include the papers presented at the 14th International Conference on Verification and Evaluation of Computer and Communication Systems (VECoS 2020), held as a virtual conference during October 26–27, 2020 (the conference was planned to be organized in Xi'an, China, during September 24–25, 2020), due to the COVID-19 pandemic. The first edition of the conference, VECoS 2007, took place in Algiers, Algeria, VECoS 2008 in Leeds, UK, VECoS 2009 in Rabat, Morocco, VECoS 2010 in Paris, France, VECoS 2011 in Tunis, Tunisia, VECoS 2012 in Paris, France, VECoS 2013 in Florence, Italy, VECoS 2014 in Béjaïa, Algeria, VECoS 2015 in Bucharest, Romania, VECoS 2016 in Tunis, Tunisia, VECoS 2017 in Montreal, Canada, VECoS 2018 in Grenoble, France, and VECoS 2019 in Portugal. The aim of the VECoS conference is to bring together researchers and practitioners in the areas of verification, control, performance, and dependability evaluation in order to discuss the state of the art and challenges in modern computer and communication systems in which functional and extra-functional properties are strongly interrelated. Thus, the main motivation for VECoS is to encourage the cross-fertilization between various formal verification and evaluation approaches, methods, and techniques, and especially those developed for concurrent and distributed hardware/software systems. The Program Committee of VECoS 2020 included 83 researchers from 20 countries. We received 60 full submissions from 11 countries. After a thorough and lively discussion phase, the committee decided to accept 20 papers. We are grateful to the Program and Organizing Committee members, to the reviewers for their cooperation, and to Springer for their professional support during the production phase of the proceedings. We are also thankful to all authors of submitted papers and to all participants of the conference. Their interest in this conference and contributions are greatly appreciated.

October 2020

Belgacem Ben Hedia
Yu-Fang Chen
Gaiyun Liu
Zhenhua Yu

Organization

VECoS 2020 was planned to be organized in Xi'an, China, during September 24–25, 2020. Due to COVID-19 pandemic, VECoS 2020 is held as a virtual conference during October 26–27, 2020.

Executive Committee

Program Co-chairs

Belgacem Ben Hedia	CEA-LIST, Saclay, France
Yu-Fang Chen	Academia Sinica, Taiwan
Gaiyun Liu	Xidian University, China

General Chair

Zhanli Li	Xi'an University of Science and Technology, China

General Co-chairs

Kamel Barkaoui	Conservatoire National des Arts et Métiers, France
Zhenhua Yu	Xi'an University of Science and Technology, China

Local Chair

Ailing Feng	Xi'an University of Science and Technology, China

Organizing Committee Co-chairs

Tian Ma	Xi'an University of Science Technology, China
Weibo Chen	Xi'an University of Science and Technology, China

Publicity Co-chairs

Yufeng Chen	Xidian University, China
Belgacem Ben Hedia	CEA-LIST, Saclay, France
Vladimir-Alexandru Paun	ENSTA ParisTech, France

Steering Committee

Djamil Aissani	LAMOS, Université de Bejaia, Algeria
Mohamed Faouzi Atig	Uppsala University, Sweden
Kamel Barkaoui (Chair)	CEDRIC CNAM, France
Hanifa Boucheneb	Veriform, Polytechnique Montreal, Canada
Francesco Flammini	Ansaldo STS, Milano, Italy
Belgacem Ben Hedia	CEA-LIST, Saclay, France
Mohamed Kaâniche	LAAS-CNRS, France

Bruno Monsuez	ENSTA UIIS, France
Nihal Pekergin	LACL, Université Paris-Est Créteil, France
Tayssir Touili	LIPN, CNRS, Université Paris Nord, France

Referees

D. Aissani
Y. A. Ameur
I. Alshabani
M. Asavoae
M. F. Atig
E. Badouel
K. Barkaoui
F. Belala
I. Ben Hafaiedh
B. Ben Hedia
A. Benzina
S. Bliudze
P. Bonhomme
Y-F. Chen
Z. Chen
F. Chu
G. Ciobanu
J-M. Couvreur
H. D. Van
M. Debbabi
I. Demongodin
A. Fantechi
S. Genaim

A. Geniet
M. Ghazel
X. Guo
S. Haddad
Y-F. Huang
P. Höfner
A. Idani
R. Iosif
M. Jmaiel
J. Julvez
M. Khalgui
K. Klai
M. Krichen
S. Lahaye
A. Legay
O. Lengal
S-W. Lin
A. Lisitsa
G. Liu
D. Liu
Z. Ma
E. Madelaine
R. Meyer

A. Mili
B. Monsuez
M. Mosbah
A. Nouri
C. Palamidessi
V-A. Paun
G. Pu
S. Qin
A-B. Rabéa
S. Renaud
A. Riesco
R. Robbana
R. J. Rodríguez
A. Sangnier
L. Sliman
F. Song
L. Su
T. Touili
K. Wolter
W. Wu
X. Yin

Additional Reviewers

R. Ammour	F. Duchene	L. Metongnon
D. Basile	F. Furbach	C. Pacheco
M. Becker	L. Gao	O. Sankur
Z. Benzadri	M. He	A. Sobral
B. Binder	Z. He	C. Wen
R. Boucebsi	J. He	W. Wu
A. Boudguiga	Y. Hou	M. Yan
L. Brenner	T. Jron	Y. Zhou
Y. Chen	J. Luo	
T. Dreossi	H. Marchand	

Sponsoring Institutions

Contents

Formal Modeling and Verification, Testing

Artificial Intelligence and Machine Learning

Petri-Net, Simulation, and Scheduling

An Approach for Supervisor Reduction of Discrete-Event Systems

Huimin Zhang[1,2], Lei Feng[3(✉)], Wu Xie[2(✉)], and Feng Yu[1]

[1] College of Computer Science and Information Engineering,
Guangxi Normal University, Guilin 541004, China
cchuiminzhang@outlook.com

[2] Guangxi Key Laboratory of Trusted Software, Guilin University of Electronic
Technology, Guilin 541004, China
xiewu588@126.com

[3] Department of Machine Design, KTH Royal Institute of Technology,
100 44 Stockholm, Sweden
lfeng@kth.se

Abstract. In the supervisory control theory of discrete event systems, since the supremal supervisor incorporates transition constrains of both the plant and the specification, the state size of the supervisor is usually large and the control logic is difficult to understand. Thus, computing for reduced supervisors with small sizes is meaningful both for designing and implementation. We propose an algorithm that a reduced supervisor can be separated from the supremal supervisor if a sufficient condition is satisfied. The algorithm for checking the sufficient condition is also presented. In the case that the sufficient condition is satisfied, a reduced supervisor can be computed in a complexity of $O(m.n)$, where the integers m and n are the state number of the supremal supervisor and the cardinality of the event set, respectively. And the state size of the reduced supervisor is equal or less than that of the specification. Some examples are presented to illustrate the proposed approach.

Keywords: Supervisory Control Theory · Discrete-Event System · Supervisor reduction

1 Introduction

In the supervisory control theory (SCT) of discrete-event systems (DES), the control problem is to enforce the logical behavior of the plant that is admissible

This work was supported in part by the National Natural Science Foundation of China under Grant 61966008, National Natural Science Foundation of GuangXi Province, China under Grant Nos. 2018GXNSFAA294052, 2020GXNSFBA159069 and 2019JJA170060, Guangxi Science and Technology Planning Project Grant No. AB18126063, Guangxi Key Laboratory of Trusted Software (Nokx201926), and Young and Middle-aged Scientific Research Basic Ability Promotion Project of Guangxi under Grant No. 2020KY02031.

© Springer Nature Switzerland AG 2020
B. Ben Hedia et al. (Eds.): VECoS 2020, LNCS 12519, pp. 3–14, 2020.
https://doi.org/10.1007/978-3-030-65955-4_1

by the specification. Both the plant and the specification are formalized by deterministic finite automata (DFA). SCT synthesizes a supremal supervisor that is controllable and nonblocking with respect to the plant with maximum permissive [1,2]. The supremal supervisor is also specified by a DFA, which has the same order of the state size as the synchronization of the plant and specifications [3,4].

However, when SCT is applied to real-world systems, the size of the supervisor is extremely large and the control logic becomes difficult to understand [5]. Since the supremal supervisor contains redundant transition constrains that already exist in the plant itself, it is possible to reduce the state size of the supervisor by removing the transition constrains of the plant while preserving control actions [6]. Computing for reduced supervisors with small sizes is meaningful both for implementation and designing. The reduction problem was first investigated by Wonham and Vaz in 1986 using control cover theory [6]. Let the state size of the supremal supervisor be an integer n. The proposed algorithm in [6] can compute reduced supervisors with minimal states but has exponential time complexity with respect to n. Su and Wonham present a polynomial-time algorithm for supervisor reduction using control congruence theory [7]. Although the reduced supervisor may not have the minimal state, the computing complexity is $O(m^4)$ [7,8]. Recently, the algorithm for supervisor reduction has been applied to supervisor localization, which aims to create a control-equivalent distributed implementation of a given centralized supervisor [9,10].

As we know, the marked language of the supremal supervisor is a subset of the intersection of the marked languages of the plant and the specification. The supremal supervisor can be obtained by removing some states and transitions from the synchronization automaton of the plant and the specification. States of the supervisor can be denoted by two-tuples (x, y), where x and y are state labels of the plant and the specifications, respectively. Using this information, we propose an approach that computes a reduced supervisor by projecting state labels of the supervisor on that of the specification. However, the proposed approach cannot work in all the DES. To alleviate this limitation, we present a sufficient condition for the algorithm that works well. If the condition is satisfied, the computing complexity of the algorithm is $O(m.n)$, where the integers m and n are the state number of the supremal supervisor and the cardinality of the event set, respectively. And the state size of the reduced supervisor is equal or less than that of the specification. Furthermore, it is explicit for the designer to understand the meaning of the control actions of the supervisor. In some cases, the reduced supervisor has less states than that obtained by Su's algorithm in reference [7].

The paper is organized as follows. Section 2 describes the preliminaries of automata and briefly reviews the SCT. Section 3 elaborates the proposed algorithm. Illustrated examples are given in Sect. 4. Finally, we conclude the paper and propose some further research directions in Sect. 5.

2 Preliminaries

2.1 Basics of Automata

Basic concepts of languages and automaton theory are introduced in this section [11]. An alphabet Σ is a non-empty and finite set of symbols. A string defined on the alphabet Σ is a sequence of symbols in Σ. The set Σ^* consists of the empty string ϵ and all finite strings defined on Σ.

A deterministic finite automaton (DFA), denoted by $\mathbf{G}$, is a quintuple

$$\mathbf{G} = (Q, \Sigma, \delta, q_0, \Gamma, Q_m),$$

where

- Q is the state set,
- Σ is the alphabet,
- δ: $Q \times \Sigma \to Q$ is the transition function,
- $q_0 \in Q$ is the initial state,
- $\Gamma : Q \to 2^E$ is the active event function; $\Gamma(q)$ is the set of all events σ for which $\delta(q, \sigma)!$,
- $Q_m \subseteq Q$ is the set of marker states.

For σ in Σ and q, q' in Q, $\delta(q, \sigma) = q'$ represents that there is a transition labeled by event σ from state q to state q'. In the following sections, δ is considered as a *partial* function, i.e., $\delta(q, \sigma)$ is not always defined. If $\delta(q, \sigma)$ is defined, it is denoted by $\delta(q, \sigma)!$. Function δ is extended from domain $Q \times \Sigma$ to domain $Q \times \Sigma^*$ in the following recursive way: $\delta(q, \varepsilon) = q$; $\delta(q, s\sigma)$ is defined if $\delta(q, s)$ is defined and reaches state q' and $\delta(q', \sigma)$ is also defined, for $q, q' \in Q$, $s \in \Sigma^*$ and $\sigma \in \Sigma$. The cardinality of the state set Q is denoted by $|Q|$. In this paper, the word *automaton* refers to a DFA. The set $L(\mathbf{G}) = \{s \in \Sigma^* | \delta(q_0, s)!\}$ is the generated language of an automaton $\mathbf{G}$, and the set $L_m(\mathbf{G}) = \{s \in L(\mathbf{G}) | \delta(q_0, s) \in Q_m\}$ denotes the accepted language of an automaton $\mathbf{G}$. $\overline{L} = \{s \in \Sigma^* | (\exists t \in \Sigma^*) st \in L\}$ is the *prefix closure* of a language L. L is *prefix-closed* if and only if $L = \overline{L}$. An automaton $\mathbf{G}$ is *nonblocking* if and only if $\overline{L_m(\mathbf{G})} = L(\mathbf{G})$.

Let $G_1 = (Q_1, \Sigma_1, \delta_1, q_{01}, \Gamma_1, Q_{m1})$ and $G_2 = (Q_2, \Sigma, \delta_2, q_{02}, \Gamma_2, Q_{m2})$ be two automata. The synchronous product of automata G_1 and G_2 is defined as $G_1 \parallel G_2 = Ac(Q_1 \times Q_2, \Sigma, \delta, (q_{01}, q_{02}), \Gamma_{1\parallel2}, Q_{m1} \times Q_{m2})$ where

$$\delta((q_1, q_2), \sigma) = \begin{cases} (\delta_1(q_1, \sigma), \delta_2(q_2, \sigma)) & \text{if } \sigma \in \Gamma_1(x_1) \cap \Gamma_2(x_2) \\ (\delta_1(x_1, \sigma), x_2) & \text{if } \sigma \in \Gamma_1(x_1) \setminus \Sigma_2 \\ (x_1, \delta_2(x_2, \sigma),) & \text{if } \sigma \in \Gamma_2(x_2) \setminus \Sigma_1 \\ \text{undefined} & \text{otherwise} \end{cases}$$

and $\Gamma_{1\parallel2} = [\Gamma_1(q_1) \cap \Gamma_2(q_2)] \cup [\Gamma_1(q_1) \setminus \Sigma_2] \cup [\Gamma_2(q_2) \setminus \Sigma_1]$.

Semantically, if $\Sigma_1 = \Sigma_2$, we have $L(G_1 \parallel G_2) = L(G_1) \cap L(G_2)$ and $L_m(G_1 \parallel G_2) = L_m(G_1) \cap L_m(G_2)$.

2.2 Supervisory Control Theory

In DES, SCT is used to enforce a plant to satisfy its specification [1,2]. A plant is modeled by a DFA $\mathbf{G}$ over an alphabet Σ. The alphabet Σ is partitioned into a controllable event set Σ_c and an uncontrollable event set Σ_u, i.e., $\Sigma = \Sigma_c \bigcup \Sigma_u$. In SCT, the specification that the plant must satisfy is modeled by a finite automaton $\mathbf{H}$, which has the same alphabet as $\mathbf{G}$.

The premise is that the behavior of $\mathbf{G}$ may violate the specification and must be modified through feedback control. In order to alter the behavior of $\mathbf{G}$, a supervisor S implements the control function $V : L(\mathbf{G}) \to \Gamma$ is introduced, where $\Gamma = \{\gamma \in 2^{\Sigma} | \Sigma_u \subseteq \gamma \subseteq \Sigma\}$ is the set of *control patterns*. The *controllable language* is a fundamental concept in SCT [2].

Definition 1. *A language $K \subseteq \Sigma^*$ is* controllable *with respect to a prefix-closed language $L \subseteq \Sigma^*$ and an uncontrollable event subset $\Sigma_u \subseteq \Sigma$ if*

$$(\forall s \in \Sigma^*, \sigma \in \Sigma_u)\, s \in \overline{K}\, \&\, s\sigma \in L \Rightarrow s\sigma \in \overline{K}.$$

Given a language $M \subseteq L$, the set of all sub-languages of M that are controllable with respect to L is

$$C(M, L) = \{K \subseteq M | K \text{ is controllable w.r.t } L\}.$$

The supremal element of $C(M, L)$ is $supC(M, L)$ [2].

A standard algorithm for $supC(M, K)$ is described in Algorithm [1,11].

Let $G = (X, \Sigma, f, \Gamma_G, x_0, X_m)$ and $H = (Y, \Sigma, g, \Gamma_H, y_0, Y)$ be the plant and the specification, respectively.

3 Proposed Approach

3.1 Problem Formalization

Given a plant $G = (X, \Sigma, f, \Gamma_G, x_0, X_m)$, a specification $H = (Y, \Sigma, g, \Gamma_H, y_0, Y)$ and the supremal supervisor $\mathbf{S} = (Q, \Sigma, \delta, q_0, \Gamma_S, Q_m)$ with respect to G and H, compute a reduced supervisor $R = (Y', \Sigma, \xi, y_0, \Gamma_R, Y'_m)$ such that

$$L_m(G \parallel R) = L_m(S),$$
$$L(G \parallel R) = L(S),$$
$$\text{and } |R| \leq |S|.$$

The supremal supervisor with respect to G and H is computed by Algorithm 1, where $Q \subseteq X \times Y$, $Q_m \subseteq X_m \times Y$ and $q_0 = (x_0, y_0)$. Each state q of the supervisor S is denoted by a two-tuples (x, y), where $x \in X$ and $y \in Y$. For each string $s \in L(S)$, if $\delta(q_0, s) = (x, y)$, then we have $f(x_0, s) = x$ and $g(y_0, s) = y$. For brevity, the same symbols of the definitions of the plant G, the specification H, the supervisor S and the automaton R are used in the following sections.

Algorithm 1: A standard algorithm for computing the supremal supervisor

---

Input: The plant G and the specification H
Output: The supremal supervisor with respect to G and H
1   Let $H_0 = H \parallel G = (Y_0, \Sigma, g_0, \Gamma_{H_0}, (x_0, y_0), Y_{0,m})$;
2   Set $i = 0$;
3   Set $Y'_i = \{(x, y) \in Y_i : \Gamma_G(x) \cap \Sigma_u \subseteq \Gamma_{H_i}(x, y)\}$;
4   Set $g'_i = g_i | Y'_i$ (The notation | stands for "restricted to");
5   Set $Y'_{i,m} = Y_{i,m} \cap Y'_i$;
6   $H_{i+1} = Trim(Y'_i, \Sigma, g'_i, (x_0, y_0), Y'_{i,m})$;
7   **while** $H_{i+1} \neq H_i$ and H_{i+1} is not an empty automaton **do**
8   Set $H_{i+1} = (Y_{i+1}, \Sigma, g_{i+1}, (x_0, y_0), Y_{i+1,m})$;
9   set $i = i + 1$;
10   Set $Y'_i = \{(x, y) \in Y_i : \Gamma_G(x) \cap \Sigma_{uc} \subseteq \Gamma_{H'}(x, y)\}$;
11   Set $g'_i = g_i | Y'_i$;
12   Set $Y'_{i,m} = Y_{i,m} \cap Y'_i$;
13     $H_{i+1} = Trim(Y'_i, \Sigma, g'_i, (x_0, y_0), Y'_{i,m})$;
14   **end**
15   Output H_{i+1};

3.2   Framework of the Proposed Approach

The framework of the proposed approach for computing a reduced supervisor for the plant is described in Fig. 1. Firstly, given a plant G and a specification H, a supremal supervisor S is computed by Algorithm 1. Secondly, a reduced supervisor R is derived from S by the proposed algorithm. Next, equations $L_m(G \parallel R) = L_m(S)$ and $L(G \parallel R) = L(S)$ are checked. If $L_m(G \parallel R) = L_m(S)$ and $L(G \parallel R) = L(S)$ hold, R is a reduced supervisor. Otherwise, the algorithm proposed by Su in ref. [7] can be used to compute a reduced supervisor for the plant.

3.3   Derivation of R from the Supremal Supervisor S

Since the supremal supervisor contains redundant transition constrains that already exist in the plant itself, it is possible to reduce the state size of the supervisor by removing the transition constrains of the plant while preserving control actions. Therefore, Algorithm 2 is proposed to derive an automaton R such that the structural information of the plant is wiped off from the supremal supervisor S.

Proposition 1. *In Algorithm 2, the output automaton $R = (Y', \Sigma, \xi, y_0, -, Y'_m)$ is closed and deterministic.*

Proof. We first show that ξ is closed in the automaton R. For $y \in Y'$ and $\sigma \in \Sigma$, if $\xi(y, \sigma)!$, we must show that $\xi(y, \sigma) \in Y'$. By Algorithm 2, if $\xi(y, \sigma)!$, there must exist states (x, y) and (x', y') in Q such that $\delta((x, y), \sigma)!$ and $\delta((x, y), \sigma) = (x', y')$

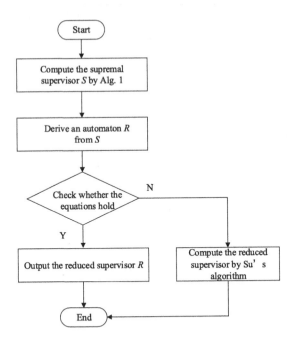

Fig. 1. Framework for computing a reduced supervisor.

Algorithm 2: Derivation of R from the supremal supervisor S

Input: The plant $G = (X, \Sigma, f, -, x_0, X_m)$ and the specification
$\qquad H = (Y, \Sigma, g, -, y_0, Y_m)$

Output: An automaton R

1　Compute the supremal supervisor $\mathbf{S} = (Q, \Sigma, \delta, q_0, -, Q_m)$ with respect to G and H by Algorithm 1 where $Q \subseteq X \times Y$, $Q_m \subseteq X_m \times Y_m$ and $q_0 = (x_0, y_0)$;

2　Let $Y' = \emptyset$ and $Y'_m = \emptyset$;

3　**for** *each state* $(x, y) \in Q$ **do**

4　　　Set $Y' = Y' \cup \{y | (x, y) \in Q\}$;

5　　　**if** $(x, y) \in Q_m$ **then**

6　　　　　Set $Y'_m = Y'_m \cup \{y | (x, y) \in Q_m\}$;

7　　　**end**

8　　　**for** *each event* $\sigma \in \Sigma$ **do**

9　　　　　**if** $\delta((x, y), \sigma) = (x', y')$ **then**

10　　　　　　　$\xi(y, \sigma) = y'$;

11　　　　　**end**

12　　　**end**

13　**end**

14　Output $R = (Y', \Sigma, \xi, y_0, -, Y'_m)$;

---

in the supervisor S. By line 4 of Algorithm 2, we have $\xi(y, \sigma) = y'$ and $y' \in Y'$. Thus, the transition function ξ of the automaton R is closed.

To show the determinism of ξ, let us consider two different states (x_1, y) and (x_2, y) in Q. For $\sigma \in \Sigma$, if $\delta((x_1, y), \sigma)!$ and $\delta((x_2, y), \sigma)!$, we have $\delta((x_1, y), \sigma) = (f(x_1, \sigma), g(y, \sigma))$ and $\delta((x_2, y), \sigma) = (f(x_2, \sigma), g(y, \sigma))$ by Algorithm 1. Let $g(y, \sigma) = y'$. Then, $\xi(y, \sigma) = y'$ holds. Therefore, ξ is deterministic.

Proposition 2. $|R| \leq |H|$.

Proof. By Algorithm 2, we have $Y' \subseteq Y$. Therefore, $|R| \leq |H|$ is true.

Let $T = G \parallel R = (Z, \Sigma, \zeta, \Gamma_Z, z_0, Z_m)$, where $Z = X \times Y'$, $z_0 = (x_0, y_0)$ and $Z_m = X_m \times Y'_m$.

Proposition 3. *If T is isomorphic to S, R is a reduced supervisor with respect to the plant G and the specification H.*

Proof. If T is isomorphic to S, $L_m(G \parallel R) = L_m(S)$ and $L(G \parallel R) = L(S)$ are true. By Proposition 2, we have $|R| \leq |H|$. Therefore, R is a reduced supervisor with respect to the plant G and the specification H.

In the supremal supervisor S, let $|G| = m$ and $|\Sigma| = n$. The computing complexity of Algorithm 2 is $O(mn)$. If Proposition 3 holds, we obtain a reduced supervisor by Algorithm 2. If T is not isomorphic to the supremal supervisor S, the algorithm proposed by Su in reference [7] is used to compute a reduced supervisor with a complexity of $O(m^4)$.

3.4   A Sufficient Condition for Guaranteeing T Isomorphic to S

Definition 2. *The supremal supervisor S is called inseparable with respect to the plant if there exist a string $w \in \Sigma^*$, states $q \in Q$, $x \in X$ and $y \in Y'$ such that the following conditions hold:*

(a) $\delta(q_0, w) = q$ and $\sigma \notin \Gamma_S(q)$ in the supremal supervisor S;
(b) $f(x_0, w) = x$ and $\sigma \in \Gamma_G(x)$ in the plant G;
(c) $\xi(y_0, w) = y$ and $\sigma \in \Gamma_R(y)$ in the automaton R obtained from Algorithm 2;

Otherwise, S is called separable with respect to the plant.

That is, there do not exist a state $(x, y) \in T$ and an event $\sigma \in \Sigma$, such that $f(x, \sigma)!$ in the plant G and $\xi(y, \sigma)!$ in the automaton R, but $\delta((x, y), \sigma)$ is not defined in the supremal supervisor S. The reason that makes S inseparable is the state merging in the line 4 of Algorithm 2. Event σ is enabled at states x and y in the plant G and the specification H, but disabled at state (x, y) in the supervisor S. At state (x, y) of the automaton T, σ is enabled again. This makes T not isomorphic to S. Algorithm 3 is presented to check whether the supervisor S is separable or not.

The computation complexity of Algorithm 3 is $O(|Q|.|\Sigma|)$. Since both sets Q and Σ are finite, Algorithm 3 can terminate.

Algorithm 3: Check whether the supervisor S is separable

    Input: The supervisor S, the plant G and the automaton R obtained by
          Algorithm 2
    Output: The supervisor S is separable or not

1 **for** *each state $q \in Q$ in the plant G* **do**
2   Choose a string s in Σ^* such that $\delta(q_0, s) = q$ in S;
3   Denote $f(x_0, s) = x$ in G and $\xi(y_0, s) = y$ in R;
4   **for** *each $\sigma \in \Gamma_G(x)$* **do**
5   **if** $\sigma \in \Gamma_R(y)$ *and* $\sigma \notin \Gamma_S(q)$ **then**
6   **return** *False*;
7   **end**
8   **end**
9 **end**
10 **return** *True*;

Proposition 4. *If S is separable with respect to the plant G, we have $L(S) = L(G \parallel R)$ and $L_m(S) = L_m(G \parallel R)$.*

Proof. (1)($\subseteq$) For every string $w \in L(S)$, $w \in L(G)$ is true. There is a state $(x, y) \in Q$ such that $\delta((x_0, y_0), w) = (x, y)$ in the supremal supervisor S. By Algorithm 2, we have $\xi(x_0, w) = x$. That is, $w \in L(R)$ holds. We have $w \in L(G) \cap L(R)$. Thus, $L(S) \subseteq L(G \parallel R)$ hods. Similarly, $L_m(S) \subseteq L_m(G \parallel R)$ is true.

    (2)($\supseteq$) This proposition is proved by induction on the length of the string w. When $|w|=0$, $w = \epsilon$. If $S \neq \emptyset$, $\epsilon \in L(S)$. Otherwise, $R = \emptyset$. Suppose that $w \in L(S)$ is true for any $w' \in L(G \parallel R)$ of length at most k $(k \geq 0)$. Let $w \in L(S)$ be a string of length $k+1$. String w can be written as $w = w'\sigma$, where $|w'| = k$ and $\sigma \in \Sigma$. There exist states $x \in X$ in G and $y \in Y'$ in R such that $\xi((x_0, y_0), w') = (x, y)$ and $\delta((x_0, y_0), w') = (x, y)$ hold. When $w'\sigma \in L(G \parallel R)$, denote $f(x, \sigma) = x_1 \in X$ and $g(y, \sigma) = y_1 \in Y'$. Assume that $w' \notin L(S)$. That is, event σ is disabled at state (x, y) of S, which is contradict with the assumption of the proposition that R is separable. Therefore, if $w \in L(G \parallel R)$, $L \in L(S)$ holds. Similarly, we have $L_m(G \parallel R) \subseteq L_m(S)$.

4   Illustrative Examples

4.1   Example 1

The plant G is the synchronization of M_1 with event set $\Sigma_1 = \{10, 11, 12, 13, 15\}$ and M_2 with event set $\Sigma_2 = \{20, 21, 22, 23\}$ shown in Figs. 2 and 3. The controllable event sets of M_1 and M_2 are $\{11, 13, 15\}$ and $\{21, 23\}$, respectively. The specification H is formalized as Fig. 4.

    Firstly, a supremal supervisor S is synthesized by Algorithm 1. An automaton R shown in Fig. 5 is derived from S by Algorithm 2. Let $T = G \| R$. The automaton T is isomorphic to the supervisor S. Thus, R is a reduced supervisor

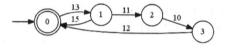

Fig. 2. M_1.

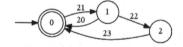

Fig. 3. M_2.

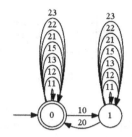

Fig. 4. The specification H.

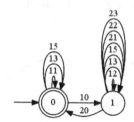

Fig. 5. Automaton R derived by Algorithm 2.

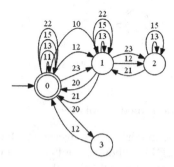

Fig. 6. The model of the reduced supervisor computed by TCT.

with respect to G. It can be checked that the supremal supervisor S is separable by Algorithm 2. Another reduced supervisor is computed by TCT given in Fig. 6 which has four states. In this example, the reduced supervisor computed by the proposed algorithm has fewer states than that obtained by using TCT [12].

4.2   Example 2

We study the example **Small Factory** taken from [2] to compare the results computed by the proposed approach using TCT. The **Small Factory** system shown in Fig. 7 consists of two machines **MACH₁**, **MACH₂** and a buffer **BUF**.

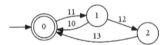

Fig. 7. Small factory.

The model of **MACH₁** is specified as Fig. 8. The model of **MACH₂** is the same as M_2 described in Fig. 3. The controllable events set is $\{11, 13, 21, 23\}$. The plant model for **Small Factory** is the synchronization of **MACH₁** and **MACH₂**. **MACH₂** is described in Fig. 3, which is the same as Fig. 9.

Fig. 8. MACH₁

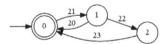

Fig. 9. MACH₂

The specifications are described as follows.

– The buffer should neither underflow nor overflow, which means that **MACH₂** cannot take a part from **BUF** when the buffer is empty and **MACH₁** cannot put a part into **BUF** when the buffer is full.
– If both machines break down, **MACH₂** must be repaired ahead of **MACH₁**.

In a further step, we perform experiments by increasing the buffer size from 2 to 12 with a step size of 2 to show the results obtained by the proposed algorithm. The results is shown in Table 1. The size of an automaton is represented by (a, b), where a denotes the number of the states and b denotes the number of the transitions. The numbers of the states and the transitions of the specification are shown in the column 'specification'. The sizes of the reduced supervisors computed by TCT and the proposed approach are shown in columns 'TCT' and 'proposed approach', respectively. If the synchronization of the automaton derived by Algorithm 2 and the plant is isomorphic to the supremal supervisor, 'Y' is filled in the corresponding entity of the column 'isomorphic'; Otherwise, 'N' is recorded.

Table 1. Size of the reduced supervisors computed by TCT and the proposed approach

Buffer size	Specification	Supervisor	TCT	Proposed approach	Isomorphic
2	(6, 35)	(21, 47)	**(5, 18)**	**(6, 25)**	Y
4	(10, 61)	(39, 96)	**(9, 34)**	**(10, 47)**	Y
6	(14, 87)	(57, 139)	**(13, 50)**	**(14, 69)**	Y
8	(18,113)	(75, 185)	**(17, 66)**	**(18, 91)**	Y
10	(22, 139)	(93, 231)	**(21, 82)**	**(22, 113)**	Y
12	(26, 165)	(93, 231)	**(25, 98)**	**(26, 134)**	N

By analyzing the data in Table 1, we conclude that the automata obtained by Algorithm 2 are indeed reduced supervisors when the numbers of slots are 2, 4, 6, 8, and 10. In this example, the number of states and transitions of the reduced supervisors obtained by the proposed algorithms is a bit bigger than that computed by TCT. When the number of slots is 12, Algorithm 2 cannot derive a reduced supervisor directly. Then, TCT can be used to compute a derived supervisor in further steps.

In this example, the size of the reduced supervisors obtained by the proposed algorithm is a little bit bigger than that computed by TCT. However, in Example 1, the size of the reduced supervisor obtained by the proposed algorithm is smaller than that computed by TCT. The complexity for computing a reduced supervisor for the plant with minimal size is exponential [1]. Generally, since Algorithm 2 adopts a simpler idea than that of Su's algorithm, the size of the reduced supervisor computed by the proposed algorithm is bigger than that obtained by TCT. However, the proposed algorithm has a lower time complexity.

5   Conclusion and Future Work

We propose an approach for computing a reduced supervisor by projecting state labels of the supremal supervisor on that of the specification. However, the proposed approach cannot work in all the DES. To alleviate this limitation, a sufficient condition for the algorithm is presented that makes the proposed algorithm work well. If the condition is satisfied, the computing complexity of the algorithm is $O(m.n)$, where the integers m and n are the state number of the supremal supervisor and the cardinality of the event set, respectively. And the state size of the reduced supervisor is equal or less than that of the specification. Since the reduced supervisor is a subset of the specification, it is explicit for the designer to understand the meaning of the control actions of the supervisor. In the future work, we will explore more general sufficient conditions for the proposed algorithm to compute reduced supervisors.

References

1. Ramadge, P.J., Wonham, W.M.: Supervisory control of a class of discrete event processes. SIAM J. Control Optim. **25**(1), 206–230 (1987)
2. Wonham, W.M., Cai, K.: Supervisory Control of Discrete-Event Systems. Springer, Cham (2019)
3. Zhang, H., Feng, L., Li, Z.: A learning-based synthesis approach to the supremal nonblocking supervisor of discrete-event systems. IEEE Trans. Autom. Control **63**(10), 3345–3360 (2018)
4. Zhang, H., Feng, L., Li, Z.: Control of black-box embedded systems by integrating automaton learning and supervisory control theory of discrete-event systems. IEEE Trans. Autom. Sci. Eng. **17**(1), 361–374 (2018)
5. Zhang, H., Feng, L., Wu, N., Li, Z.: Control of black-box embedded systems by integrating automaton learning and supervisory control theory of discrete-event systems. IEEE Trans. Autom. Sci. Eng. **15**(1), 2–15 (2018)
6. Vaz, A.F., Wonham, W.M.: On supervisor reduction in discrete-event systems. Int. J. Control **44**(2), 475–491 (1986)
7. Su, R., Wonham, W.M.: Supervisor reduction for discrete-event systems. Discrete Event Dyn. Syst. **14**(1), 31–53 (2004)
8. Su, R., Wonham, W.M.: What information really matters in supervisor reduction? Automatica **95**, 368–377 (2018)
9. Cai, K., Wonham, W.M.: Supervisor localization: a top-down approach to distributed control of discrete-event systems. IEEE Trans. Autom. Control **55**(3), 605–618 (2010)
10. Liu, Y., Cai, K., Li, Z.: On scalable supervisory control of multi-agent discrete-event systems. Automatica **108**, 605–618 (2019)
11. Cassandras, C.G., Lafortune, S.: Introduction to Discrete Event Systems. Springer Science & Business Media, New York (2008)
12. Feng, L., Wonham, W. M.: TCT: a computation tool for supervisory control synthesis. In: Proceedings of the 8th International Workshop on Discrete Event Systems, pp. 388–389. IEEE, Ann Arbor (2006)

Multi-robot Path Planning Using Petri Nets

Hongbin Zhang[1], Jiliang Luo[1]($\boxtimes$), Jinjun Long[2], Yisheng Huang[3], and Weimin Wu[4]

[1] Department of Control Science and Engineering, Huaqiao University, Xiamen 361021, China
hopingzhang@foxmail.com, jlluo@hqu.edu.cn
[2] KENGIC Intelligent Equipment Co., Ltd., No. 2 Workshop No. 321 Jinrong Road High-Tech Zone, Qingdao 266000, China
jinjun.long@kengic.com
[3] Department of Electrical Engineering, National Ilan University, Yilan 26047, Taiwan, People's Republic of China
huang.ccit@gmail.com
[4] State Key Laboratory of Industrial Control Technology, Institute of Cyber-Systems and Control, Zhejiang University, Hangzhou 310027, China
wmwu@iipc.zju.edu.cn

Abstract. A formal approach is proposed for planning a team of mobile robots such that no collision occurs and tasks represented by a Boolean specification are satisfied. It should be specially noted that the order, by which the given tasks are executed, is taken into account by the specification. First, a team of mobile robots and their environment are modeled as a Petri net (PN). Second, a method is presented to design place nodes enforcing a given specification on the PN model. Consequently, the resultant PN can be used to model the robot team's behaviors that satisfy the specification. Third, an optimal problem, minimizing the total traveling distance that the robots take to perform given tasks, is formulated as an integer linear programming (ILP) problem via the PN model. By solving the ILP problem, the optimal action sequence is obtained, which actually means an optimal strategy to schedule robots.

Keywords: Multi-robot · Discrete event systems · Path planning · Optimization · Petri Nets

1 Introduction

Since a team of mobile robots can efficiently perform complex delivery tasks, multi-robot systems are intensively utilized in modern industries. The path planning problem for a multi-robot system draws more and more interest in the last decade.

This work was supported in part by National Science Foundation of China under Grant No. 61973130, 61573158 and 61773343, and Natural Science Foundation of FuJian Province of China under Grant 2014J01241.

B. Ben Hedia et al. (Eds.): VECoS 2020, LNCS 12519, pp. 15–26, 2020.
https://doi.org/10.1007/978-3-030-65955-4_2

Among the various tasks that a multi-robot system can accomplish, the most basic is a team of robots reaching multiple destinations and avoiding obstacles and collisions. In [17–19], genetic algorithm, particle swarm optimization algorithm and artificial potential field method are used to accomplish the task mentioned above.

Such tasks are often called MAPF (multi-agent path-finding), where tasks that do not assign targets to each agent in advance are called annoymous MAPF or TAPF (combined target-assignment and path-finding) [12]. The current state-of-the-art approach to solve the annoymous MAPF problem uses the min-cost max-flow algorithm to obtain the optimal solution in polynomial time [20], but the complexity of this algorithm is related to the number of robots, and this modeling approach makes it difficult to propose more complex tasks for robot formations that might be used in real life.

In [2,4,7,15], a more expressive specification is used to describe the tasks of the robots. The transition system and buchi automaton are used to achieve these tasks, but suffers from the state explosion issue if the number of robots increases.

[14] model robots and their environment with Petri nets (PN), and construct an integer linear programming (ILP) problem that is to minimize the total cost of robots. The use of PN effectively solves the problem of state space explosion as the number of robots increases. Compared to the direct construction of the ILP problem, the intermediate construction based on the PN model can effectively utilize the PN firing rules to simplify the complex and numerous constraints. The specification can be added more intuitively and flexibly by adding PN structures, and the use of PN modeling also facilitates subsequent analysis of system properties such as deadlock and liveness.

A great progress has been made on the synthesis of PN [3,6,8] to represent complex dynamical behavior of discrete event systems. [9] presents an optimal PN controller to prevent automated guided vehicles (AGV) from colliding for AGV systems with uncontrollable and indistinguishable events. Fruitful works are developed on the PN control method for linear constraints [1,10,11,13,16]. These motivate us to explore how to design place nodes enforcing a given specification on the PN models of multi-robot systems, and to in turn develop methods for the path planning of robots by PNs.

In this work, an approach is proposed to the path planning problem of a multi-robot system based on PNs. First, the environment of robots is modeled as an ordinary PN. Second, a method is presented to design place nodes to enforce requirements on trajectories of robots, which are expressed by a Boolean specification, on the PN model. Third, a path planning problem is formulated as an ILP problem via the augmented PN model. Fourth, the optimal trajectories are derived by efficiently solving the ILP problem by IBM ILOG CPLEX [5].

Compared with the methods in [14], not only the environment of robots but also complex requirements, such as collision avoidance, and that some region should be or not be visited, are modeled by a PN. As a result, the constraints of an ILP problem can be generated only according to the PN model. This greatly simplify and ease the work of designing the ILP problem. Furthermore, the requirements on the order of actions of robots can also be expressed by the PN model, which are not taken into consideration by [14].

The rest of this paper consists of the following sections. Section 2 presents basic concepts about PNs. Section 3 describes the path planning problem of robots. Section 4 presents the PN synthesis methods. Section 5 describes how to compute optimal trajectories of robots. Simulation experiments are made in Sect. 6. Section 7 concludes this work.

2 Preliminaries

A Petri net (PN) structure is a four-tuple $N = (\mathcal{P}, \mathcal{T}, \mathcal{F}, \mathcal{W})$, where $\mathcal{P}$ is a finite set of places, $\mathcal{T}$ is a finite set of transitions, and a relation $\mathcal{F} \subseteq (\mathcal{P} \times \mathcal{T}) \cup (\mathcal{T} \times \mathcal{P})$ is a collection of directed arcs that connect to places and transitions, $\mathcal{W} : \mathcal{F} \to \mathbb{Z}^+$ is a mapping that assigns each arc with a positive integer weight. It is assumed that PNs considered in this paper are ordinary, i.e., the weight of each arc is 1.

A PN structure is represented by a graph where places, transitions and relations are denoted by circles, rectangles and directed arcs, respectively. The marking is a distribution of tokens, denoted by dots, in places, and we say that a place has an empty token if it is empty. The pre-incidence matrix C^- is a $|\mathcal{P}| \times |\mathcal{T}|$ matrix such that, for all $(p, t) \in \mathcal{P} \times \mathcal{T}$, $C^-(p, t) = 1$ if $(p, t) \in \mathcal{F}$, $C^-(p, t) = 0$ otherwise. The post-incidence matrix C^+ is a $|\mathcal{P}| \times |\mathcal{T}|$ matrix such that, for all $(p, t) \in \mathcal{P} \times \mathcal{T}$, $C^+(p, t) = 1$ if $(t, p) \in \mathcal{F}$, $C^+(p, t) = 0$ otherwise. $C = C^+ - C^-$ is called the incidence matrix, which represents the flow of tokens between places caused by the firing of transition. A marking of a place represents the number of tokens in it, and each place can have a non-negative integer of tokens. A marking m of a PN structure means a distribution of tokens in places, and $m(p)$ is the marking of $p \in \mathcal{P}$. m_0 is the marking at the initial instant, and a PN is a PN structure with an initial marking denoted by $\langle N, m_0 \rangle$. $\forall \alpha, \beta \in \mathcal{P} \cup \mathcal{T}$, α is an input of β if $(\alpha, \beta) \in \mathcal{F}$, and α is an output one of β if $(\beta, \alpha) \in \mathcal{F}$. ${}^\bullet\alpha (\alpha^\bullet)$ denotes the set of inputs (outputs) of α.

A transition t is enabled at m if each input place of it contains at least one token, i.e., $\forall p \in {}^\bullet t, m(p) \geq 1$. If an enabled transition t is fired, the PN reaches the marking $\tilde{m}$ from m, where $m[t\rangle \tilde{m}$: $\tilde{m} = m + C(:, t)$.

3 Problem Descriptions

Given a team of identical robots, its path planning problem is how to determine the path trajectory for each robot such that all traveling tasks are completed with a minimal total distance on the condition that requirements or specifications on path trajectories are satisfied.

In order to formalize the problem, the environment where robots move is discretized as a set of region cells, and $\Omega = \{\Pi_1, \Pi_2, \cdots, \Pi_{|\Omega|}\}$ is a set of regions of interest, where the region Π denotes a set of region cells. Consequently, a path trajectory means a sequence of region cells. Further, specifications mean that some regions should be or not be visited, that at most one robot can enter a region cell in the same time such that any collision is avoided, and so on. Boolean specification are utilized to formally represent specifications on path trajectories, and a basic unit, which constitutes a Boolean specification, is an atomic proposition in P_t and P_f, where $P_t = \{\Pi_1, \Pi_2, \cdots, \Pi_{|\Omega|}\}$, and $P_f = \{\pi_1, \pi_2, \cdots, \pi_{|\Omega|}\}$. Every $\Pi \in P_t$ means that the region $\boldsymbol{\Pi}$ should be visited at least one time, and every $\pi \in P_f$ means that there should be a robot finally staying in the region $\boldsymbol{\Pi}$. Such atomic propositions can be connected by $\neg$, $\wedge$ and $\vee$ to make up a complex specification.

φ is used to represent the total specification, and consists of four sub-specifications b, G, V and τ, i.e., $\varphi = b \wedge G \wedge V \wedge \tau$ means the traveling tasks of robots which are described as follows.

$$b = \bigwedge_{\Pi \in P_b} (\neg \Pi) = \neg(\bigvee_{\Pi \in P_b} \Pi), \tag{1}$$

where $P_b \subseteq P_t$, means the set of regions corresponding to atomic propositions in P_b should not be visited.

$G = g_1 \wedge g_2 \wedge \cdots$, where

$$g_i = \bigvee_{\pi \in P_{g_i}} \pi, \tag{2}$$

where $P_{g_i} \subseteq P_f$, means that there should be at least one robot staying in one of the regions corresponding to atomic propositions in P_{g_i} eventually.

$V = v_1 \wedge v_2 \wedge \cdots$ where

$$v_j = \bigvee_{\Pi \in P_{v_j}} \Pi, \tag{3}$$

where $P_{v_j} \subseteq P_t$, means one of the regions corresponding to atomic propositions in P_{v_j} should be visited at least one time.

$$\tau = \bigwedge_{(P_v, P_{v'}) \in P_\tau} (\neg v \, \mathbf{u} \, v'), \tag{4}$$

where $P_\tau \subseteq 2^{P_t} \times 2^{P_t}$, means that there is a region cell defined by Π' that should be visited before any one defined by Π is visited.

Example 1. Let's consider the case where there is the total specification on the robot system shown in Fig. 1, as follows:

$$\varphi = (\neg(\Pi_1 \vee \Pi_2)) \wedge (\pi_2 \vee \pi_3) \wedge (\Pi_4 \wedge \Pi_5) \wedge (\neg \Pi_4 \, \mathbf{u} \, \Pi_5), \tag{5}$$

where $\boldsymbol{\Pi}_1 = \{c_0\}$, $\boldsymbol{\Pi}_2 = \{c_3\}$, $\boldsymbol{\Pi}_3 = \{c_2\}$, $\boldsymbol{\Pi}_4 = \{c_5\}$, $\boldsymbol{\Pi}_5 = \{c_0, c_1\}$.

$$b = \neg(\Pi_1 \vee \Pi_2) \tag{6}$$

means that any robot is forbidden to pass the region cell c_0 or c_3.

$$G = g_1 = \pi_2 \vee \pi_3 \tag{7}$$

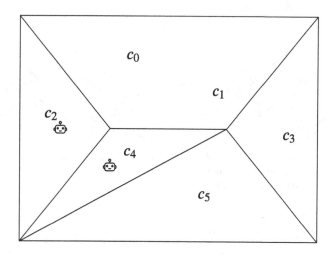

Fig. 1. A multi-robot system

means that there should be at least one robot in c_3 or c_2 eventually.

$$V = v_1 \wedge v_2 = \Pi_4 \wedge \Pi_5 \tag{8}$$

means that c_5 should be visited, and that c_0 or c_1 should be visited by robots in the planned trajectories.

$$\tau = \neg \Pi_4 \, \mathbf{u} \, \Pi_5 \tag{9}$$

means that c_5 should not be visited before that c_0 or c_1 is visited.

4 PN Synthesis Method for Modeling Multi-robot Systems

In this section, a method is proposed on how to design a PN model for a multi-robot system and its given specifications.

4.1 PN Model for a Multi-robot System

Suppose the environment of robots is decomposed into a set of small triangular cells, and a set of such cells of interest can be represented by Π corresponding to an atomic proposition Π.

Definition 1. *Given the environment of a multi-robot system, its PN is a 4-tuple* $\mathcal{N}_e = \langle \mathcal{P}_e, \mathcal{T}_e, \mathcal{F}_e, m_{e,0} \rangle$ *such that iff there is a region cell* c_i, *then* $\mathcal{P}_e$ *contains a place* p_i *representing it, and iff there is an edge between two adjacent cells* c_i *and* c_j, *then* $\mathcal{T}$ *contains two transitions* t *and* t', *and* $\mathcal{F}$ *contains four directed arcs* (p_i, t), (t, p_j), (p_j, t'), *and* (t', p_i), *and iff a region cell* c_i *has* $k \in \{0, 1, 2 \cdots\}$ *robots at the initial instant, then* $m_{e,0}(p_i) = k$.

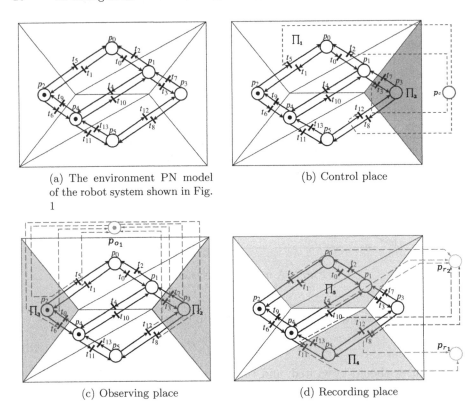

(a) The environment PN model of the robot system shown in Fig. 1

(b) Control place

(c) Observing place

(d) Recording place

Fig. 2. PN model and places enforcing requirements on path trajectories of robots

Example 2. Take the multi-robot system, as shown in Fig. 1, as example. The environment PN model can be determined according to Definition 1, and is shown in Fig. 2(a). There are six places representing six region cells, and two tokens for two robots.

4.2   Synthesis Method of Control Places Enforcing the Sub-specification b

According to (1), the sub-specification b require that a set of region cells not be visited. This implies that a set of corresponding places should not be marked no matter how a PN model runs. To do so, we can design a control place enforcing b on an environment PN.

Definition 2. *Given a sub-specification b and the environment PN of a multi-robot system, b's forbidden set, denoted by $\mathcal{P}_b$, is defined as the set of places that represent the region cells in $\cup_{\Pi \in P_b} \Pi$.*

According to Definition 2, the sub-specification b is actually to prevent any place in its forbidden set $\mathcal{P}_b$ from being marked. Hence, Algorithm 1 is obtained to design a control place enforcing b.

Algorithm 1. Control places enforcing b

Require: $\mathcal{N}_e$, sub-specification b
Ensure: a control place p_c
 1: $m_0(p_c) = 0$
 2: **for** $t \in T_e$ **do**
 3:    **if** $t^\bullet \subseteq \mathcal{P}_b \wedge {}^\bullet t \not\subseteq \mathcal{P}_b$ **then**
 4:       $\mathcal{F}_e := \mathcal{F}_e \cup \{(p_c, t)\}$
 5:    **end if**
 6: **end for**

Example 3. For the sub-specification b in (6) on the environment PN shown in Fig. 2(a), the control place p_c is designed by Algorithm 1, as shown in Fig. 2(b).

4.3   Synthesis Method of Observing Places Enforcing the Sub-specification G

According to (2), g_i(one item of G) requires that eventually there be robots in one of some region cells.

Definition 3. *Given a sub-specification $G = g_1 \wedge g_2 \wedge \cdots$ and the environment PN of a multi-robot system, $\mathcal{P}_{g_j}$ is defined as the set of places representing the region cells in $\bigcup_{\Pi \in P_{g_j}} \Pi$.*

Hence, Algorithm 2 is obtained to design observing places for G according to Definition 3.

Algorithm 2. Synthesis of observing places for G

Require: $\mathcal{N}_e$, a sub-specification G
Ensure: a set of observing places $\mathcal{P}_o$
 1: $\mathcal{P}_o = \emptyset$
 2: **for** $g_i \in G$ **do**
 3:    $\mathcal{P}_o := \mathcal{P}_o \cup \{p_{o_i}\}$
 4:    $m_0(p_{o_i}) = \sum\limits_{p \in \mathcal{P}_{g_i}} m_{e,0}(p)$
 5:    **for** $t \in T_e$ **do**
 6:       **if** ${}^\bullet t \subseteq \mathcal{P}_{g_i} \wedge t^\bullet \not\subseteq \mathcal{P}_{g_i}$ **then**
 7:          $\mathcal{F}_e := \mathcal{F}_e \cup \{(p_{o_i}, t)\}$
 8:       **end if**
 9:       **if** $t^\bullet \subseteq \mathcal{P}_{g_i} \wedge {}^\bullet t \not\subseteq \mathcal{P}_{g_i}$ **then**
10:          $\mathcal{F}_e := \mathcal{F}_e \cup \{(t, p_{o_i})\}$
11:       **end if**
12:    **end for**
13: **end for**

Example 4. Consider G in (7) for the environment PN in Fig. 2(a). $P_{g_1} = \{\pi_2, \pi_3\}$. Hence, $\mathcal{P}_{g_1} = \{p_2, p_3\}$ according to Definition 3. The observing place p_{o_1} is designed, as shown in Fig. 2(c).

4.4 Synthesis Method of Recording Places Enforcing the Sub-specification V and τ

Definition 4. *Given a sub-specification $V = v_1 \wedge v_2 \wedge \cdots$ and the environment PN of a multi-robot system, $\mathcal{P}_{v_j}$ is defined as the set of places that represent the region cells in $\bigcup_{\Pi \in P_{v_j}} \Pi$.*

According to Definition 4, Algorithm 3 is obtained to design observing places for V.

Example 5. Consider the sub-specifications V in (8) and τ in (9) for the environment PN, as shown in Fig. 2(a). $\mathcal{P}_{v_1} = \{p_5\}$ and $\mathcal{P}_{v_2} = \{p_0, p_1\}$. By Algorithm 3, the recording places p_{r_1} and p_{r_2} are designed, as shown in Fig. 2(d).

5 ILP Problem Obtained by PNs

So far, we have designed a PN for a multi-robot system. Since the PN expresses both the environment and specifications on path trajectories, an ILP to formulate the path planning problem is obtained by the PN as follows.

Algorithm 3. Synthesis of recording places for V and τ

Require: $\mathcal{N}_e$, the sub-specifications V and τ
Ensure: a set of recording places $\mathcal{P}_r$
1: $\mathcal{P}_r = \emptyset$
2: **for** v_j in V **do**
3:      $\mathcal{P}_r := \mathcal{P}_r \cup \{p_{r_j}\}$
4:      $m_0(p_{r_j}) = \sum\limits_{p \in \mathcal{P}_{v_j}} m_{e,0}(p)$
5:      **for** $t \in T_e$ **do**
6:          **if** ${}^{\bullet}t \not\subseteq \mathcal{P}_{v_j} \wedge t^{\bullet} \subseteq \mathcal{P}_{v_j}$ **then**
7:              $\mathcal{F}_e := \mathcal{F}_e \cup \{(t, p_{r_j})\}$
8:          **end if**
9:      **end for**
10: **end for**
11: **for** $(\neg v_i \, \mathbf{u} \, v_j)$ in τ **do**
12:     **for** $t \in {}^{\bullet}p_{r_i}$ **do**
13:         $\mathcal{F}_e := \mathcal{F}_e \cup \{(t, p_{r_j}), (p_{r_j}, t)\}$
14:     **end for**
15: **end for**

$$\min : \ w^T \cdot \sum_{i=1}^{k} \sigma_i + \sum_{i=1}^{k} i \cdot \sum_{t \in \mathcal{T}} (\sigma_i(t))$$

$$\forall 1 \leq i \leq k, \sigma_i \in \{0,1\}^{|\mathcal{T}|},$$
$$\forall 1 \leq i \leq k, \forall p \in \mathcal{P}_e \cup \mathcal{P}_c, m_i(p) \in \{0,1\},$$
$$\forall 1 \leq i \leq k, \forall p \in \mathcal{P}_o \cup \mathcal{P}_r, m_i(p) \in \mathbb{N}, \tag{10}$$
$$\forall 1 \leq i \leq k, m_i = m_{i-1} + C \cdot \sigma_i,$$
$$\forall 1 \leq i \leq k, m_{i-1} - C^- \cdot \sigma_i \geq 0,$$
$$\forall p \in \mathcal{P}_o \cup \mathcal{P}_r, m_k(p) \geq 1.$$

where w is a column vector with $|\mathcal{T}|$ elements representing the traveling distances for transitions, σ_i means the set transitions that are fired at the i-th firing instant, and k denotes the number of transition fires that should be taken to ensure all robots reach their target regions.

Once the ILP problem (10) is obtained, it can be efficiently solved by the ILP solver in [5]. The solution obtained by our method can give not only the motion path of robots, but also the sequence of motions. According to this motion sequence, the planned trajectories of a team of robots can satisfy the specifications, and are with a total traveling distance that may be minimal. The complexity of the ILP problem is related to the number of unknown variables and constraints. The number of unknown variables in (10) is: $(k \times |\mathcal{P} \cup \mathcal{T}| + \mathcal{P})$; the number of constraints is: $((3k+2) \times |\mathcal{P}| + k \times |\mathcal{T}| + |\mathcal{P}_o| + |\mathcal{P}_r|)$.

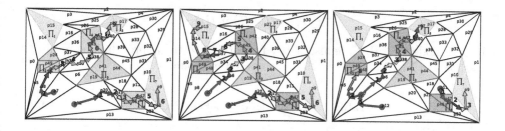

Fig. 3. A numerical experiment with two robots and the total specification (11)

Fig. 4. A numerical experiment with two robots and the total specification (12)

Fig. 5. A numerical experiment with three robots and the total specification (12)

6 Numerical Experiments

We develop a set of Python programs to implement the proposed approach, and conduct several numerical experiments on a computer with an i7-7700 CPU, which are summarized as follows. The considered robot system is shown in Fig. 3, where there are 7 regions of interest. Since a cell is represented by a place, a place p_i also denotes a cell c_i in the experiments for simplicity. By the proposed approach, the environment PN is obtained with 50 environment places, and 146 transitions.

In the first numerical experiment, the total specification is

$$\varphi_1 = \neg \Pi_3 \wedge \Pi_2 \wedge \Pi_4 \wedge \Pi_6 \wedge (\pi_1 \vee \pi_5) \wedge \pi_7, \tag{11}$$

which requires that Π_3 should not be visited by any robot, Π_2, Π_4 and Π_6 should be passed through by robots, at least one robot stay in Π_7 finally, and at least one robot stay in Π_1 or Π_5 finally. By the proposed approach, four recording places, two observing places, and one control place are added into the environment PN. By setting $k = 25$, the path trajectories of the robots are calculated, as shown in Fig. 3 and Table 1, the solution was obtained in around 0.446 s.

Table 1. Trajectories of robots in Fig. 3

R_1	p_7	$\to p_8$	$\to p_{48}$	$\to p_{34}$	$\to p_{35}$	$\to p_{38}$	$\to p_{43}$	$\to p_{42}$	$\to p_{40}$	$\to p_{27}$
R_2	p_{12}	$\to p_{20}$	$\to p_{21}$	$\to p_{46}$	$\to p_{47}$	$\to p_{22}$	$\to p_9$			

In the second numerical experiment, the specification $\tau = \neg \Pi_2 \mathbf{u} \Pi_4$ (Π_4 should be visited before Π_2) is taken into account besides (11), and its total requirement is consequently.

$$\varphi_2 = \varphi_1 \wedge (\neg \Pi_2 \mathbf{u} \Pi_4). \tag{12}$$

Its PN model and ILP are also derived according to the approach. The results, as shown in Fig. 4 and Table 2, are computed in around 0.71 s.

Table 2. Trajectories of robots in Fig. 4

R_1	p_7	$\to p_6$	$\to p_{35}$	$\to p_{38}$	$\to p_{43}$	$\to p_{36}$	$\to p_{37}$	$\to p_{24}$	$\to p_{14}$	$\to p_{15}$
R_2	p_{12}	$\to p_{20}$	$\to p_{21}$	$\to p_{46}$	$\to p_{47}$	$\to p_{22}$	$\to p_9$			

In the third numerical experiment, another robot is added into the place p_{11}, as shown in Fig. 5. Similarly, the path trajectories of three robots are computed in around 0.14 s, and the results are shown in Fig. 5 and Table 3. It is worth noting that R_2 stays at p_8 and waits until R_1 reaches p_{43} (Π_4 is visited), and then R_2 starts to move into p_{48} (Π_2 is visited).

In the ILP problems for the multi-robot systems, as respectively shown in Fig. 3, Fig. 4 and Fig. 5, the numbers of unknown variables are all equal to 5106, and them of linear constraints are all equal to 2861.

Table 3. Trajectories of robots in Fig. 5

R_1	p_7	$\rightarrow p_6$	$\rightarrow p_{35}$	$\rightarrow p_{38}$	$\rightarrow p_{43}$	$\rightarrow p_{42}$	$\rightarrow p_{40}$	$\rightarrow p_{27}$
R_2	p_{12}	$\rightarrow p_7$	$\rightarrow p_8$	stay	stay	$\rightarrow p_{48}$		
R_3	p_{11}	$\rightarrow p_{47}$	$\rightarrow p_{22}$	$\rightarrow p_9$				

7 Conclusion

An approach is proposed to address the path planning problem of a team of robots. Its central idea is to model a multi-robot system by a PN. Via the PN model, an ILP problem is constructed, and is resolved to obtain the path trajectories of robots.

Due to the use of PN, our algorithm does not suffer from state space explosion as the number of robots increases (the number of ILP variables is independent of the number of robots), and allows specification for robot formations that certain areas should be traversed with limited sequential requirements. Since the problem is ultimately converted to an ILP problem, different objective functions are allowed to be customized to meet different planning requirements (e.g.., shortest completion time, shortest total travel distance). However, since the number of ILP variables is related to the number of places (region cells), the algorithm may have the problem of state space explosion due to the excessive number of regions.

In the future, we will explore how to efficiently resolve more complex optimal problems generated by PNs, and how to formally represent multi-robot systems with more complex specifications by PNs.

References

1. Basile, F., Cordone, R., Piroddi, L.: A branch and bound approach for the design of decentralized supervisors in Petri net models. Automatica **52**, 322–333 (2015)
2. Ding, X., Kloetzer, M., Chen, Y., Belta, C.: Automatic deployment of robotic teams. IEEE Robot. Autom. Mag. **18**(3), 75–86 (2011)
3. Giua, A., Silva, M.: Petri nets and automatic control: a historical perspective. Annu. Rev. Control **45**, 223–239 (2018)
4. Guo, M., Tumova, J., Dimarogonas, D.V.: Cooperative decentralized multi-agent control under local LTL tasks and connectivity constraints. In: Proceedings of the 53rd IEEE Conference on Decision and Control, Los Angeles, CA, USA, pp. 75–80 (2014)
5. IBM: IBM ILOG CPLEX optimization studio (2016). https://www.ibm.com/analytics/cplex-optimizer
6. Iordache, M.V., Antsaklis, P.J.: Petri nets and programming: a survey. In: Proceedings of American Control Conference, St. Louis, MO, USA, pp. 4994–4999, 10–12 June 2009. https://doi.org/10.1109/ACC.2009.5159987
7. Kloetzer, M., Mahulea, C.: LTL-based planning in environments with probabilistic observations. IEEE Trans. Autom. Sci. Eng. **12**(4), 1407–1420 (2015)

8. Li, Z.W., Zhou, M.C., Wu, N.Q.: A survey and comparision on Petri net-based deadlock prevention control policies for flexible manufacturing systems. IEEE Trans. Syst. Man Cybern. Part C Appl. Rev. **38**(2), 173–188 (2008)
9. Luo, J., Wan, Y., Wu, W., Li, Z.: Optimal Petri-net controller for avoiding collisions in a class of automated guided vehicle systems. IEEE Trans. Intell. Transp. Syst. **21**, 1–12 (2019). https://doi.org/10.1109/TITS.2019.2937058
10. Luo, J., Wu, W., Zhou, M., Shao, H., Nonami, K., Su, H.: Structural controller for logical expression of linear constraints on Petri nets. IEEE Trans. Autom. Control **65**, 1 (2019). https://doi.org/10.1109/TAC.2019.2918052
11. Luo, J., Zhou, M.: Petri-net controller synthesis for partially controllable and observable discrete event systems. IEEE Trans. Autom. Control **62**(3), 1301–1313 (2017). https://doi.org/10.1109/TAC.2016.2586604
12. Ma, H., Koenig, S.: Optimal target assignment and path finding for teams of agents (2016)
13. Ma, Z., Li, Z., Giua, A.: Design of optimal Petri net controllers for disjunctive generalized mutual exclusion constraints. IEEE Trans. Autom. Control **60**, 1774–1785 (2015)
14. Mahulea, C., Kloetzer, M.: Robot planning based on Boolean specifications using Petri net models. IEEE Trans. Autom. Control **63**(7), 2218–2225 (2018)
15. Kloetzer, M., Mahulea, C.: A Petri net based approach for multi-robot path planning. Discrete Event Dyn. Syst. **24**(4), 417–445 (2013). https://doi.org/10.1007/s10626-013-0162-6
16. Moody, J.O., Antsaklis, P.J.: Petri net supervisors for DES with uncontrollable and unobservable transitions. IEEE Trans. Autom. Control **45**(3), 462–476 (2000)
17. Purcaru, C., Precup, R., Iercan, D., Fedorovici, L., Petriu, E.M., Voisan, E.: Multi-robot GSA- and PSO-based optimal path planning in static environments. In: Proceedings of the 9th International Workshop on Robot Motion and Control, pp. 197–202 (2013). https://doi.org/10.1109/RoMoCo.2013.6614608
18. Shibata, T., Fukuda, T.: Coordinative behavior in evolutionary multi-agent system by genetic algorithm. In: IEEE International Conference on Neural Networks, vol. 1, pp. 209–214 (1993). https://doi.org/10.1109/ICNN.1993.298558
19. Warren, C.W.: Multiple robot path coordination using artificial potential fields. In: IEEE International Conference on Robotics and Automation, Cincinnati, OH, USA, pp. 500–505 (1990)
20. Yu, J., LaValle, S.M.: Multi-agent path planning and network flow (2012)

Deadlock Avoidance of Flexible Manufacturing Systems by Colored Resource-Oriented Petri Nets with Novel Colored Capacity

ZhaoYu Xiang$^{(\boxtimes)}$ (iD)

Institute of System Engineering, Macau University of Science and Technology,
Taipa 999078, Macao, Special Administrative Region of China
1909853gmi30002@student.must.edu.mo

Abstract. A variety of colored resource-oriented Petri nets (CROPN)-based control method to forbid deadlock in flexible manufacturing systems (FMS) are to add control places to the original net, which makes the net being complex. This paper proposes a novel concept in colored resource-oriented Petri nets (CROPN) called colored capacity. Firstly, the foraml definition of colored capacity in a CROPN is given. Based on this concept, the new execution rule of the transitions is proposed. Then, a procedure is developed such that the colored capacity function of each place in a CROPN can be obtained. By colored capacity function, all control places that are used to forbid illegal markings in CROPN are displaced by the colored capacity and the deadlock can be avoided by the new execution rule, which makes the structure of the net much simpler than the net with control places. Finally, an FMS example is used to illustrate the proposed method.

Keywords: Deadlock avoidance · Discrete event systems · Flexibility manufacturing systems · Petri nets · Colored resource-oriented Petri net

1 Introduction

Flexible manufacturing systems (FMS) have been widely used in industrial fields [1–7]. However, since a large number of jobs have to share same resource in an FMS, deadlocks may arise, which leads to serious consequences. To deal with deadlock issue in FMSs, many control policies based on Petri net models have been established.

In these control policies, a deadlock avoidance strategy to prevent the FMS modeled by Petri nets from being deadlock is to add some constraints to the targeted FMS such that the system is deadlock-free. The work in [9] proposes

This work was supported in part by the Science and Technology Development Fund, Macau SAR (File Nos. 0012/2019/A1).

B. Ben Hedia et al. (Eds.): VECoS 2020, LNCS 12519, pp. 27–40, 2020.
https://doi.org/10.1007/978-3-030-65955-4_3

a colored resource-oriented Petri net (CROPN) model to analyze the deadlock problem in FMS. The CROPN is considered to be more powerful than other Petri nets such as the resource place-based Petri nets [5].

The work in [5] establishes the deadlock-free operations in CROPN. However, to use the deadlock avoidance policy proposed in [5], we need to add control places to the original net, which makes the net being complex.

In this paper, a novel CROPN concept called colored capacity is proposed. We define the colored capacity of the place, which represents the biggest number of tokens with the same color that the place can retain simultaneously. It is shown that a place in CROPN may have different colored capacity corresponding to different color. Based on the concept of colored capacity, the new transitions firing rule for CROPN is presented. Then, after we obtain all deadlock and impending deadlock markings by using the method proposed in [5], a method is presented to determine the colored capacity for each place in CROPN by simple calculation. It is shown that the colored capacity of a place in CROPN will change along with the marking change. Based on the dynamically changing color capacity, combined with the new transitions firing rule proposed in this paper, all deadlock markings and impending deadlock markings in a flexible manufacturing systems (FMS) modeled by CROPN are forbidden, therefore, this FMS is deadlock-free and we do not need to add control places to the net.

The contributions in this paper are as follows:

1) In this paper, the colored concept for FMS modeled by CROPN is proposed, which is not considered in [5].
2) Based on the dynamically changing color capacity, combined with the new transition firing rule proposed in this paper, all deadlock markings and impending deadlock markings in a flexible manufacturing systems (FMS) modeled by CROPN are forbidden without the need to add control places to the net.

This paper is organized as follows. Section 1 is the introduction. Section 2 defines the concept of colored capacity for CROPN. Section 2 proposes a method to determine the colored capacity for each place in all interactive subnets in CROPN by simple calculation. Section 3 presents a FMS example to explain the application of the proposed method. We conclude in Sect. 4.

2   Realization of Control Policies for Interactive Subnets

In this section, a method is developed to determine the colored capacity for each place in CROPN such that the deadlock-avoidance policy $U(v^k)$ is realized by using the new transitions firing rule (Fig. 1).

Consider CROPN in Fig. 2 [5], there are three part types A, B, and C, where $p_0 \rightarrow t_2 \rightarrow p_1 \rightarrow t_4 \rightarrow p_2 \rightarrow t_5 \rightarrow p_3 \rightarrow t_7 \rightarrow p_4 \rightarrow t_9 \rightarrow p_0$ for the operating of part-A and $p_0 \rightarrow t_{10} \rightarrow p_4 \rightarrow t_8 \rightarrow p_3 \rightarrow t_6 \rightarrow p_2 \rightarrow t_1 \rightarrow p_0$ for the operating of part-B and $p_0 \rightarrow t_2 \rightarrow p_1 \rightarrow t_3 \rightarrow p_5 \rightarrow t_{11} \rightarrow p_0$ for the operating of part-C with $K(p_1) = K(p_3) = K(p_5) = 1$ and $K(p_2) = K(p_4) = 2$. Hence, we have

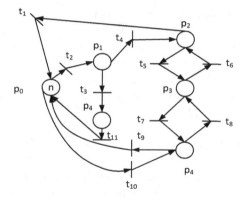

Fig. 1. A CROPN with one interactive subnet formed by two PPCs [5]

$C(p_0) = \{b_2, b_{10}\}, C(p_1) = \{b_3, b_4\}, C(p_2) = \{b_1, b_5\}, C(p_3) = \{b_6, b_7\}, C(p_4) = \{b_8, b_9\}, C(p_5) = \{b_{11}\}$. In this net, there is only one interactive subnet $v^2 = \{p_2, t_5, t_6, p_3, t_7, t_8, p_4\}$ formed by two PPCs:

$$v_1 = \{p_2, t_5, p_3, t_6\}, v_2 = \{p_3, t_7, p_4, t_8\} \tag{1}$$

$$\text{Let } \Theta(v^2) = (M(p_2)(b_5), \ M(p_2)(b_1), \ M(p_3)(b_7), \ M(p_3)(b_6) \tag{2}$$

Then, by the algorithm proposed in [5], we have $M_{FBM} ** (v^2) = \{M_1, M_2, M_3\}$ such that

$$M_1(v^2) = \Theta_1(v^2) = (0, 0, 1, 0, 0, 2)^T$$
$$M_2(v^2) = \Theta_2(v^2) = (2, 0, 0, 1, 0, 0)^T \tag{3}$$
$$M_3(v^2) = \Theta_3(v^2) = (2, 0, 0, 0, 0, 2)^T$$

Hence, to prevent deadlock in this net, we have three control policies in $U(v^2)$ used to prohibit $M_{FBM} ** (v^2)$ as follows:

$$C(M_1) : M(p_3)(b_7) + M(p_4)(b_8) \leq M_1(p_3)(b_7)$$
$$+ M_1(p_4)(b_8) - 1 = 2; \tag{4}$$

$$C(M_2) : M(p_2)(b_5) + M(p_3)(b_6) \leq M_2(p_2)(b_5)$$
$$+ M_2(p_3)(b_6) - 1 = 2; \tag{5}$$

$$C(M_3)) : M(p_2)(b_5) + M(p_4)(b_8) \leq M_3(p_2)(b_5)$$
$$+ M_3(p_4)(b_8) - 1 = 3. \tag{6}$$

where $M \notin M_B(v^2)$.

Next, we discuss how to realize $U(v^2)$ by using colored capacity and its new execution rule. First, let us present some notations as follows:

1. In a CROPN, given a marking M. For a place $p \in P$, let $RM(M(p)(b_i)) = K(p)(b_i)(M) - M(p)(b_i)$ denote the number of free room available to tokens with color b_i in p at marking M, where $b_i \in C(p)$.

2. Given a CROPN, a subnet v^k in the net with $M_i \in M_{FBM} * *(v^k)$, let $S_i(v^k) = \Lambda_i(v^k) - 1$, where $\Lambda_i(v^k) = \sum_{D(v^k)(i,l,j)} M_i(p_l)(b_j)$.

3. In a CROPN, given a mrking M, for a place $p \in P$, we use $C(p,z)$ to denote a set of colores of $b_z \in C(p)$. Hence, let $\sum_{C(p,z)} M(p)(b_z)$ denote the number of tokens corresponding to $C(p,z)$ for marking M.

4. In a CROPN, we use $A(L,J)$ to denote the set of combinations of $p_L \in P$ and $b_J \in C(P)$, where $C(P) = \{C(p_u)|u = 0,1,...,|P|-1\}$.

5. Given a CROPN, a subnet v^k in the net, we use $C(v^k)(L,J)$ to denote the set of combinations of $p_L \in P(v^k)$ and $b_J \in C(v^k)$, where $C(v^k) = \{C(p)|p \in P(v^k)\}$.

6. Given a CROPN, a subnet v^k in the net. When $M_{FBM} * *(v^k)$ have been found, let us rearrange the markings in $M_{FBM} * *(v^k)$ as $M_1, M_2, ...,$. Then, define $D(v^k)(A,l,j) = \{D(v^k)(1,l,j) \cup D(v^k)(2,l,j)\cup ...\}$

7. Given a CROPN, a subnet v^k in the net. Define $ND(v^k)(L,J) = C(v^k)(L,J) - D(v^k)(A,l,j)$ such that for any combination $(p_L, b_J) \in ND(v^k)(L,J)$, we have $(p_L, b_J) \in C(v^k)(L,J)$ and $(p_L, b_J) \notin D(v^k)(A,l,j)$.

In view of the above-mentioned definitions, given an interactive subnet v^k in a CROPN, we classify the colored capacity $K(p)(b_i)(M), p \in P(v^k), b_i \in C(p), M \notin M_B(v^k)$ into two categories:(1) $K(p)(b_i)(M)$ for $(p,b_i) \in D(v^k)(A,l,j)$. (2) $K(p)(b_i)(M)$ for $(p,b_i) \in ND(v^k)(L,J)$.

To realize $U(v^2)$ for subnet v^2 in the net shown in Fig. 2, we need to consider all items in $D(v^2)(A,l,j)$, where $D(v^2)(A,l,j) = D(v^2)(1,l,j) \cup D(v^2)(2,l,j) \cup D(v^2)(3,l,j)$ and $U(v^2) = \{C(M_1), C(M_2), C(M_3)\}$.

We first determine $K(p)(b_i)(M), (p,b_i) \in D(v^2)(1,j,l)$ for realizing $C(M_1)$ by using the new execution rule. Consider two combinations (p_3, b_7) and (p_4, b_8) from $D(v^2)(1,l,j)$. Suppose

$$RM(M(p_3)(b_7)) = \left\{ S_1(v^2) - \sum_{D(v^2)(1,l,j)} M(p_l)(b_j) \right\} \tag{7}$$

$$RM(M(p_4)(b_8)) = \left\{ S_1(v^2) - \sum_{D(v^2)(1,l,j)} M(p_l)(b_j) \right\} \tag{8}$$

where $M \notin M_B(v^2)$, $S_1(v^2) = 2$, $D(v^2)(1,l,j) = \{(p_3, b_7),(p_4, b_8)\}$. Then, we have $K(p_3)(b_7)(M) = 2 - M(p_4)(b_8)$ and $K(p_4)(b_8)(M) = 2 - M(p_3)(b_7)$. It is clear that for any $M \notin M_B(v^2)$, the constraint is realized by using the colored capacity and new transitions firing rule. For example, assume that $M(p_3)(b_7) = 0, M(p_4)(b_8) = 2$, we have $K(p_3)(b_7)(M) = 0 < M(p_3)(b_7) - I(p_3, t_5)(b_5) + O(p_3, t_5)(b_5) = 1$. Hence, transition t_5 can not fire at marking M. Similarly, the transition t_{10} can not fire at marking M. Hence, $\nexists t \in T, M[t > M'$ such that $M'(p_3)(b_7) + M'(p_4)(b_8) > 2$. Therefore, with $K(p_3)(b_7)(M)$ and $k(p_4)(b_8)(M)$ being determined for any reachable marking $M \notin M_B(v^2)$, we have $M(p_3)(b_7) + M(p_4)(b_8) \leq 2$, imply that $C(M_1)$ is realized.

Note that, for any marking $M \notin F_B(v^2)$, we have $K(p_3)(b_7)(M) \leq K(p_3) = 1$ and $K(p_4)(b_8)(M) \leq K(p_4) = 2$. But, assume that $M(p_4)(b_8) = M(p_4)(b_9) = 0$

and $M(p_3)(b_7) = M(p_3)(b_6) = 0$, by Eqs. (11) and (12) we have $K(p_3)(b_7)(M) = 2$, which is a contradiction. Hence, Eq. (11) and (12) must be rewritten as follows:

$$RM(M(p_3)(b_7)) = \min \left\{ S_1(v^2) - \sum_{D(v^2)(1,l,j)} M(p_l)(b_j), \right.$$
$$\left. K(p_3) - \sum_{C(p_3,z)} M(p_3)(b_z) \right\} \tag{9}$$

$$RM(M(p_4)(b_8)) = \min \left\{ S_1(v^2) - \sum_{D(v^2)(1,l,j)} M(p_l)(b_j), \right.$$
$$\left. K(p_4) - \sum_{C(p_4,z)} M(p_4)(b_z) \right\} \tag{10}$$

where $C(p3, z) = \{b_7, b_6\}$ and $C(p4, z) = \{b_9, b_8\}$. For example, assume that $M(p_4)(b_8) = M(p_4)(b_9) = 0$ and $M(p_3)(b_7) = M(p_3)(b_6) = 0$, by Eq. (10) and Eq. (11), we have $RM(M(p_3)(b_7)) = 1$ and $RM(M(p_4)(b_8)) = 2$. That is $K(p_3)(b_7)(M) = 1$ and $K(p_4)(b_8)(M) = 2$ do not contradict $K(p_3)(b_7)(M) \leq K(p_3) = 1$ and $K(p_4)(b_8)(M) \leq K(p_4) = 2$.

Furthermore, assume that $M(p_2)(b_5) = 2$, $M(p_3)(b_7) = 0, M(p_3)(b_6) = 0$ and $M(p_4)(b_8) = 1, M(p_4)(b_9) = 0$, by Eq. (13) and Eq. (14), we have $K(p_3)(b_7)(M) = 1$ and $K(p_4)(b_8)(M) = 2$. By new transitions firing rule, the transition t_{10} can fire, i.e., $M[t_{10} > M'$ such that $M'(p_4)(b_8) = 2, M'(p_2)(b_5) = 2$, and $M'(p_3)(b_7) = 0$. Note that, at marking M', constraint is satisfied. But, some constraint is not satisfied, since $M'(p_4)(b_8) + M'(p_2)(b_5) = 4 > 3$. Hence, Eq. (13) and (14) must be rewritten as follows:

$$RM(M(p_3)(b_7)) = \min \left\{ S_1(v^2) - \sum_{D(v^2)(1,l,j)} M(p_l)(b_j), \right.$$
$$\left. K(p_3) - \sum_{C(p_3,z)} M(p_3)(b_z) \right\} \tag{11}$$

$$RM(M(p_4)(b_8)) = \min \left\{ S_1(v^2) - \sum_{D(v^2)(1,l,j)} M(p_l)(b_j), \right.$$
$$S_3(v^2) - \sum_{D(v^2)(3,l,j)} M(p_l)(b_j), \tag{12}$$
$$\left. K(p_4) - \sum_{C(p_4,z)} M(p_4)(b_z) \right\}$$

where $S_3(v^2) = 3$, $D(v^2)(3, l, j) = \{(p_2, b_5), (p_4, b_8)\}$. Then, assume that

$$M(p_2)(b_5) = 2,$$
$$M(p_3)(b_7) = 0, M(p_3)(b_6) = 0, \tag{13}$$
$$M(p_4)(b_8) = 1, M(p_4)(b_9) = 0.$$

By Eq. (15) and Eq. (16), we have $K(p_3)(b_7)(M) = 1$ and $K(p_4)(b_8)(M) = 1$. Hence, by new transition firing rule, the transition t_{10} can not fire at marking M.

Similarly, to realize constraint, we select two combinations (p_2, b_5) and (p_3, b_6) from $D(v^2)(2, l, j)$ and determine the corresponding colored capacity as follows:

$$RM(M(p_2)(b_5)) = \min \left\{ S_2(v^2) - \sum_{D(v^2)(2,l,j)} M(p_l)(b_j), \right.$$

$$S_3(v^2) - \sum_{D(v^2)(3,l,j)} M(p_l)(b_j), \qquad (14)$$

$$\left. K(p_2) - \sum_{C(p_2,z)} M(p_2)(b_z) \right\}$$

$$RM(M(p_3)(b_6)) = \min \left\{ S_2(v^2) - \sum_{D(v^2)(2,l,j)} M(p_l)(b_j), \right.$$

$$\left. K(p_3) - \sum_{C(p_3,z)} M(p_3)(b_z) \right\} \qquad (15)$$

where

$$M \notin M_B(V^2)$$
$$D(v^2)(2, l, j) = \{(p_2, b_5), (p_3, b_6) \qquad (16)$$
$$C(p2, z) = \{b_1, b_5\}$$

Thus $RM(M(p_2)(b_5)) + M(p_2)(b_5) = K(p_2)(b_5)(M)$, and $RM(M(p_3)(b_6)) + M(p_3)(b_6) = K(p_3)(b_6)(M)$.

Note that, with $K(p_3)(b_7)(M), K(p_4)(b_8)(M), K(p_2)(b_5)$, and $K(p_3)(b_6)(M)$ being determined, where $M \notin M_B(v^2)$, all items in $D(1, l, j), D(2, l, j)$, and $D(3, l, j)$ are considered, the constraint set $U(v^2)$ is realized by using the excuation rule.

The next step is to determine $K(p_L)(b_J)(M)$ for $(p_L, b_J) \in ND(v^2)(L, J)$. Since every item in $ND(v^2)(L, J)$ has no contribution to realize $U(v^2)$, we do not need to think over $S_1(v^2), S_2(v^2), S_3(v^2)$ for $K(p_L)(b_J)(M)$. Therefore, we have

$$RM(M(p_0)(b_2)) = RM(M(p_0)(b_{10})) = \infty \qquad (17)$$

$$RM(M(p_1)(b_3)) = RM(M(p_1)(b_4)) = K(p_1)$$
$$- \sum_{C(p_1,z)} M(p_1)(b_z) \qquad (18)$$

$$RM(M(p_2)(b_1)) = K(p_2) - \sum_{C(p_2,z)} M(p_2)(b_z) \qquad (19)$$

$$RM(M(p_4)(b_9)) = K(p_4) - \sum_{C(p_4,z)} M(p_4)(b_z) \qquad (20)$$

$$RM(M(p_5)(b_{11})) = K(p_5) - \sum_{C(p_5,z)} M(p_5)(b_z) \qquad (21)$$

where $M \notin M_B(v^2)$ and $K(p_0)(b_2)(M) = K(p_0)(b_{10})(M) = \infty$, $K(p_1)(b_3)(M) = K(p_1) - M(p_1)(b_4)$, $K(p_1)(b_4)(M) = K(p_1) - M(p_1)(b_3)$, $K(p_2)(b_1)(M) = K(p_2) - M(p_2)(b_5)$, $K(p_4)(b_9)(M) = K(p_4) - M(p_4)(b_8)$, $K(p_5)(b_{11})(M) = K(p_5)$.

Next, we transform the above analysis into Algorithm 1 for determining the colored capacity of each place in an interactive subnet in a CROPN.

Algorithm 1. *Determine the colored capacity of every place in an interactive subnt in a CROPN.*

Given a CROPN, an interactive subnet v^k in the net. With $M_{FBM} * *(v^k)$ being found by Algorithm presented in [5], we rearrange the markings in $M_{FBM} * *(v^k)$ as $M_1, M_2, ..., M_{|M_{FBM} * *(v^k)|}$.

Input: $D(v^k)(i, l, j)$, $ND(v^k)(L, J)$, $S_i(v^k)$, $i \in \{1, 2, ..., |M_{FBM} * *(v^k)|\}$. $K(p)$, and $C(p, z), p \in P(v^k)$;

Output: colored capacity $K(p_L)(b_J)(M)$, where $M \notin M_B(v^k)$, and $(p_L, b_J) \in C(v^k)(L, J)$.

1. *Step 1:*
   For $I=1$ to $|M_{FBM} * *(v^k)|$;
   $Relay(I) \leftarrow D(v^k)(I, l, j)$; End for;
   Let $RM_t = \emptyset$;
2. *Step 2:*
   For $q=1$ to $|M_{FBM} * *(v^k)|$;
   While $D(v^k)(q, l, j) \neq \emptyset$;
   Select a combination (p_l, b_j) from $D(v^k)(q, l, j)$;
   $RM_t \leftarrow S_q(v^k) - \sum_{Relay(q)} M(p_l)(b_j)$;
   For $d= q+1$ to $|M_{FBM} * *(v^k)|$;
   If (p_l, b_j) is an item in $Relay(d)$;
   Delete the item (p_l, c_l) from $D(v^k)(d, l, j)$;
   $RM_t \leftarrow RM_t \cup \{ S_d(v^k) - \sum_{Relay(d)} M(p_l)(b_j) \}$;
   End if;
   End for;
   $RM_t \leftarrow RM_t \cup K(p_l) - \sum_{C(p_l, z)} M(p_l)(b_z)$;
   $K(p_l)(b_j)(M) = min(RM_t) + M(p_l)(b_j)$;
   Delete the item (p_l, c_l) from $D(v^k)(q, l, j)$;
   $RM_t \leftarrow \emptyset$;
   End while;
   End for;
3. *Step 3:*
   While $ND(v^k)(L, J) \neq \emptyset$;
   Select a combination (p_L, b_J) from $ND(v^k)(L, J)$;
   $K(p_L)(b_J)(M) = k(p_L) - \sum_{C(p_L, z)} M(p_L)(b_z)$
   $+M(p_L)(b_J)$;
   Delete the item (p_L, b_J) from $ND(v^k)(L, J)$;
   End while;

In Algorithm 1, we preprocess a set in step 1. Next, in step 2, we determine $K(p_l)(b_j)(M)$ for any $(p_l, b_j) \in D(v^k)(A, l, j)$, where $M \notin M_B(v^k(x))$. Finally, in step 3, we determine $K(p_L)(b_J)(M)$ for any $(p_L, b_J) \in ND(v^k)(L, J)$.

The Algorithm 1, combined with the analysis in this section, we have the following theorems.

Theorem 1. *Given a CROPN, an interactive subnet v^k in the net, if for any combination $(p_L, b_J) \in C(v^k)(L, J)$, $K(p_L)(b_J)(M)$ is determined by Algorithm 1, where $M \notin M_B(v^k)$, then the subnet v^k is deadlock-free.*

Proof: According to the analysis in this section, if for every place in subnet v^k, the colored capacity is determined by Algorithm 1, then $U(v^k)$ is realized by using the new execuation rule. Furthermore, by Algorithm 1, no control places and arcs are added such that no new bad marking will produce in that subnet. Therefore, the subnet v^k is deadlock-free. The conclusion holds.   ∎

Theorem 2. *Given a CROPN, an interactive subnet v^k in the net, and a set $M_{FBM}**(v^k)$, if $U(v^k)$ is realized by the colored capacity and the new execuation rule, such that for any reachable marking M, $\sum_{D(v^k)(i,l,j)} M(p_l)(b_j) \leq \Lambda_i(v^k) - 1$, where $M_i \in M_{FFM}**(v^k)$, then, the control policy set $U(v^k)$ is maximally permissive.*

Proof: By Theorem 1, with $U(v^k)$ being realized by using the colored capacity and the new execuation rule, the subnet v^k is deadlock-free, imply that for any marking $M_b \in M_{FBM}$, M_b is prohibited. Suppose the control policy set $U(v^k)$ is not maximally permissive. Then, $\exists M_1 \in M_L$, M_1 is prohibited, which implies that there exists $M_i \in M_{FBM}**(v^k)$ such that $\sum_{D(v^k)(i,l,j)} M_1(p_l)(b_j) > \Lambda_i(v^k) - 1$, which is a contradiction. The conclusion holds.   ∎

By Theorem 1, with $K(p_L)(b_J)(M)$, $(p_L, b_J) \in C(v^k)(L, J)$ being determined by Algorithm 1, the net is deadlock-free. The remaining problem is how to determine $K(p)(b_J)(M)$ for $(p, b_j) \in A(L, J) - C(v^k)(L, J)$. On the one hand, for any place $p \notin P(v^k)$, p has no influence to deadlock. On the other hand, we need to deliver as many products as possible into system modeled by CROPN. Therefore, we have following Algorithm.

Algorithm 2. *Determine the colored capacity of place that is not in interactive subnts in a CROPN.*
   Given a CROPN formed by l subnet $v^k(1), v^k(2), ...,$and $v^k(l)$.
   Input: $A(L, J)$, and $C(v^k(1))(L, J),...,C(v^k(l))(L, J)$.
   Output: colored capicity $K(p_L)(b_J)(M)$, where $M \in R(M_0)$, and $(p_L, b_J) \in A(L, J) - C(v^k(1))(L, J) \cup C(v^k(2))(L, J), ..., C(v^k(l))(L, J)$, and $K(p)$, $p \in P(v^k)$.

1. *Step 1:*
   $Relay = \emptyset$;
   $Relay1 = \emptyset$;
   For $I=1$ to l;
   $Relay \leftarrow C(v^k(I))(L, J) \cup Relay$;
   End for;
   $Relay1 = A(L, J) - Relay$;
2. *Step 2:*
   While $Relay1 \neq \emptyset$;
   Select a combination (p_L, b_J) from $Relay1$;

$K(p_L)(b_J)(M) = k(p_L) - \sum_{C(p_L,z)} M(p_L)(b_z)$
$+M(p_L)(b_J);$
Delete the item (p_L, b_J) *from Relay1;*
End while;

In summary, we develop a deadlock avoidance procedure as follows such that for every interactive subnet v^k in a CROPN, $U(v^k)$ is realized and the net is made deadlock-free.

Procedure 1. *Deadlock avoidance procedure for FMSs modeled by CROPN.*

- *Step 1: Given a FMS, let us construct its corresponding CROPN model by using the Procedure proposed in [5].*
- *Step 2: Find all interactive subnets in the CROPN [5].*
- *Step 3: For every interactive subnet v^k in the net, let us determine the colored capacity $K(p_L)(b_J)(M)$ for $(p_L, b_J) \in C(v^k)(L, J)$ by using Algorithm 1.*
- *Step 4: Determine the colored capacity $K(p_L)(b_J)(M)$ for $(p_L, b_J) \in A(L, J) - C(v^k(1))(L, J) \cup C(v^k(2))(L, J), ..., C(v^k(l))(L, J)$ by using Algorithm 2.*

3   FMS Example

Consider a flexible manufacturing system (FMS) cell with three machines m_1, m_2, and m_3 processing two part types A and B. A-part has three operations: $m_1 \to m_2 \to m_3$. B-part has three operations: $m_3 \to m_2 \to m_1$. The capacity of resource m_1, m_2 and m_3 is $1, 1$ and 1, respectively.

To make the FMS cell deadlock-free, by Procedure 1, the first step is to construct its corresponding CROPN model as shown in Fig. 3, where the place p_0 represents the central storage with $K(p_0) = \infty$, the place p_x represents the machine m_x with $K(p_x) = 1$ for $x \in \{1, 2, 3\}$. With the modele, we have $C(t_i) = b_i$, where $i \in \{1, 2, ..., 8\}$. Hnece, we have $C(p_0) = \{b_2, b_8\}$, $C(p_1) = \{b_1, b_3\}$, $C(p_2) = \{b_4, b_5\}$, and $C(p_3) = \{b_6, b_7\}$.

The second step is to find all interactive subnets in the net. It is clear that there is only one interactive subnet $v^2 = \{p_1, t_3, p_2, t_5, p_3, t_6, t_4\}$ formed by two PPCs: $v_1 = \{p_1, t_3, p_2, t_4\}, v_2 = \{p_2, t_5, p_3, t_6\}$. Hence, we have the following result:

$$\sum_{C(p_1,z)} M(p_1)(b_z) = M(p_1)(b_3) + M(p_1)(b_1)$$

$$\sum_{C(p_2,z)} M(p_2)(b_z) = M(p_2)(b_4) + M(p_2)(b_5)$$

$$\sum_{C(p_3,z)} M(p_3)(b_z) = M(p_3)(b_6) + M(p_3)(b_7)$$

$$A(L, J) = \{(p_0, b_2), (p_0, b_8), (p_1, b_1), (p_1, b_3), (p_2, b_4),$$
$$(p_2, b_5), (p_3, b_6), (p_3, b_7)\}$$

$$C(v^2)(L, J) = \{(p_1, b_1), (p_1, b_3), (p_2, b_4),$$
$$(p_2, b_5), (p_3, b_6), (p_3, b_7)\}$$

In the third step, we first need to do the reachability analysis for the net by using the traditional execution rules Eq. (1) and Eq. (2). The reachability graph and reachable markings of the net are shown in Fig. 4 and Table 1, respectively. Note that, in Table 1, only markings $M(p)(b_i) > 0$ are presented. It is clear that markings $M_0 - M_3$, $M_5 - M_7$, $M_{10} - M_{12}$, and $M_{15} - M_{19}$ are legal markings, while markings M_4, M_8, and M_9 are impending deadlock markings, markings M_{13}, M_{14} are deadlock markings. Hence, we have $M_{FBM} = \{M_4, M_8, M_9, M_{13}, M_{14}\}$. Further, we have $M_{FBM} * *(v^2) = \{M_4(v^2), M_8(v^2), M_9(v^2)\}$, where $M_4(v^2) = (0, 1, 0, 0, 1, 0)^T$, $M_8(v^2) = (0, 0, 0, 1, 1, 0)^T$, and $M_9(v^2) = (0, 1, 1, 0, 0, 0)^T$. Let us rearrange the markings in $M_{FBM} **(v^2)$ as M_1, M_2, M_3, where $M_1 = M_4(v^2)$, $M_2 = M_8(v^2)$, and $M_3 = M_9(v^2)$. Hence, we have the following result:

$$D(v^2)(1, l, j) = \{(p_1, b_3), (p_3, b_6)\}$$
$$D(v^2)(2, l, j) = \{(p_2, b_5), (p_3, b_6)\}$$
$$D(v^2)(3, l, j) = \{(p_1, b_3), (p_2, b_4)\}$$
$$S_1(v^2) = S_2(v^2) = S_3(v^2) = 1$$
$$C(M_1) : M(p_1)(b_3) + M(p_3)(b_6) \leq 1$$
$$C(M_2) : M(p_2)(b_5) + M(p_3)(b_6) \leq 1$$
$$C(M_3) : M(p_1)(b_3) + M(p_2)(b_4) \leq 1$$
$$D(v^2)(A, l, j) = \{(p_1, b_3), (p_2, b_4), (p_2, b_5), (p_3, b_6)\}$$
$$ND(v^2)(L, J) = \{(p_1, b_1), (p_3, b_7)\}$$

With these results, we then determine the colored capacity function of each place in interactive subnets in the net by using the Algorithm 1 as follows:

$$K(p_1)(b_1)(M) = K(p_1) - M(p_1)(b_3) \tag{22}$$

$$
\begin{aligned}
K(p_1)(b_3)(M) = \min \{&1 - M(p_1)(b_3) - M(p_3)(b_6), \\
&1 - M(p_1)(b_3) - M(p_2)(b_4), \\
K(p_1) - M(p_1)(b_3) - M(p_1)&(b_1)\} + M(p_1)(b_3)
\end{aligned}
\tag{23}
$$

$$
\begin{aligned}
K(p_2)(b_4)(M) = \min \{&1 - M(p_1)(b_3) - M(p_2)(b_4), \\
K(p_2) - M(p_2)(b_4) - M(p_2)&(b_5)\} + M(p_2)(b_4)
\end{aligned}
\tag{24}
$$

$$
\begin{aligned}
K(p_2)(b_5)(M) = \min \{&1 - M(p_2)(b_5) - M(p_3)(b_6), \\
K(p_2) - M(p_2)(b_4) - M(p_2)&(b_5)\} + M(p_2)(b_5)
\end{aligned}
\tag{25}
$$

$$
\begin{aligned}
K(p_3)(b_6)(M) = \min \{&1 - M(p_1)(b_3) - M(p_3)(b_6), \\
&1 - M(p_2)(b_5) - M(p_3)(b_6), \\
K(p_3) - M(p_3)(b_6) - M(p_3)&(b_7)\} + M(p_3)(b_6)
\end{aligned}
\tag{26}
$$

$$K(p_3)(b_7)(M) = K(p_3) - M(p_3)(b_6) \tag{27}$$

Finally, by step 4 in Procedure 1, we have

$$K(p_0)(b_2)(M) = \infty \tag{28}$$

$$K(p_0)(b_8)(M) = \infty \tag{29}$$

With the colored capacity function of each place in the net determined by Procedure 1, the non-colored capacity CROPN net becomes the colored capacity CROPN net. Let us do the reachability analysis for it by using the new execution rule for verifying whether the FMS cell modeled by colored capacity net is made deadlock-free. The result is shown in Fig. 3, where markings $\{M_4, M_8, M_9, M_{13}, M_{14}\}$ in Fig. 4 and Table 1 are forbidden and no legal marking is forbidden, imply that the FMS cell modeled by colored capacity CROPN net is made deadlock-free.

Let us analyze the mechanism of making the colored capacity net deadlock-free by using the new execution rule. We consider the following three cases.

1. If we use the non-colored capacity net (traditional CROPN) and the traditional execution rule, then By Fig. 4, the transitions t_8 and t_2 can fire at marking M_1 and M_2, respectively. This means that the FMS cell modeled by non-colored capacity net can enter bad state M_4 from state M_1 or M_2. If the colored capacity are introduced into the net, then the net becomes colored capacity net. At marking M_1 (Table 1), by Eq. (25), we have $K(p_3)(b_6)(M_1) = 0$. This implies that $K(p_3)(b_6)(M_1) = 0 < M(p_3)(b_6) - I(p_3, t_8)(b_8) + o(p_3, t_8)(b_8) = 1$. Hence, by Definition ??, the transition t_8 can not fire at marking M_1. Similarly, at marking M_2, we have $K(p_1)(b_3)(M_2) = 0$ and $K(p_1)(b_3)(M_2) = 0 < M(p_1)(b_3) - I(p_1, t_2)(b_2) + o(p_1, t_2)(b_2) = 1$. Hence, the transition t_2 can not fire at marking M_2. This means that the FMS cell modeled by colored capacity net cannot enter bad marking M_4 from state M_1 or M_2.

2. At marking M_3, we have $K(p_3)(b_6)(M_3) = 0$ and $K(p_3)(b_6)(M_3) = 0 < M(p_3)(b_6) - I(p_3, t_8)(b_8) + o(p_3, t_8)(b_8) = 1$. Hence, the transition t_8 can not fire at marking M_3. At marking M_5, we have $K(p_1)(b_3)(M_5) = 0$ and $K(p_1)(b_3)(M_5) = 0 < M(p_1)(b_3) - I(p_1, t_2)(b_2) + o(p_1, t_2)(b_2) = 1$. Hence, the transition t_2 can not fire at marking M_5. Therefore, the FMS cell modeled by the colored capacity net cannot enter bad markings M_8 and M_9 from state M_3 and M_5 while the traditional non-colored capacity net can as shown in Fig. 4.

3. At marking M_6, we have $K(p_3)(b_6)(M_6) = 0$ and $K(p_3)(b_6)(M_6) = 0 < M(p_3)(b_6) - I(p_3, t_8)(b_8) + o(p_3, t_8)(b_8) = 1$. Hence, the transition t_8 can not fire at marking M_6. At marking M_{11}, we have $K(p_1)(b_3)(M_{11}) = 0$ and $K(p_1)(b_3)(M_{11}) = 0 < M(p_1)(b_3) - I(p_1, t_2)(b_2) + o(p_1, t_2)(b_2) = 1$. Hence, the transition t_2 can not fire at marking M_{11}. Therefore, the FMS cell modeled by colored capacity net cannot enter markings M_{13} and M_{14} from markings M_6 and M_{11}, while the traditional non-colored capacity net can as shown in Fig. 4.

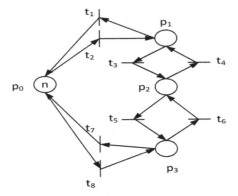

Fig. 2. The CROPN for FMS example

In terms of deadlock prevention, to make the FMS cell deadlock-free, compared the work in [5], the control policies of both proposed methods are equal. However, as shown in Table 2, 1 control place and 8 arcs are added to the non-colored capacity CROPN net for this example in [5], while no control place and arc needed to be added by using the method proposed in this paper. Thus, the method proposed in this paper is structurally simpler than the work in [5].

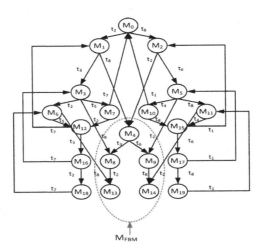

Fig. 3. Reachability graph of the non-colored capacity CROPN in Fig. 3

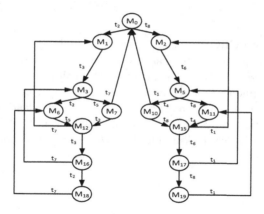

Fig. 4. Reachability graph of the colored capacity CROPN in Fig. 3

Table 1. Reachable markings of the CROPN in Fig. 3

Marking	
M_0	$M_0(p_0)(b_2) = n$ and $M_0(p_0)(b_8) = n$
M_1	$M_1(p_0)(b_2) = n$, $M_1(p_0)(b_8) = n$ and $M_1(p_1)(b_3) = 1$
M_2	$M_2(p_0)(b_2) = n$, $M_2(p_0)(b_8) = n$ and $M_2(p_3)(b_6) = 1$
M_3	$M_3(p_0)(b_2) = n$, $M_3(p_0)(b_8) = n$ and $M_3(p_2)(b_5) = 1$
M_4	$M_4(p_0)(b_2) = n$, $M_4(p_0)(b_8) = n$, $M_4(p_1)(b_3) = 1$ and $M_4(p_3)(b_6) = 1$
M_5	$M_5(p_0)(b_2) = n$, $M_5(p_0)(b_8) = n$ and $M_5(p_2)(b_4) = 1$
M_6	$M_6(p_0)(b_2) = n$, $M_6(p_0)(b_8) = n$, $M_6(p_1)(b_3) = 1$ and $M_6(p_2)(b_5) = 1$
M_7	$M_7(p_0)(b_2) = n$, $M_7(p_0)(b_8) = n$ and $M_7(p_3)(b_7) = 1$
M_8	$M_8(p_0)(b_2) = n$, $M_8(p_0)(b_8) = n$, $M_8(p_2)(b_5) = 1$ and $M_8(p_3)(b_6) = 1$
M_9	$M_9(p_0)(b_2) = n$, $M_9(p_0)(b_8) = n$, $M_9(p_1)(b_3) = 1$ and $M_9(p_2)(b_4) = 1$
M_{10}	$M_{10}(p_0)(b_2) = n$, $M_{10}(p_0)(b_8) = n$ and $M_{10}(p_1)(b_1) = 1$
M_{11}	$M_{11}(p_0)(b_2) = n$, $M_{11}(p_0)(b_8) = n$, $M_{11}(p_2)(b_4) = 1$ and $M_{11}(p_3)(b_6) = 1$
M_{12}	$M_{12}(p_0)(b_2) = n$, $M_{12}(p_0)(b_8) = n$, $M_{12}(p_1)(b_3) = 1$ and $M_{12}(p_3)(b_7) = 1$
M_{13}	$M_{13}(p_0)(b_2) = n$, $M_{13}(p_0)(b_8) = n$, $M_{13}(p_1)(b_3) = 1$, $M_{13}(p_2)(b_5) = 1$ and $M_{13}(p_3)(b_6) = 1$
M_{14}	$M_{14}(p_0)(b_2) = n$, $M_{14}(p_0)(b_8) = n$, $M_{14}(p_1)(b_3) = 1$, $M_{14}(p_2)(b_4) = 1$ and $M_{14}(p_3)(b_6) = 1$
M_{15}	$M_{15}(p_0)(b_2) = n$, $M_{15}(p_0)(b_8) = n$, $M_{15}(p_1)(b_1) = 1$ and $M_{15}(p_3)(b_6) = 1$
M_{16}	$M_{16}(p_0)(b_2) = n$, $M_{16}(p_0)(b_8) = n$, $M_{16}(p_2)(b_5) = 1$ and $M_{16}(p_3)(b_7) = 1$
M_{17}	$M_{17}(p_0)(b_2) = n$, $M_{17}(p_0)(b_8) = n$, $M_{17}(p_1)(b_1) = 1$ and $M_{17}(p_2)(b_4) = 1$
M_{18}	$M_{18}(p_0)(b_2) = n$, $M_{18}(p_0)(b_8) = n$, $M_{18}(p_1)(b_3) = 1$, $M_{18}(p_2)(b_5) = 1$ and $M_{18}(p_3)(b_7) = 1$
M_{19}	$M_{19}(p_0)(b_2) = n$, $M_{19}(p_0)(b_8) = n$, $M_{19}(p_1)(b_1) = 1$, $M_{19}(p_2)(b_4) = 1$ and $M_{19}(p_3)(b_6) = 1$

Table 2. Comparison between the supervisors obtained based on CROPN and colored capacity CPOPN.

The number of control places of the net	The number of arcs of the net
CROPN: 1	8
Colored capacity CROPN:0	0

4   Conclusion

In this paper, a novel CROPN concept called colored capacity is proposed. We define the colored capacity of the place, which represents the biggest number of tokens with the same color that the place can retain simultaneously. A method is presented to determine the colored capacity for each place in CROPN by simple calculation. Then all deadlock markings and impending deadlock markings in a flexible manufacturing systems (FMS) modeled by CROPN are forbidden, therefore, this FMS is deadlock-free and we do not need to add control places to the net.

References

1. Chen, Y., Liu, G.: Computation of minimal siphons in Petri nets by using binary decision diagrams. ACM **12**(1), 1–15 (2013)
2. Li, X., Liu, G., Li, Z., Wu, N., Barkaoui, K.: Elementary siphon-based robust control for automated manufacturing systems with multiple unreliable resources. IEEE Access **12**(1), 1–15 (2013)
3. Liu, G., Yun, C., Daniel, Y., Uzam, M.: A merging method for the siphon-based FMS maximally permissive controllers with simpler structures. IMA J. Math. Control Inf. **31**(4), 551–573 (2014)
4. Liu, G., Chao, D., Yu, F.: Control policy for a subclass of Petri nets without reachability analysis. IET Control Theory Appl. **7**(8), 1131–1141 (2013)
5. Chen, H., Wu, N.Q., Zhou, M.C.: A novel method for deadlock prevention of AMS by using resource-oriented Petri nets. Inf. Sci. **363**, 178–189 (2016)
6. Feng, Y., Xing, K., Zhou, M., Liu, H.: Liveness analysis and deadlock control for automated manufacturing systems with multiple resource requirements. IEEE Trans. Syst. Man Cybern. Syst. **50**, 1–14 (2017)
7. Li, Z.W., Wu, N.Q., Zhou, M.C.: Deadlock control of automated manufacturing systems based on Petri nets—A literature review. IEEE Trans. Syst. Man Cybern. Part C **42**(4), 437–462 (2012)

The Modeling and Simulation on SRM Drive System Using Variable-Proportional-Desaturation PI Regulator

Wei ZiHan, Zhao Mi$^{(\boxtimes)}$, and Fu ChangXin

College of Mechanical and Electrical Engineering, Shihezi University,
Shihezi 832003, China
WeiZiHan615@163.com, zhaomi530@163.com, fcx623623@163.com

Abstract. In this article, a novel variable-proportional-desaturation PI (VPDPI) speed regulator is designed to improve the dynamic performance and anti-disturbance of the drive system. Firstly, an equivalent model of three closed-loop control system for switched reluctance motor (SRM) is established based on the direct torque control (DTC) strategy. Secondly, the control parameters of VPDPI speed regulator are adjusted in terms of its equivalent mathematical model and the system response. Finally, the effectiveness of the proposed method is verified by MATLAB/Simulink platform under the different operation conditions. The simulation results indicate that the output of the controlled system has the shorter settling time and avoiding the overshoot. Consequently, the dynamic performance and its anti-disturbance ability of the system are both improved.

Keywords: Switched reluctance motor · PI regulator · Variable-proportional-desaturation · Modelling and simulation

1 Introduction

Electrical motors as the most promising electrical equipment exists in a huge scope of the proper control and high-energy efficiency in electric drives systems [1–3]. In order to obtain highly efficient operations, the variable speed control of the motor is most vital for energy-efficient and environmentally friendly technology, especially in the high-speed drives environment [4,5]. Obviously, the required output power of the drive systems would increase as the speed rises. However, not all motors are suitable for operating on the high speed in terms of their mechanical structure.

By contrast, switched reluctance motors (SRMs) provides the double salient-pole structure without any winding in rotor, whose magnetic circuit enables an

Supported by the Natural Science Foundation of China under Grant No. 61563045.

B. Ben Hedia et al. (Eds.): VECoS 2020, LNCS 12519, pp. 41–53, 2020.
https://doi.org/10.1007/978-3-030-65955-4_4

operation with very high speed [6, 7]. However, its salient-pole structure leads to the seriously nonlinear problem of the magnetic circuit, and further resulting in the flux linkage saturation phenomenon of the SRM model. Therefore, the main purpose of the precise control on electric drives system with SRM are twofold. One is to weaken essentially nonlinear characteristics of SRM. The other is to solve the wind-up phenomenon both in model features and controller design. As the former, the model predictive control has shown the clear advantage and great potential [8–10]. However, it requires more additional time to verify the correctness of the resulted model when the parameters is slow time-varying. Moreover, the fractional-order differential equations have gained significant development on controlling nonlinear objects [11]. In a practical system, the difficulty of complex system modeling is reduced and the accuracy of the system is improved by combining fractional order theory, but the fractional order differential equation is solved with too much computation and reserves [12]. For the traditional speed control, the proportional integral (PI) controller is conventional and feasible, which can also closely merge many intelligent control methods under different control requirements, such as artificial neural network algorithms [13], fuzzy algorithms [14] and particle swarm optimization (PSO) algorithms [15].

Nevertheless, the wind-up phenomenon accompanies more or less in the part of integral component and even the related simulator, which limits the function of maximum output in controller. Undoubtedly, this issue is awful when the obtained results are influenced by the integral saturation. Therefore, its harmful effect should be treated timely. To achieve it, the anti-wind-up precautions have been firstly introduced [16–18]. In order to reduce the impact of integral saturation, the anti-saturation PI speed controller are proposed based on the analysis of integral saturation in [19] and [20], and the dynamic performance of the speed regulation system is improved.

Generally, many approaches about anti-saturation have focused on how to suppress the saturation effect. With the load and related parameters quick changed, the saturation states cannot be avoided effectively. Once the resistability of the controller is failed, the drives system would be vibrational and even uncontrollable. Therefore, if the integral saturation effect of the speed regulator can be thoroughly eliminated, the system dynamic response process and even its steady-state performance would be improved greatly. Motivated by the in-depth analysis of integral saturation phenomenon, this paper presents a novel control method based on the variable-proportional-desaturation PI (VPDPI) regulator for 60kW SRM drives system. Furthermore, the design and the main parameters adjustment methods of the VPDPI speed controller is proposed explicitly. The simulation comparison among the classical PI, anti-saturation PI and VPDPI is carried out, which covers the dynamic responsiveness and interference immunity from the aspect of rise time, steady-state error, overshoot and settling time under different disturbance conditions.

2   The Overall Design of the SRM Drive System

The SRM drive system based on the direct torque control (DTC) strategy [21] is mainly composed of speed regulator, DTC controller, SRM and position detection module, whose structural diagram of the whole drive system is described in Fig. 1. Where the speed regulator is one of the most core modules in the controlled system, which can ensure the controllability of the output. The speed error signal $\triangle n$ and the system reference torque T^* as the input and output of the module, respectively. In addition, T^* as well as the actual torque T and feedback signals of the flux linkage ψ are processed using the DTC controller. And the module generates trigger signals to control the switched reluctance motor, and ensure the system torque and flux linkage smoothly changing [21]. Moreover, the speed, torque and flux linkage of switched reluctance motor can be measured by the position detection module.

In order to achieve the precise control of the whole drive system, particularly in the complement of speed controller, the control specification and its related parameters of the controller should be optimized based on the equivalent math model except the speed regulator. In this paper, the synthesization of the improved PI controller and its corresponding parameters is formalized in the following section.

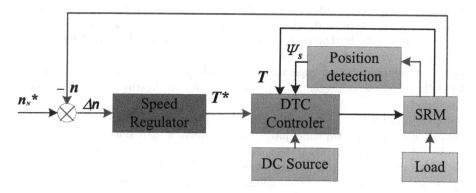

Fig. 1. The structure block diagram of SRM drive system.

3   The Optimized Design of Speed Regulator for SRM Drive System

3.1   Integral Saturation of PI Regulator

As is well known, the traditional PI regulator is linearly composed of the proportion element and the integration element, whose dynamical differential equation is shown as Eq. (1):

$$u(t) = K_p e + K_i \int edt \qquad (1)$$

Where K_p (resp. K_i) is the proportional coefficient (resp. integral coefficient).

In real controlled system, PI regulator can realize the rapid response and zero steady-state error by properly proportional unit and integrator part. However, it is worth to noting that if the integral accumulation time of PI regulator is too long, the output value of PI regulator will reach the upper limit and remain unchanged, which is called integral saturation phenomenon. Obviously, the saturation phenomenon can lead to the fact that the input and output of the regulator no longer meet the control specification and cause worse results, and further resulting in the slower dynamic response and the larger overshoot. In order to weaken the influence of integral saturation phenomenon in the SRM controlled system, anti-saturation PI (ASPI) regulator has been proposed in the literatures. The deviation signal of speed Δn is usually utilized as the control input to select whether the integral unit is applied to the system or not. Although the anti-saturation PI regulator can improve the dynamic performance partially by properly switching integral element, it does not fundamentally eliminate the integral saturation effect. To achieve it, the concept of the desaturation is introduced for the speed regulator such that its control effect is more perfect.

3.2   The Design of Variable-proportional-desaturation Speed PI Regulator

In this subsection, an optimized method based on desaturated PI (DPI for short) regulator is firstly proposed, whose structural diagram is subsequently depicted as shown in Fig. 2. Furthermore, the variable-proportional control method is combined to further improve the dynamic response of the regulator. Consequently, the computation algorithm of variable-proportional-desaturation PI (VPDPI) regulator is presented, and its control specification is also discussed in detailed.

According to Fig. 2, c and φ are the threshold values of the variable-proportional and desaturation parts, respectively. γ is defined as the reverse compensation coefficient. Moreover, K_{p1} and K_i denote the basic proportional coefficient and the integral coefficient of the speed regulator, respectively. And K_{p2} is the enhanced proportional coefficient of VPDPI regulator.

It can be seen that if the integral saturation occurs, desaturated PI regulator makes integral positive accumulates amount to reduce rapidly by introducing $\gamma(\gamma < 0)$. Therefore,the regulator is always in an unsaturated state, which obviously weakens the influence of saturation phenomenon on the control performance of the regulator.

Note that if the deviation signal of speed e exceeds the set threshold c, the regulator can control the dynamic response time of the system by changing the proportional coefficient while ensuring the steady-state performance of the system, whose corresponding control specification are explained in different cases as follows:

* **Unsaturated state** ($e > c$): The regulator utilizes the traditional PI control mode, and the proportional coefficient equals to the basic coefficient K_{p1}, the regulator outputs $u(t) = K_{p1}e + K_i \int edt$.

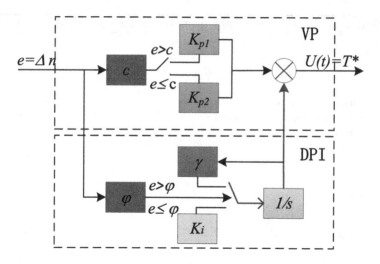

Fig. 2. The structure diagram of VPDPI regulator control algorithm.

* **Critical saturation state** ($\varphi < e \leq c$): The regulator is still in the traditional PI control mode, but the integral is about to be saturated. In order to better improve the corresponding dynamic performance of the system, the variable-proportional regulation effect starts to strengthen at this stage, with the proportional coefficient of K_{p2} and the regulator output $u(t) = K_{p2}e + K_i \int edt$.
* **Saturation state** ($e \leq \varphi$): The regulator introduces desaturation compensation coefficient γ. At the same time, in order to improve the effect of desaturation compensation coefficient γ on the dynamic response performance of the system, the variable-proportional regulation effect is continuously strengthened, with the proportional coefficient of K_{p2} and the regulator output $u(t) = K_{p2}e + \gamma K_i \int edt$.

Secondly, the control conception of variable proportion coefficient is converged to continue perfect the response speed of the SRM drive system while maintaining its stability performance. Next, the implementation process of VPDPI regulator is formalized as shown in Algorithm 1.

According to Algorithm 1, the general formula of VPDPI regulator can be obtained as shown in Eq. (2):

$$u(t) = [\rho(K_{p2} - K_{p1}) + K_{p1}]e + \gamma K_i \int edt \qquad (2)$$

In Eq. (2), let ρ be the selection coefficient, and if $e > c$ then $\rho = 0$, otherwise $\rho = 1$. In conclusion, the VPDPI regulator can enhance the proportional control effect in advance and improve the response speed of the regulator before the integral saturation phenomenon occurs by properly adjusting the proportional selection coefficient ρ and saturation compensation coefficient γ. When the satu-

---

Algorithm 1. The computation algorithm of $VPDPI$ for SRM drive system

---
Input: The given speed of SRM drive system n_N^*
Output: The referenced torque T_e^*

1: begin;
2: Setting n_N^*, c, φ
3: Identify and Sampling n at t time.
4: Calculating the error value of speed $e = n_N^* - n$
5: **if** $e \leq c$ **then**
6:     $\rho := 1$
7:       $K_p = K_{p2}$
8: **else**
9:     $\rho := 0$
10:       $K_p = K_{p1}$
11: **end if**
12: **if** $e \leq \varphi$ **then**
13:     $\gamma := 1$
14: **else**
15:     $\gamma := -2$
16: **end if**
17: Calculate $u(t) = [\rho(K_{p2} - K_{p1}) + K_{p1}]e + \gamma K_i \int edt$. Actually, $u(t)$ is the referenced torque T_e^* at t moment.
18: Output the referenced torque T_e^{*}.
19: end.

---

ration phenomenon occurs, it can desaturate the system in time and reduce the steady state deviation of the system.

3.3 VPDPI Regulator Parameter Adjusting

Assuming that hysteresis loss, eddy current loss and coupling relationship between motor windings are all ignored from the perspective of energy conversion, the input energy of SRM is mainly consumed through kinetic energy and internal energy storage after internal conversion [22]. Let W_i, W_f and W_o denote the input energy, internal energy storage and kinetic energy, respectively. Based on the energy conversation, the electromechanical equation is established as shown in Eq. (3):

$$W_o = W_i - W_f = \int T_e d\theta = \int L(\theta, i)idi \tag{3}$$

Where T_e is the electric torque of the motor, L is the winding inductance, and i is phase current. According to Eq. (3), when the current is given and the inductance is equivalent to a linear model [23,24], the torque can be simplified to an univariate function of position angle, which can be derived as follows:

$$\begin{cases} L(\theta) = \psi_m(a + b\sin(N_r\theta)) \\ T_e = \frac{1}{2}i^2\frac{dL}{d\theta} = \frac{1}{2}i^2(bN_r\psi_m\cos(N_r\theta)) \end{cases} \tag{4}$$

In Eq. (4), ψ_m and θ denote the flux saturation value and position angle, respectively. In addition, a and b are both undetermined coefficients, which can be calculated through obtaining the motor parameters. N_r represent the number of rotor poles.

If the viscosity coefficient of the system is ignored, its motion equation can be derived in terms of the input-output relationship of the SRM system.

$$\frac{d\omega}{dt} = \frac{1}{J}(T_e - T_L) = \frac{1}{J}(\frac{1}{2}i^2 bN_r\psi_m \cos(N_r\theta) - T_L) \tag{5}$$

Where $d\omega$ is the motor rotation angular velocity, J is instantaneous value of inertia, T_L is the load torque.

According to the Kirchhoff's voltage law (KVL) equation satisfied by the electrical system and the relation between flux linkage and position angle and current, the voltage equation of SRM is deduced as shown in Eq. (6):

$$U - iR_s = L\frac{di}{dt} + i\frac{dL}{d\theta}\frac{d\theta}{dt} \tag{6}$$

Where U is DC voltage and R_s is stator resistance. By combining Eq. (4) with Eq. (6), the relationship between single-phase winding current and time can be obtained as shown in Eq. (7):

$$\frac{di}{dt} = \frac{U - iR_s - i\omega bN_r\psi_m \cos(N_r\theta)}{\psi_m(a + b\sin(N_r\theta))} \tag{7}$$

When the speed is constant, Eq. (7) is simplified as follows:

$$i = \frac{U}{\omega bN_r\psi_m \frac{\int_0^T \cos(N_r\theta)dt}{T}} \tag{8}$$

Next, Eq. (8) is substituted into Eq. (5):

$$\frac{d\omega}{dt} = \frac{1}{J}(\frac{1}{2}\frac{U^2}{\omega^2 bN_r\psi_m \frac{\int_0^T \cos(N_r\theta)dt}{T}} - T_L) \tag{9}$$

When the motor reaches the steady state value, the speed variation rate tends to zero, and its average speed $\overline{\omega}$ is written as follows:

$$\overline{\omega} = \sqrt{\frac{U^2}{2T_L bN_r\psi_m \frac{\int_0^T \cos(N_r\theta)dt}{T}}} \tag{10}$$

Equation (9) is discretized at the average speed $\overline{\omega}$ and the average control voltage $\overline{U}$, and the linear equation of the rate of change of speed is obtained as follows:

$$\frac{d\omega}{dt} = \frac{d\alpha}{dU}|_{\overline{\omega},\overline{U}}\Delta U + \frac{d\alpha}{d\omega}|_{\overline{\omega},\overline{U}}\Delta\omega \tag{11}$$

After the Laplace transform, the transfer function of control voltage and speed is derived as shown in Eq. (12):

$$
\frac{\Delta\omega(s)}{\Delta u(s)} = \frac{\frac{2T_L}{J\overline{U}}}{s + \frac{2T_L}{J\overline{U}}\sqrt{2T_L b N_r \psi_m \frac{\int_0^T \cos(N_r\theta)dt}{T}}}
$$

$$
= \frac{\frac{\overline{\omega}}{\overline{u}}}{J b N_r \psi_m \frac{\int_0^T \cos(N_r\theta)dt}{T} \frac{\overline{\omega^3}}{\overline{u^2}} s + 1} = \frac{k_p}{sT_p + 1}
\tag{12}
$$

According to Algorithm 1, the transfer function of the VPDPI controller is computed:

$$
W_{pi}(s) = \frac{\Delta u}{\Delta n} = \rho(K_{p2} - K_{p1}) + K_{p1} + \gamma\frac{K_i}{s}
\tag{13}
$$

Furthermore, on the basis of the equivalent first-order transfer function of the motor in Eq. (12), in order to make the output closely track the given speed, the relationship between the SRM motor and the regulator can be established. Hence, the main parameters can be obtained in Eq. (14):

$$
\begin{cases}
K_{p1} = K_a\frac{\overline{u}}{\overline{\omega}} \\
K_{p2} = K_b\frac{\overline{u}}{\overline{\omega}} \\
K_i = K_c(J b N_r \psi_m \cos(N_r\theta)\frac{\overline{u^3}}{\overline{\omega^2}})^{-1}
\end{cases}
\tag{14}
$$

In Eq. (14), K_a, K_b and K_c are all system correction coefficients.

4 The Simulation Analysis and Performace Analysis of SRM Drive System

In order to verify the control performance of VPDPI regulator, the simulation model of the 6/4 SRM drive system is constructed in MATLAB/Simulink platform, which is shown in Fig. 3, whose main parameters are also depicted in Table 1.

Moreover, the control performance of SRM drive system is further analyzed through comparing three different regulators among the traditional PI, anti-saturation PI and the proposed VPDPI method under variable operation conditions.

4.1 The Performance Analysis of the Drive System Under Rated Speed with No-Load

Assuming that the SRM drive system runs at rated speed with no-load, the waveform of the speed control effect of the three PI regulators is shown in Fig. 4, whose zooming-in waveform of the steady-state value from 0.05 s to 0.25 s is also given in Fig. 5.

According to the simulation results of the SRM drive system, the performance criteria of three regulators under rated speed with no load are shown in Table 2.

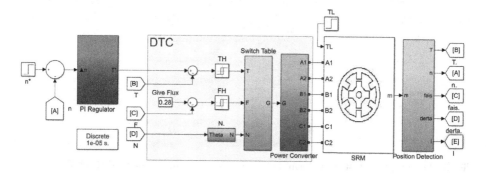

Fig. 3. The simulation model of SRM drive system.

Table 1. The main parameters of SRM drive system model.

Parameters (symbol)	Value	Parameters (symbol)	Value
Rated power (P_N)	60 (KW)	Aligned inductance (L_{min})	0.67 (mH)
Rated speed (n_N^*)	1000 (r/min)	Unaligned inductance (L_{max})	23.62 (mH)
Rated flux link (Ψ^*)	0.28 (Wb)	Saturation inductance (Ψ_m)	0.15 (mH)
DC voltage (DC)	240 (V)	Stator resistance (R_s)	0.05 (Ω)
Inertia (J)	0.05 ($Kg \cdot m^2$)	Friction (f)	0.02 ($N \cdot m \cdot s$)
γ	-2	K_{p1}	0.1
K_{p2}	1	K_i	0.1
c	230	φ	100

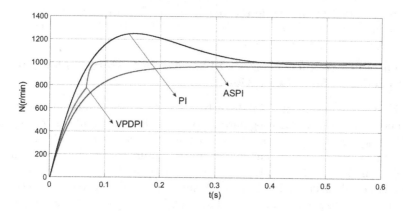

Fig. 4. The waveform of the speed control effect of three PI regulators under constant speed and no load.

From Table 2, it can be concluded that the traditional PI regulator can ultimately reach the steady-state without any error. However, the integral saturation phenomenon leads to the longer adjusting-time and the larger overshoot, the dynamic performance should be greatly improved. By contrast, the ASPI regulator can resist the saturation phenomenon to some extent, which greatly avoids the system overshoot. However, the system response would become slow

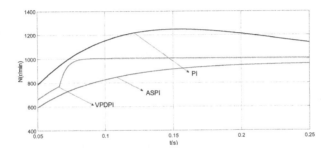

Fig. 5. The amplification waveform diagram of speed control effect of three PI regulators under rated speed with no load.

Table 2. The performance comparison of three PI regulator under rated speed with no-load.

Speed regulators	Rise time (s)	Peak time (s)	Settling time (s)	Overshoot (%)	Steady-state error (r/min)
PI	0.08	0.16	0.35	25	0
ASPI	0.525	0.525	0.203	0	32
VPDPI	0.09	0.09	0.077	0	0

as the saturation status aggravates. Most of all, the controlled system exists the steady-state error. In contrast, the results indicate that the proposed VPDPI regulator can realize the moderate requirements for static errorless, fast and non-overshoot, which ensures the precise control for SRM drive system.

4.2 The Performance Analysis of the Drive System Under Variable Speed and Load Operation

When the system has no external disturbance, the system simulation has verified the superiority of the VPDPI regulator. However, the speed and load of SRM are variable timely in the actual condition. Therefore, the feasibility and effectiveness of the controller under different disturbances should be demonstrated through the digital simulation. Firstly, the dynamic tracking performance of three regulators under speed variation is analyzed. Assuming that the speed sudden changes from 1000 r/min to 1200 r/min at t = 0.6 s. Furthermore, in order to compare the immunity performance of three regulators under load disturbance, assuming that the load changes from 0 N · m to 10 N · m at $t = 1$ s. Their comparison waveform of speed controller under variable speed and load conditions is shown in Fig. 6.

To make a more intuitive comparison of the three regulators performance during in speed and load saltation, whose zooming-in waveform of the steady-state value under variable speed(from 0.6 s to 0.8 s) and variable load (1 s to 1.2 s) are also given in Fig. 7(a) and (b), respectively.

According to the simulation results of the SRM drive system, the dynamic performance and the steady-state performance criteria of three regulators under variable speed operation are presented in Table 3.

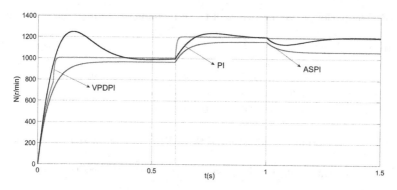

Fig. 6. The waveform of speed control effect of three PI regulators under variable speed $(n = 1200 \, \text{r/min})$ and load $(T_L = 10 \, N \cdot m)$.

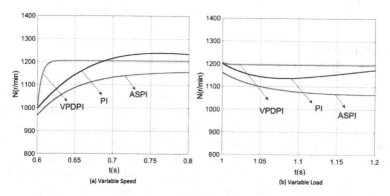

Fig. 7. The magnifies waveform of speed control effect of three PI regulators at variable speed.

After that, the anti-disturbance performance criteria of SRM drive system under variable load are also discussed in Table 4 through starkly compare the load disturbance resistance of three regulators.

According to the aforementioned results, a huge overshoot happen in traditional PI regulator and a lower response speed and steady-state error occur in ASPI regulator. However, the VPDPI regulator under variable speed condition can instantly track to the steady-state value without overshoot and steady-state error. In a word, the dynamic performance and steady-state performance criteria of the VPDPI regulator are superior in the three regulators.

Secondly, when the system load is suddenly changed, the dynamic landing and recovery time of the VPDPI regulator is smaller than the other two regulators. Therefore, the VPDPI regulator has a reliable disturbance immunity under having a preferable following performance.

Table 3. The performance indicators of three PI regulators under variable speeds.

Speed regulators	Rise time (s)	Peak time (s)	Settling time (s)	Overshoot (%)	Steady-state error (r/min)
PI	0.688	0.750	0.650	3.3	0
ASPI	0.935	0.935	>1	0	29
VPDPI	0.62	0.62	0.608	0	0

Table 4. The performance indicators of three PI regulators under variable loads.

Speed regulators	Landing dynamic (r/min)	Recovery time (s)
PI	65	1.21
ASPI	99	>2
VPDPI	5	1

5 Conclusion

In order to better address the integral saturation problem occurred in drive system of PI controller, the variable-proportional-desaturation PI speed control for the 6/4 SRM is designed in this paper. The parameters setting of the proposed controller are also constructed from its mathematical model and the response of the given system. Finally, the simulating experiment for the whole drive system is carried out through the direct torque control strategy under various operating circumstances. According to the verification results and the detailed comparative analysis on three different speed regulators, the proposed PI controller is more proficient against speed variation and load torque disruption. Furthermore, the VPDPI regulator possesses superior dynamic and steady performance than other controllers, which nominate it for various velocity modulation systems.

References

1. Zhang, H., Zhao, W.Z.: Decoupling control of steering and driving system for in-wheel-motor-drive electric vehicle. Mech. Syst. Signal Process. **101**, 389–404 (2018)
2. Hu, J.J., Yang, Y., Jia, M.X., et al.: A novel energy optimization control strategy for electric drive system based on current angle. Appl. Sci. **10**(3778), 3778 (2020)
3. Xin, X.S., Shan, H.C., Yang, Y.: Control strategy for motor drive system of electric vehicle. IEEJ Trans. Electr. Electron. Eng. **11**(3), 374–383 (2016)
4. Chaoui, H., Khayamy, M., Okoye, O., et al.: Simplified speed control of permanent magnet synchronous motors using genetic algorithms. IEEE Trans. Power Electron. **34**(4), 3563–3574 (2019)
5. Hu, H.S., Wang, T.T., Zhao, S.Y., et al.: Speed control of brushless direct current motor using a genetic algorithm—optimized fuzzy proportional integral differential controller. Adv. Mech. Eng. **11**(11), 1–13 (2019)
6. Tang, Y., He, Y.J., Wang, F.X., Lee, D.-H., Ahn, J.-W., Kennel, R.: Back-EMF-based sensorless control system of hybrid SRM for high-speed operation. Adv. Mech. Eng. **112**(6), 867–873 (2018)

7. Kohei, A., Kan, A.: High speed SRM using vector control for electric vehicle. CES Trans. Electr. Mach. Syst. **4**(1), 61–68 (2020)
8. Mayne, D., Falugi, P.: Stabilizing conditions for model predictive control. Int. J. Robust Nonlinear Control **29**(4), 894–903 (2019)
9. Li, C.H., Wang, G.F., Li, Y.: An improved finite-state predictive torque control for switched reluctance motor drive. IET Electr. Power Appl. **12**(1), 144–151 (2018)
10. Xin, L., Shamsi, P.: Model predictive current control of switched reluctance motors with inductance auto-calibration. IEEE Trans. Ind. Electron. **63**(6), 3934–3941 (2016)
11. Abdolrahman, R., Behrouz, S.: A modified fractional-order unscented Kalman filter for nonlinear fractional-order systems. Circ. Syst. Signal Process. **37**(9), 3756–3784 (2018)
12. Pandey, S., Junghare, A.S., Dwivedi, P., et al.: A newborn hybrid anti-windup scheme for fractional order proportional integral controller. Electr. Eng. **43**(6), 3049–3063 (2018)
13. Li, C.H., Wang, G.F., Li, Y., et al.: Direct adaptive neural network control for switched reluctance motors with input saturation. IEEJ Trans. Electr. Electron. Eng. **13**(12), 1804–1814 (2018)
14. Ben, J.A., Chaker, N., Aloui, H., et al.: Adaptive fuzzy-PI control for speed's tracking in a leisure electrical vehicle. J. Electr. Syst. **12**(3), 475–489 (2016)
15. Kalai Selvi, R., Suja Mani Malar, R.: A bridgeless Luo converter based speed control of switched reluctance motor using Particle Swarm Optimization tuned proportional integral controller. Microprocess. Microsyst. **75**, 103039 (2020)
16. Stanislav, V., Bohumil, S.: Saturation in engineering simulation. IFAC Proc. Volumes **46**(17), 102–107 (2013)
17. Azar, A.T., Serrano, F.E.: Design and modeling of anti wind up PID controllers. In: Zhu, Q., Azar, A. (eds.) Complex System Modelling and Control Through Intelligent Soft Computations. Studies in Fuzziness and Soft Computing, vol. 319, pp. 1–44. Springer, Cham (2015). https://doi.org/10.1007/978-3-319-12883-2_1
18. Srikanth, M., Sreedhar, L., Suresh Kumar, M.V.: Digital PI controller using anti-wind-up mechanism for a speed controlled electric drive system. Int. J. Innovative Technol. Exploring Eng. **3**(1), 239–242 (2013)
19. Hoo, C.L., Chung, E.C.Y., Haris, S.M., Mohamed, N.A.N.: Steady-state integral proportional integral controller for PI motor speed controllers: a theoretical approach. ICIC Express Lett. **9**(6), 1777–1782 (2015)
20. Mohammad, R.S., Esmaeel, K., Siamak, G.: An anti-windup rate-varying integral control applied to electromechanical actuator. Proc. Inst. Mech. Eng. Part C J. Mech. Eng. Sci. **229**(4), 692–702 (2015)
21. Hafeezul, H., Halil, I.O.: FLC-DTC method for torque ripples minimization of 8/6 switched reluctance motors drive. J. Adv. Res. Electr. Eng. **4**(1), 51–55 (2020)
22. Memon, A.A., Shaikh, M.M.: Input data for mathematical modeling and numerical simulation of switched reluctance machines. Data in Brief **14**, 138–142 (2017)
23. Zeng, M.: Research on the torque mathematical model of 12/8 pole switched reluctance motor. Mech. Electron. Inf. **3**, 20–21 (2019)
24. Song, S.C., Qu, B.N., Song, J.C.: Application of variable gain PI controller in switched reluctance motor. Miniature Spec. Mot. **45**(8), 61–64, 67 (2017)

Vehicle Scheduling Problem in Terminals: A Review

Pei Wang[✉]

China Merchants Holdings (International) Information Technology Co., Ltd.,
Shenzhen, Guangdong 518054, People's Republic of China
wangpei@cmhk.com

Abstract. This paper presents a review on vehicle scheduling problem in terminal. First of all, the current status of vehicle scheduling in terminal is introduced. The introduction includes the main types of current terminals and the main operating machinery in those terminals. Then, three main issues in vehicle scheduling problems are discussed and clarified in this paper, they are fleet sizing problem, vehicle dispatching problem and path planning problem. Research on these issues is divided into multiple types, and the development of these studies is introduced in the article. At last, this article explores the advantages and disadvantages of these studies. By comparing various types of research, this article presents the future research directions of these issues.

Keywords: Vehicle scheduling problem · Terminal · Literature overview

1 Introduction

1.1 Background

The twenty-first century is the century of large volume of trade. Due to the expensive cost and limited transportation volume of air travel, railways and other modes of transport, oceans and rivers have become the busiest trading places. Therefore, ports and terminals play an increasingly important role in the modern international production, trade and transportation system. As we move into the era of automation, the automatic terminal has gradually become a new norm.

The development of the world's ports has roughly gone through three phases. In the first phase, ports serve as a pure transport center, providing ship docking, loading, unloading, transshipment and storage of goods. As industrial and commercial activities boomed, ports have become not only transport centers, but also service centers, which add value to the goods by providing extra service. Nowadays, ports have gone into the third phase, in which they serve as international logistics centers, integrating goods, technology, capital and information in a single place. Automatic ports and terminals arise at this historic moment. With the emergence of automated terminals, the vehicle scheduling problem has received increasing attention.

B. Ben Hedia et al. (Eds.): VECoS 2020, LNCS 12519, pp. 54–67, 2020.
https://doi.org/10.1007/978-3-030-65955-4_5

1.2   Introduction to Terminal Operations

In general, the seaport is composed of various types of terminals, which import and export containers, bulk cargoes and other types of goods. In every type of terminal, goods are transported by vehicles like truck, automated guided vehicle (AGV), etc. Every terminal has two interfaces: the quayside with loading and unloading of ships, and the landside where goods are loaded and unloaded on/off vehicles from/to yards. There is a horizontal transportation zone between the quayside and the landside, in which the transportation vehicle runs. The loading and unloading equipment of quayside is generally quay cranes (QC), bulk cargo ship unloader, etc. The loading and unloading equipment of landside is generally rail mounted gantry cranes (RMG), rubber tyred gantries (RTG) and overhead bridge cranes (OBC), etc. Horizontal transport vehicles are generally trucks, AGVs (Automated Guided Vehicles), etc (Figs. 1 and 2).

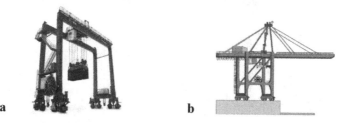

Fig. 1. a,b Stacking cranes and quay crane. a Stacking crane. b Quay crane.

Fig. 2. a,b Container truck and AGV. a Container truck. b AGV.

There are three types of terminal transportation operations: ship loading operations, ship unloading operations and yard transport operations. Transport vehicles travel differently in these operations (Table 1).

Table 1. Transportation operations.

Operation type	Transport vehicle action
Ship unloading operations	Transport vehicles transport goods from quayside to yard.
Ship loading operations	Transport vehicles transport goods from yard to quayside.
Yard transport operations	Transport vehicles transport goods from one area of the yard to another area of the yard.

2 Vehicle Scheduling Problem in Terminal

2.1 Introduction to Vehicle Scheduling Problem in Terminal

As shown in Fig. 3, the working area of vehicle is between yards and vessels, where vehicles transport containers. In order to make vessels leave port earlier, containers should be loaded and unloaded as quickly as possible. To achieve this goal, QC must work with maximum efficiency. According to the actual situation of the port, the efficiency of QC is limited by the efficiency of vehicles transporting containers. How to manage vehicle's transportation tasks determines the efficiency of QC.

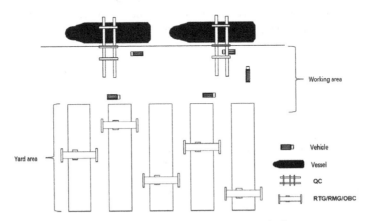

Fig. 3. Layout of Port Container Terminal

To maximize the efficiency of vehicle, two problems need to be solved, vehicle dispatching problem and path planning problem. Vehicle dispatching problem considers the optimal matching between transportation tasks and vehicles. The best vehicle dispatching algorithm can ensure QC's continuous operation and keep the vehicle's empty driving distance at a low level. Path planning problem considers which path the vehicle takes to complete the current task. A good path planning algorithm can find the lowest-cost path for vehicles.

To maximize the efficiency of QC, it seems plausible to increase the number of vehicles. However, this method will bring new problems — the cost of vehicles and the congestion in terminals will increase. So, another problem that needs to be solved is vehicle fleet sizing problem. Fleet sizing problem focuses on how many vehicles is the most suitable for transportation tasks.

2.2 Fleet Sizing Problem

The problem of vehicle fleet sizing refers to the analysis of the optimal number of vehicles according to the transportation demand in a certain operation scenario. In the seaport, trucks, AGVs or other vehicles are responsible for the transportation of containers and other goods. A small number of vehicles will lead to the delay of goods transportation and cause delays in the operation of the entire terminal. A large number of vehicles will increase vehicle waiting time and road congestion. Configuring the right number of vehicles is of great significance to the efficient operation of ports. In recent years, more and more scholars are studying the problem of determining the size of fleets.

As early as the 1970s, many scholars conducted researches on the problem of fleet sizing [1, 2]. With the increasing demand for trucks, AGVs, etc. At the port, more and more studies have been conducted on the size of the fleet in the port. Some scholars analyze the mathematical model to obtain the optimal fleet size, while others determine it through simulation. This section will classify and explain according to the different ways to solve fleet sizing problem.

Mathematical Model Analysis

Some research studies the problem of fleet size in a static environment. Koo et al. [3] solved the problem of the fleet operation planning for the static freight container transport system. In this plan, all transportation needs are pre-determined at the beginning of the planning phase. The best fleet planning model is used to determine the minimum number of vehicles required. And then in 2004 [4] they designed a procedure for a static environment in a seaport. The procedure is divided into two phases: in the first phase, the procedure determines the lower bound of the fleet size based on the shortest travel time. In the second phase, the tabu search algorithms are used to calculate the vehicle's route with the least number of vehicles.

Vis et al. [5] studied the problem of container terminal fleet size minimization within time-window [release time, due time] constraints. Vehicles can only transport containers from any time within the time window. The author developed a linear integer programming model to analyze the minimum fleet size, and used simulation to verify the analysis results. In their another study [6], the problem of the number of trucks in the port is solved using the minimum flow algorithm. In this model, the literature assumes that the waiting time of all cranes is 0, and the cranes' loading and unloading operation time is fixed, and the travel time of the trucks between any two fixed facilities is known. They abstracted each loading and unloading task into a node. If two tasks can be completed by a truck, the weight between the two nodes is set to 1, otherwise it is set to 0. And they determined the minimum fleet size by calculating the minimum flow between the start and end points.

Most of these studies have made some idealized assumptions, such as constant vehicle speed and zero shore bridge operating time. However, the actual operating environment will be disturbed by many random events which lead to the optimal fleet size being unsuitable for the real operating environment.

Simulation Analysis

For the determination of the fleet size, some studies have combined it with the researches of dispatching and routing to establish a mathematical model and analyze the optimal fleet size. Gobal and Kasilingam [7] uses a SIMUN-based simulation model to determine the size of the fleet. The authors proposed determining the number of vehicles based on the waiting time and idle time of vehicles and other machines. As the number of vehicles increases, the waiting time and idle time of the loading and unloading machines become shorter, but the idle time of the vehicles becomes longer. Wibisono et al. [8] believed that the most optimized fleet size configuration is to achieve the shortest operating time and the highest production capacity. They solved fleet sizing problem with multi objectives in using a simulation approach based on ARENA Software. Kuo-Hao et al. [9] studied the problem of vehicle fleet sizing in semiconductor manufacturing, and proposed a solution method called Simulation Sequential Element Model (SSM) to help determine the optimal vehicle fleet size while meeting time constraints reduce vehicle costs. The proposed method is to construct a series of meta-models in sequence and evaluate the quality of the resulting solution.

Many studies used data envelopment analysis (DEA) to analyze transportation process and then used simulation to determine the fleet size. Pjevcevic et al. [10] proposed a decision-making method based on DEA for determining effective container handling processes. They defined the performance indicators of the proposed container handling process and analyzed the impact of changing the number of vehicles. The value of the performance indicator is collected and calculated from the simulation. Fauadi et al. [11] proposed a vehicle demand estimation method based on discrete event simulation and DEA. They developed a transportation simulation with SIMU8 simulation software, based on the specified goals and performance indicators, a set of solutions can be obtained, and the decision maker can decide the necessary fleet size. Chang et al. [12] believed that vehicle fleet sizing problem is a big challenge when facing multiple objectives. They proposed a framework that integrates simulation optimization and data envelopment analysis techniques to determine the optimal vehicle fleet size under multiple objectives of automated materials handling systems.

Simulating the real working environment makes the research results more suitable for the real working environment. However, this method's versatility is weak, and it is often necessary to rebuild a new simulation environment when changing the scenarios.

2.3 Vehicle Dispatching Problem

In terms of assigning tasks to vehicles, there are many methods to solve them. These methods can be divided into 3 categories by the methods they used. They are classic rules strategy, heuristic strategy and machine learning methods.

Classic rule strategy
To deal with vehicle dispatching problem in terminal, many researches have been done. Samuel and Raff discussed vehicle dispatching problems in different cases. Vehicle dispatching problem in terminal was introduced in chapter 4 [13].

Chin-I Liul and P.A. Ioannou made a comparison for different dispatching rules in Automated Container Terminal [14]. Several simple rule dispatching strategies are involved in this article, they are Random Vehicle (RV), Shortest Travel Distance (STD), Longest Travel Distance (LTD) and Minimum Queue (MQ). B.Mahadevan and T.T.Narendran proposed a method to combine track layout design and vehicle dispatching [15]. Their simulation results proved the effectiveness of their method. Cui Wanan and Wei Bin's study also concerns the influence of distribution center when dispatching vehicles [16]. They proposed a mathematical modeling method to solve the vehicle scheduling problem between the distribution center and the port. To deal with congestion problem in a large fleet situation, Yong-Leong Cheng et al. proposed a network flow formulation [17]. According to simulation results, their method can outperform other methods in throughput. In container terminal, there are different types of vehicles. Martin Grunow et al. studied the multi-load vehicle dispatching problem in his article [18].

Dispatching strategy using fuzzy theory is one of those classic rule strategies, and it is also a research hotspot. Seiichi Kagaya et al. used fuzzy theory to group transportation tasks based on trip similarity [19]. Their method can be combined with different scheduling models. And the method permits the selection of different similarity criteria. Umashankar N and V N Karthik proposed a multi-criteria dispatching control method to dispatch AGVs [20]. In order to better merge multiple attributes, they used fuzzy theory to construct a two-layer decision model. In order to improve the effect of fuzzy control, lots of scholars combine fuzzy theory with other algorithms. For example, Xianhui Song et al. combine Ant Colony algorithm (ACA) with fuzzy theory [20]. Their research results have been applied to a port terminal in California, USA.

Heuristic strategy

Due to the complexity of the terminal environment, classic rule strategies are difficult to solve some situations in the terminal. So many scholars turned their attention to heuristic algorithms.

E. Kozan and P. Preston used genetic algorithm to schedule container transfer [21]. And his research indicated that genetic algorithm can be used to evaluate the result of different strategies. In Kap Hwan Kim and Jong Wook Bae's study, they proposed a mixed-integer programming model, and a heuristic algorithm was used to overcome the excessive computational time needed for solving the mathematical model [22]. To test the performance of their heuristic algorithm, a simulation study was conducted to compare the performance with other algorithms. Jeongmin Kim et al. proposed a Multi-objective optimization of dispatching strategies with a multi-objective evolutionary algorithm (MOEA) [23]. With their strategy, two objectives can be achieved. The first one is productivity of the terminal and the other one is the reduction of CO_2. N.M.Lee et al. proposed a new heuristic algorithm in their study [24]. This new heuristic algorithm is inspired by the human immune system. N.M.Lee et al. compared their new algorithm with Genetic Algorithm (GA), the results showed Immune Algorithm's superiority. Pyung-Hoi Koo proposed an algorithm aiming to improve vehicle turnover efficiency [25]. The heuristic is based on a network flow optimization and look-ahead concept. Fuzzy theory is widely used in the field of vehicle scheduling, Ferani E. Zulvia et al. combined fuzzy theory with a hybrid Ant Colony Optimization and Genetic Algorithm (HACOGA) [26]. They conducted simulation experiments, the results showed that their improved algorithm outperforms Ant Colony Optimization (AGO) and Genetic Algorithm.

One key indicator of performance in terminal is the efficiency of QC. But with classic strategies, due to the calculation time, it is hard to take QC's performance into account. With the advent of heuristic algorithms, more and more scholars have considered QC's performance in scheduling algorithms. The study of Yao Xing et al. investigated the synchronization scheduling problem between the automated guided vehicles and these new QCs [27]. And two heuristic algorithms were used to overcome the computation complexity and balance computation time and the quality of the solution. S.M. Homayouni et al. also did a research on scheduling AGVs with QCs [28]. And in his research, a hybrid genetic-heuristic algorithm was proposed, which was proved to be efficient. J.

Bose's study aimed to minimize the delay time of container transports [29]. An evolutionary algorithm was used to achieve that goal. J.W. Bae and K.H. Kim proposed a new form to dispatch vehicles [30]. In their study, vehicles are shared among multiple cranes. They adopted the concept of vehicle pool to complete their algorithm. And also, a heuristic algorithm was used to reduce computation time. Lijun Yue et al. take QC's performance, the stability of vessel and maximum endurance of vehicle into account [31]. And a genetic algorithm was used in their study.

Machine learning strategy

The advantages of heuristic strategy are its efficiency and simplicity. Due to the variability of the port environment, it is difficult for the heuristic strategy to keep up with changes in the terminal environment. In order to overcome this defect, scholars have tried to use machine learning technology to solve vehicle dispatching problem.

Bian Zhicheng et al. proposed a simulation-based vehicle dispatching strategy [32]. They used online calculation of trip time to make the solution more accurate. Their simulation result showed the superiority of the proposed method against other methods. In the study of Jae Kook Lim et al., a Q-learning technique was used to design guide-path network for automated guided vehicle [33]. A comparison between the proposed algorithm with Kim and Tanchoco's algorithm was conducted, the result showed their superiority to Kim's algorithm. A dispatching method combining reinforcement learning and simulation is proposed by Qingcheng Zeng et al. [34]. This method can include multiple devices to coordinate scheduling, including yard cranes and yard trailers. Jean-Yves Potvin et al. proposed a neural networks method to dispatch vehicles [35]. They believe that the flexibility of neural network methods will be important in vehicle dispatching problem. The study of Ri Choe et al. proves this view. In response to the changing environment of the terminal, they proposed an online preference learning method [36]. And lots of simulation experiments were conducted by them, all results showed their proposed method has better adaptability to changing environments.

2.4   Path Planning Problem

Path planning is an important branch in the field of truck dispatching problem. The path planning problem is to find an optimal path that has the minimum combined cost of truck moving from the initial location to the target location. The basic algorithm of path planning stays the same regardless of its application scenarios, whether in terminals or warehouses. In general, path planning methods can be divided into two categories: traditional methods and intelligent methods. Intelligent methods can make corresponding adjustments according to the environment while traditional methods cannot.

Traditional Methods

Graph search strategy A directed graph, consisting of vertices and edges, is built based on the environment of the actual workspace. The graph is usually abbreviated as G=<V, E>, where V denotes nodes (intersections) and E denotes path between two nodes. A virtual weight is always given to the edge, indicating the cost of choosing this particular road.

Once generating the directed graph, a lot of algorithms are invented and upgraded to solve the shortest path planning or k-shortest path planning problem. Dijkstra and A* algorithms are the most basic of its kind. S Wei et al. improved the A* algorithm's heuristic function to adjust to the factory path but did not consider the path inflection point and may cost a long search time [37]. R J Yang et al. upgraded A* algorithm by updating the edges' weight in real time [38]. T Zheng et al. used characteristics of jump point search to improve the node search mode and the search speed, and added the angle evaluation to the cost function of A* algorithm to find the path with the least inflection point [39]. S Sedighi et al. combined Hybrid A* search engine with the Visibility Diagram planning to find the shortest possible non-holonomic path, which proves to be 40% faster than Hybrid A* algorithm [40]. C B Wang et al. introduced factors of turning, and edge removal to improve A* algorithm in k-shortest path problem, which generates k times shorter path by deleting edges [41].

Configuration space algorithm (CSA) In the configuration space algorithm, the truck is reduced to a point, and the obstacles and boundary around it are scaled up accordingly, so that the truck can move to any point in the obstacle space without colliding with the obstacles or moving out of the boundary. Therefore, CSA ensures the safety of truck movement in the path planning phase. The advantage of CSA is its flexibility. Changes of the starting point and the target point of the truck will not cause the reconstruction of the connected graph. The disadvantage is that when obstacles increase, the computational complexity will increase correspondingly, and the optimal path may not be found in a limited time.

J H Chuang et al. proposed an analytically tractable potential field model of configuration space, which assumes that the border of every two-dimensional region is uniformly charged [42]. T C Manjunath et al. simulated and implemented configuration space algorithm in an autonomous system, proving the effectiveness of the method [43].

Artificial potential field (APF) The traditional artificial potential field method regards the movement of vehicles in the environment as a kind of movement in the abstract artificial force field. The target points generate "gravitation" to the vehicle, and the obstacles generate "repulsion" to the vehicle. Finally, the movement of vehicles is controlled by seeking the resultant force. However, because the potential field method compresses all information into a single resultant force, it will sometimes discard valuable information about the distribution of obstacles and fall into the local minimum easily.

M Zhang et al. constructed a dynamic artificial potential field based on the local information to solve the problem of the path planning. The problem of global minimum point is solved by adjusting the trucks' movement direction within the escaping area [44]. L Chen et al. redefined the gravitational force function and the repulsive force function, therefore the kinetic characteristic of vehicles in the artificial potential field is analyzed and the obstacles' impact is neutralized [45]. Q Chen et al. added speed factor to the potential field function of the traditional artificial potential field method, so that the truck can avoid dynamic obstacles and reach the dynamic target [46]. J J Yun et al. used "virtual obstacle" and "associate target" methods to improve APF, and applied it to robot path planning successfully [47]. L Zhou et al. proposed an improved obstacle potential field function model considering the size of the vehicle and the obstacles and changes the weight of the obstacle potential field function adaptively to guide the vehicle away from the local minima [48].

Intelligent methods

In recent years, with the wide application of intelligent methods such as genetic algorithm, path planning methods have also made great progress. Many researchers focus on the research of path planning based on intelligent methods. Among them, fuzzy method, neural network and genetic algorithm are the most mentioned.

Fuzzy logic control By simulating the driver's thoughts, the fuzzy logic algorithm combines the robustness of fuzzy theory with the physiology-based "perception-action" behavior, providing a new direction for path planning of vehicles in unknown or uncertain environments. The advantage of fuzzy logic control is that the algorithm is intuitive, easy to implement, can facilitate the integration of human experience into the algorithm, does not need huge computation, and can meet the real-time requirements. The disadvantages are: when the environment is very complex, the summarized rules are difficult to cover all aspects.

Y. Makita et al. proposed a hybrid method of fuzzy control rules combined with a potential field for vehicle control in the presence of obstacles, which considers the operator's technique and needs only the map's information [49]. J J Yun et al. proposed a method based on particle swarm optimization for fuzzy controller, which utilizes the PSO algorithm to optimize the threshold of the fuzzy functions thus the parameters can be adjusted in different environments [50]. Q Li et al. developed control rules respectively for u-slot and maze environments by modifying and optimizing some rules of the universal fuzzy controller [51]. M Wang et al. proposed a minimum-risk approach that could escape from the local minimum in unknown environments. The approach is theoretically proved to guarantee global convergence even in extreme environments [52]. M B Motamedinejad et al. proposed a fuzzy model for path planning. The proposed model is based on virtual force field technique, which minimizes tracking error by considering both static and dynamic obstacles and vehicles' relative priorities [53].

Genetic algorithm (GA) Genetic algorithm is a highly parallel, random and adaptive search algorithm based on natural selection and evolution in the biological world. The genetic algorithm can search the solution space in different regions constantly, avoid the situation of falling into the local minimum solution, and get the global optimal path. However, when traditional genetic algorithms are used to deal with complex, multi-objective and multi-variable optimization problems, they tend to be premature or slow in convergence.

S D Li et al. proposed a new GA based on path network, in which a feasible path is represented by a chromosome. The upgraded algorithm converges quickly and works efficiently [54]. Z Q Yang et al. proposed a new adaptive multi-object genetic algorithm (AMOGA) and applied it to the flexible inspection path planning. The encoding mechanism, crossover operator and mutation operator are designed to be adjusted online [55]. R K Panda et al. proposed a different kind of approach to the dynamic motion planning problems of mobile robots in uncertain dynamic environments based on the behavior dynamics. They transformed the problem into an optimization problem in the robot's acceleration space [56]. Y G Sun et al. proposed a path planning method for mobile robot based on quantum genetic algorithm. Their team used the quantum-bit with the superposition state to encode a path and introduced quantum genetic algorithm to optimize the cost function of path planning [57]. M M Ali et al. improved GA to decrease the number of iteration and processing power by using predefined priorities for parent initial path generation rather than creating parent paths randomly [58].

Neural network algorithm The basic principle of the neural network method in path planning is to take environmental obstacles and other information as the input layer of the neural network, and process them in parallel through the neural network. The output layer of the neural network outputs the desired steering angle and speed, etc., to guide the vehicle to avoid obstacles until it reaches the destination. The advantages of neural network algorithm include its high parallel processing efficiency and its quick convergence to the optimal path. However, when there are many obstacles, the network scale is often large, which makes it difficult to apply in practice.

M Luo et al. proposed an improved 3D path planning method based on multi-scale map method to reduce the time cost. By introducing this multi-scale map method into the multi-robot bioinspired neural network algorithm, the time cost and mathematical complexity of the path planning algorithm can be greatly reduced [59]. A. A. Kassim et al. proposed a general neural network architecture called the wave expansion neural network (WENN). The neural network takes the workspace which needs to be navigated as input and generates artificial potential fields which are then subsequently used to find a path [60]. Y M Li et al. combined traditional quadtrees and framed-quadtrees with the shunting equation based neural network model to improve the efficiency of path planning. Their method can generate shorter collision free trajectory and has bigger representation capabilities of maps [61]. H Y Yuan et al. proposed a path planning algorithm using a four-layer neural network structure to realize the planning of the collision path of smart vehicles in unknown environment [62]. Z Y Lv et al. proposed a new neural networks vehicle path planning algorithm by rearranging neurons in the configuration space [63].

3 Conclusions and Outlook

Although many scholars have studied the problem of vehicle sizing, and have obtained many exciting results. However, there are still many problems to be solved:(1) The vehicle fleet sizing problem is often based on a specific model and scheduling strategy, but the assumptions in most studies are too ideal, and the research results of the studies are difficult to apply to the real operating environment. (2) The versatility of the methods proposed in the study is very low, and these methods can only be applied to solve the problem of fleet size in a certain scenario. So, it is an important direction to find the theoretical basis of the fleet sizing problem and combine it with simulation to verify in multiple scenarios, so as to obtain a universal and achievable method.

For vehicle dispatching problem, numerous research results from scholars have accelerated the development of vehicle transportation systems in terminals. But there are still some problems to be overcome: (1) There are few studies that directly optimize the QC idle time, which is the most important goal of vehicle dispatching problem in terminal. (2) Until now, no research considers the uncertainty of manned vehicles, which have a great impact on the scheduling results.

To overcome those problems, there are two directions. The first one is data-driven method. With the development of automated ports, more and more operational information is retained. These data can be used to analyze the efficiency of vehicle scheduling and build a model of vehicle scheduling. The other one is simulation method. Although many studies are based on simulation, simulation is only a means to verify the effectiveness of their method. Simulation can be used to make dispatching decisions and forecast future conditions. However, as mentioned before, this method still has some problems to be solved, one of them is robustness of simulation.

As for path planning problem, AGV path planning has always been a key issue in terminal automation. In the last decade, with the fast development and implementation of AGVs in nearly every field of automation, the number of publications in path planning has increased dramatically, bringing a lot of new insights. However, there are still problems regarding the path planning problem in automatic terminals. First, path planning process could be more dynamic, especially in the place where changes could happen all the time. Researches on online path changes are still lacking. It would be more efficient if AGVs' paths are not static. Second, the balance between getting the optimal solution and reducing the computational time is not achieved well. There's still a lot of room to improve the algorithm with less computational power.

References

1. Salzborn, F.J.M, Buckley, D J.: Minimum fleet size models for transportation systems. In: Buckley, D.J., (Ed.), Proceedings of the 6th International Symposium on Transportation & Traffic Theory (ISTTT6), pp. 607–624. Elsevier, Sydney (1974)
2. Proll, L.G.: Letter to the editor—a note on the minimum fleetsize for a suburban railway system. Transp. Sci. **6**(2), 204–207 (1972)
3. Koo, P.H., Jang, D.W., Lee, W.S.: Fleet sizing and vehicle routing for static freight container transportation. IE Interfaces **16**(2), 174–184 (2003)
4. Koo, P.H., Lee, W.S., Jang, D.W.: Fleet sizing and vehicle routing for container transportation in a static environment. OR Spectr. **26**(2), 193–209 (2004)
5. Vis, I.F.A., René (M.)B.M., de Koster, M.W.P.: Savelsbergh minimum vehicle fleet size under time-window constraints at a container terminal. Transp. Sci. **39**(2), 249–260 (2005)
6. Vis, I.F.A., De Koster, R., Roodbergen, K.J., Peeters, L.W.: Determination of the number of automated guided vehicles required at a semi-automated container termina. J. Oper. Res. Soc. **52**(4), 409–417 (2001)
7. Gobal, S.L., Kasilingam, R.G.: A simulation model for estimating vehicle requirements in automated guided vehicle systems. Comput. Ind. Eng. **21**(1–4), 623–627 (1991)
8. Wibisono, R., Ai, T.J., Yuniartha, D.R.: Fleet sizing of automated material handling using simulation approach. Mater. Sci. Eng. Conf. Ser. **319**, 012030 (2018)
9. Chang, K.H., Huang, Y.H., Yang, S.P.: Vehicle fleet sizing for automated material handling systems to minimize cost subject to time constraints. Lie Trans. **46**(3), 301–312 (2014)
10. Pjevcevic, D., Nikolic, M., Vidic, N., Vukadinovic, K.: Data envelopment analysis of AGV fleet sizing at a port container terminal. Int. J. Prod. Res. **55**(14), 4021–4034 (2016)

11. bin Md Fauadi, M.H.F., Li, W.-L., Murata, T., Prabuwono, A.S.: Vehicle requirement analysis of an AGV system using discrete-event simulation and data envelopment analysis. In: 2012 8th International Conference on Computing Technology and Information Management (NCM and ICNIT), pp. 819–823. IEEE, Seoul (2012)

12. Chang, K.H., Chang, A.L., Kuo, C.Y.: A simulation-based framework for multi-objective vehicle fleet sizing of automated material handling systems: an empirical study. J. Simul. 8(4), 271–280 (2014)

13. Samuel, R.: Routing and scheduling of vehicles and crews: the state of the art. Comput. Oper. Res. 10(2), 63–67 (1983)

14. Liu, C.I., Ioannou, P.A.: A comparison of different AGV dispatching rules in an automated container terminal. In: IEEE International Conference on Intelligent Transportation Systems, pp. 880–885. IEEE, Singapore (2002)

15. Mahadevan, B., Narendran, T.T.: Design of an automated guided vehicle-based material handling system for a flexible manufacturing system. Int. J. Prod. Res. 28(9), 1611–1622 (1990)

16. Wanan, C.C., Bin, W.: Vehicle dispatching under the shortest path and port centralization. In: Fourth International Symposium on Knowledge Acquisition & Modeling, pp. 139–142. IEEE, Sanya (2012)

17. Cheng, Y.L., Sen, H.C., Natarajan, K., Teo, C.P., Tan, K.C.: Dispatching automated guided vehicles in a container terminal. Supply Chain Optimization, 355–389 (2006)

18. Günther, H.-O., Kim, K.H.: Dispatching multi-load AGVs in highly automated seaport container terminals. Container Terminals and Automated Transport Systems. 231–255(2005)

19. Kagaya, S., Kikuchi, S., Donnelly, R.A.: Use of a fuzzy theory technique for grouping of trips in the vehicle routing and scheduling problem. Eur. J. Oper. Res. 76(1), 143–154 (1994)

20. Umashankar, N., Karthik, V.N.: Multi-criteria intelligent dispatching control of automated guided vehicles in FMS. In: 2006 IEEE Conference on Cybernetics and Intelligent Systems, pp. 1–6. IEEE, Bangkok (2006)

21. Kozan, E., Preston, P.: Genetic algorithms to schedule container transfers at multimodal terminals. Int. Trans. Oper. Res. 6(3), 311–329 (2010)

22. Kim, K.H., Bae, J.W.: A look-ahead dispatching method for automated guided vehicles in automated port container terminals. Transp. Sci. 38(2), 224–234 (2004)

23. Kim, J., Choe, R., Kwang, R.R.: Multi-objective optimization of dispatching strategies for situation-adaptive AGV operation in an automated container terminal. In: Proceedings of the 2013 Research in Adaptive and Convergent Systems (RACS 2013), pp. 1–6. Association for Computing Machinery, New York (2013)

24. Lee, N.M.Y., Lau, H.Y.K., Ko, A.W.Y.: An immune inspired algorithm for solving dynamic vehicle dispatching problem in a port container terminal. In: Andrews, P.S., Timmis, J., Owens, N.D.L., Aickelin, U., Hart, E., Hone, A., Tyrrell, A.M. (eds.) ICARIS 2009. LNCS, vol. 5666, pp. 329–342. Springer, Heidelberg (2009). https://doi.org/10.1007/978-3-642-03246-2_30

25. Koo, P.H.: Dispatching transport vehicles in maritime container terminals. Int. J. Bus. Tour. Appl. Sci. 1, 90–97 (2013)

26. Zulvia, F.E., Kuo, R.J., Hu, T.L.: Solving CVRP with time window, fuzzy travel time and demand via a hybrid ant colony optimization and genetic algorithm. In: 2012 IEEE Congress on Evolutionary Computation, pp. 1–8. IEEE, Brisbane (2012)

27. Xing, Y., Yin, K., Quadrifoglio, L., Wang, B.X.: Dispatch problem of automated guided vehicles for serving tandem lift quay crane. Transp. Res. Rec. 2273(1), 79–86 (2012)

28. Homayouni, S.M., Tang, S.H., Ismail, N., Ariffin, M.K.A.M., Samin, R.: A hybrid genetic-heuristic algorithm for scheduling of automated guided vehicles and quay cranes in automated container terminals. In: 2009 International Conference on Computers & Industrial Engineering, pp. 96–101. IEEE, Troyes (2009)

29. Bose, J., Reiners, T., Steenken, D., Voß, S.: Vehicle dispatching at seaport container terminals using evolutionary algorithms. In: Proceedings of the 33rd annual Hawaii international conference on system sciences, pp. 10-pp. IEEE, Maui (2000)

30. Bae, J.W., Kim, K.H.: A pooled dispatching strategy for automated guided vehicles in port container terminals. Int. J. Manage. Sci. **6**(2), 47–60 (2000)

31. Yue, L., Fan, H., Zhai, C.: Joint configuration and scheduling optimization of a dual-trolley quay crane and automatic guided vehicles with consideration of vessel stability. Sustainability **12**(1), 1–16 (2019)

32. Zhicheng, B., Yaozhou, Z., Xuemin, Z., Yansong, X., Jiaqi, C., Weijian, M.: Simulation-based AGV dispatching in automated container terminal. In: 2019 International Conference on Advances in Construction Machinery and Vehicle Engineering (ICACMVE), pp. 414–42. IEEE, Changsha (2019)

33. Lim, J.K., Lim, J.M., Yoshimoto, K., Kim, K.H., Takahashi, T.: Designing guide-path networks for automated guided vehicle system by using the Q-learning technique. Comput. Ind. Eng. **44**(1), 1–17 (2003)

34. Zeng, Q., Yang, Z., Hu, X.: A method integrating simulation and reinforcement learning for operation scheduling in container terminals. Transport **26**(4), 383–393 (2011)

35. Potvin, J.Y., Shen, Y., Rousseau, J.M.: Neural networks for automated vehicle dispatching. Comput. Oper. Res. **19**(3–4), 267–276 (1992)

36. Choe, R., Kim, J., Ryu, K.R.: Online preference learning for adaptive dispatching of AGVs in an automated container terminal. Appl. Soft Comput. **38**, 647–660 (2016)

37. Wei, S., Wang, L., Wang, B.R., Ren, H.J., Yang, Y.S., Liu, X.L., Ding, Y.C.: Improvement of A-star algorithm and its application in AGV path planning. Autom. Instrument. **38**, 51–54 (2017)

38. Yang, R., Cheng, L.: Path Planning of restaurant service robot based on a-star algorithms with updated weights. In: 2019 12th International Symposium on Computational Intelligence and Design (ISCID), pp. 292–295. IEEE, Hangzhou (2019)

39. Zheng, T., Xu, Y., Zheng, D.: AGV path planning based on improved A-star algorithm. In: 2019 IEEE 3rd Advanced Information Management, Communicates, Electronic and Automation Control Conference (IMCEC), pp. 1534–1538. IEEE, Chongqing (2019)

40. Sedighi, S., Nguyen, D., Kuhnert, K.: Guided hybrid A-star path planning algorithm for valet parking applications. In: 2019 5th International Conference on Control, Automation and Robotics (ICCAR), pp. 570–575. IEEE, Beijing (2019)

41. Wang, C., Wang, L., Qin, J., Wu, Z., Duan, L., Li, Z.,... & Lu, Z.: Path planning of automated guided vehicles based on improved A-Star algorithm. In: 2015 IEEE International Conference on Information and Automation, pp. 2071–2076. IEEE, Lijiang (2015)

42. Chuang, J.H., Ahuja, N.: An analytically tractable potential field model of free space and its application in obstacle avoidance. IEEE Trans. Syst. Man, Cybern. Part B (Cybern.), **28**(5), 729–736 (1998)

43. Manjunath, T.C., Nagaraja, B.G., Kusagur, A.: Simulation & implementation of shortest path algorithm with a mobile robot using configuration space approach. In: 2009 International Conference on Advanced Computer Control, pp. 197–201. IEEE, Singapore (2009)

44. Zhang, M., Shen, Y., Wang, Q., Wang, Y.: Dynamic artificial potential field based multi-robot formation control. In: 2010 IEEE Instrumentation & Measurement Technology Conference Proceedings, pp. 1530–1534. IEEE, Austin (2010)

45. Chen, L., Liu, C., Shi, H., Gao, B.: New robot planning algorithm based on improved artificial potential field. In: 2013 Third International Conference on Instrumentation, Measurement, Computer, Communication and Control, pp. 228–232. IEEE, Shenyang (2013)

46. Qian, C., Qisong, Z., Li, H.: Improved artificial potential field method for dynamic target path planning in LBS. In: 2018 Chinese Control And Decision Conference (CCDC), pp. 2710–2714. IEEE, Shenyang (2018)

47. Jianjun, Y., Hongwei, D., Guanwei, W., Lu, Z.: Research about local path planning of moving robot based on improved artificial potential field. In: 2013 25th Chinese Control and Decision Conference (CCDC), pp. 2861–2865. IEEE, Guiyang (2013)
48. Zhou, L., Li, W.: Adaptive artificial potential field approach for obstacle avoidance path planning. In: 2014 Seventh International Symposium on Computational Intelligence and Design Vol. 2, pp. 429–432. IEEE, Hangzhou (2014)
49. Makita, Y., Hagiwara, M., Nakagawa, M.: A simple path planning system using fuzzy rules and a potential field. In: Proceedings of 1994 IEEE 3rd International Fuzzy Systems Conference, pp. 994–999. IEEE, Orlando (1994)
50. Yu, J., Sun, Y., Ruan, X., Zhang, Y.: Research on path planning for robots based on PSO optimization for fuzzy controller. In: Proceeding of the 11th World Congress on Intelligent Control and Automation, pp. 5293–5298. IEEE, Shenyang (2014)
51. Li, Q., Zhang, C., Han, C., Xu, Y., Yin, Y., Zhang, W.: Path planning based on fuzzy logic algorithm for mobile robots in static environment. In: 2013 25th Chinese Control and Decision Conference (CCDC), pp. 2866–2871. IEEE, Guiyang (2013)
52. Wang, M.: Fuzzy logic based robot path planning in unknown environment. In: 2005 International Conference on Machine Learning and Cybernetics Vol. 2, pp. 813–818. IEEE, Guangzhou (2005)
53. Motamedinejad, M.B., Barzamini, R., Jouzdani, J., Khosravi, A.: A new fuzzy path planning for multiple robots. In: 2006 International Conference on Information and Automation, pp. 295–300. IEEE, Shandong (2006)
54. Li, S., Ding, M., Cai, C., Jiang, L.: Efficient path planning method based on genetic algorithm combining path network. In: 2010 Fourth International Conference on Genetic and Evolutionary Computing, pp. 194–197. IEEE, Shenzhen (2010)
55. Zeqing, Y., Libing, L., Zhihong, T., Weiling, L.: Application of adaptive genetic algorithm in flexible inspection path planning. In: 2008 27th Chinese Control Conference, pp. 75–80. IEEE, Kunming (2008)
56. Panda, R.K., Choudhury, B.B.: An effective path planning of mobile robot using genetic algorithm. In: 2015 IEEE International Conference on Computational Intelligence & Communication Technology, pp. 287–291. IEEE, Ghaziabad (2015)
57. Sun, Y., Ding, M.: Quantum genetic algorithm for mobile robot path planning. In: 2010 Fourth International Conference on Genetic and Evolutionary Computing, pp. 206–209. IEEE, Shenzhen (2010)
58. Ali, M.M., Farooq, O., Khan, M.H., Haxha, S.: Hardware implementation of compact genetic algorithm for robot path planning in globally static environment in 8-bit microcontroller. In: 2019 5th International Conference on Information Management (ICIM), pp. 242–247. IEEE, Cambridge (2019)
59. Luo, M., Hou, X., Yang, J.: Multi-Robot one-target 3D path planning based on improved bio-inspired neural network. In: 2019 16th International Computer Conference on Wavelet Active Media Technology and Information Processing, pp. 410–413. IEEE, Chendu (2019)
60. Kassim, A. A., Kumar, B. V.: A neural network architecture for path planning. In: Proceedings 1992 of IJCNN International Joint Conference on Neural Networks Vol. 2, pp. 787–792. IEEE, Baltimore (1992)
61. Li, Y., Meng, M.Q.H., Li, S., Chen, W., You, Z., Guo, Q., Liang, H.: A quad tree based neural network approach to real-time path planning. In: 2007 IEEE International Conference on Robotics and Biomimetics (ROBIO), pp. 1350–1354. IEEE, Sanya (2007)
62. Yuan, H., Zhang, G., Li, Y., Liu, K., Yu, J.: Research and implementation of intelligent vehicle path planning based on four-layer neural network. In: 2019 Chinese Automation Congress (CAC), pp. 578–582. IEEE, Hangzhou (2019)
63. Lv, Z., Cao, J.: Path planning methods of mobile robot based on new neural network. In: Proceedings of the 32nd Chinese Control Conference, pp. 3222–3226. IEEE, Xi'an (2013)

Coverage Analysis of Net Inscriptions in Coloured Petri Net Models

Faustin Ahishakiye[✉], José Ignacio Requeno Jarabo, Lars Michael Kristensen, and Volker Stolz

Department of Computer Science, Electrical Engineering, and Mathematical Sciences, Faculty of Engineering and Natural Science, Western Norway University of Applied Sciences, Bergen, Norway
{fahi,jirj,lmkr,vsto}@hvl.no

Abstract. High-level Petri nets such as Coloured Petri Nets (CPNs) are characterised by the combination of Petri nets and a high-level programming language. In the context of CPNs and CPN Tools, the inscriptions (e.g., arc expressions and guards) are specified using Standard ML (SML). The application of simulation and state space exploration (SSE) for validating CPN models traditionally focusses on behavioural properties related to net structure, i.e., places and transitions. This means that the net inscriptions are only implicitly validated, and the extent to which their sub-expressions have been covered is not made explicit. The contribution of this paper is an approach that establishes a link between coverage analysis known from programming languages and net inscriptions of CPN models. Specifically, we consider Modified Condition/Decision Coverage (MC/DC) of Boolean SML decisions, which cannot be measured within CPN Tools neither through state space exploration nor model checking directly. We have implemented our approach in a library for CPN Tools comprised of an annotation and instrumentation mechanism that transparently intercepts and collects evaluation of boolean conditions, and a post-processing tool that, for a set of model executions (runs), determines whether each decision is MC/DC-covered. We evaluate our approach on four large publicly available CPN models.

1 Introduction

Coverage analysis is important for programs in relation to fault detection. Structural coverage criteria are required for software safety and quality design assurance [6], and low coverage indicates that the software product has not been extensively tested. Two common metrics are statement- and branch coverage [16], where low coverage concretely indicates that certain instructions have never actually been executed. Coloured Petri Nets [9] and CPN Tools [8] have been widely used for constructing models of concurrent systems with simulation and state

This work was partially supported by the European Horizon 2020 project COEMS under grant agreement no. 732016 (https://www.coems.eu/). We thank Svetlana Jakšić for discussions on this work.

B. Ben Hedia et al. (Eds.): VECoS 2020, LNCS 12519, pp. 68–83, 2020.
https://doi.org/10.1007/978-3-030-65955-4_6

space exploration (SSE) being the two main techniques for dynamic analysis. CPN model analysis is generally concerned with behavioural properties related to boundedness, reachability, liveness, and fairness properties. This means that the main focus is on structural elements such as places, tokens, markings (states), transitions and transition bindings. Arc expressions and guards are only implicitly considered via the evaluation of these net inscriptions taking place as part of the computation of transition enabling and occurrence during model execution. This means that design errors in net inscriptions may not be detected as we do not obtain explicit information on for instance whether both branches of an if-then-else expression on an arc have been covered.

We argue that from a software engineering perspective, it is important to be explicitly concerned with quantitative and qualitative analysis of the extent to which net inscriptions have been covered. Our hypothesis is that the coverage criteria used for traditional source code can also be applied to the net inscriptions of CPN models. Specifically, we consider the modified condition decision coverage (MC/DC) criterion. MC/DC is a well-established coverage criteria for safety-critical systems, and is required by certification standards, such as the DO-178C [15] in the domain of avionic software systems. In the context of MC/DC, a *decision* is a Boolean expression composed of sub-expressions and Boolean connectives (such as logical conjunction). A *condition* is an atomic (Boolean) expression. According to the definition of MC/DC [11, 16], each condition in a decision has to show an independent effect on that decision's outcome by: (1) varying just that condition while holding fixed all other possible conditions; or (2) varying just that condition while holding fixed all other possible conditions that could affect the outcome. MC/DC is a coverage criterion at the condition level and is recommended due to its advantages of being sensitive to code structure, requiring few test cases ($n + 1$ for n conditions), and it is the only criterion that considers the independence effect of each condition.

Coverage analysis for software is usually provided through dedicated instrumentation of the software under test, either by the compiler, or additional tooling, such as binary instrumentation. Transferring this to a CPN model under test, our aim is to combine the execution of a CPN model (by simulation or SSE) with coverage analysis of SML guard and arc expressions. Within CPN Tools, there is no coverage analysis of the SML expressions in a CPN model. This means that to record coverage data for a CPN model under test, it is necessary to instrument the Boolean expressions such that the truth-values of individual conditions are logged in addition to the overall outcome of the decision. Our approach to instrumentation makes use of side-effects by outputting intermediate results of conditions and decisions, which we then process to obtain the coverage verdict. No modifications to the net structure of the CPN model are necessary. Furthermore, the instrumentation has little impact on model execution so that it does not delay the simulation and SSE.

The remainder of this paper is organised as follows. In Sect. 2, we introduce the MC/DC coverage criterion in more detail. In Sect. 3, we present our approach to deriving coverage data and show how to instrument guard and arc expressions to collect the required coverage data. In Sect. 4 we consider the post-processing of coverage data. We demonstrate the application of our library for coverage analysis on publicly available 3rd-party CPN models in Sect. 5. In this section, we also evaluate our approach with respect to overhead in execution, and discuss our findings. Section 6 discusses related work, and we present our conclusions including directions for future work in Sect. 7. Our coverage analysis library, the example models, the Python code to produce reports and graphs, and documentation is available at htttps://github.com/selabhvl/cpnmcdctesting.

2   Coverage Analysis and MC/DC

When considering CPN models, we will be concerned with coverage analysis of guard and arc expressions. A guard expression is a list of Boolean expressions all of which are required to evaluate to true in a given transition binding for the transition to be enabled. We refer to such Boolean expressions as *decisions*. Similarly, an if-then-else expression on an arc will have a decision determining whether the then- or the else-branch will be taken. Decisions are constructed from *conditions* and Boolean operators.

Definition 1 (Condition, Decision). *A **condition** is a Boolean expression containing no Boolean operators except for the unary operator NOT.*

*A **decision** is a Boolean expression composed of conditions and zero or more Boolean operators. It is denoted by $D(c_1, c_2, \cdots, c_i, \cdots, c_n)$, where c_i, $1 \leq i \leq n$ are conditions.*

As an example, we may have a guard (or an arc expression) containing a decision of the form $D = (a \wedge b) \vee c$, where a, b, and c are conditions. These conditions may in turn refer to the values bound to the variables of the transition. The evaluation of a decision requires a *test case* assigning a *value* $\in \{0, 1, ?\}$ to the conditions of the decision, where ? means that a condition was not evaluated due to short-circuiting. Short-circuit means that the right operand of the *and*-operator ($\&\&/\wedge$) is not evaluated if the left operand is false, and the right operand of the *or*-operator ($||/\vee$) is not evaluated if the left operand is true.

Depending on the software safety level (A-D) which is assessed by examining the effects of a failure in the system, different structure coverage criteria are required. *Statement* coverage for software levels A-C, *branch/decision* coverage for software levels A-B, and MC/DC for software level A [16]. Statement coverage is considered inadequate because it is insensitive to some control structures. Both statement- and branch coverage are completely insensitive to the logical operators ($\vee$ and $\wedge$) [4]. The criteria taking logical expressions into consideration have been defined [6]. These are *condition coverage* (CC), where each condition in a decision takes on each possible outcome at least once true and once false during testing; *decision coverage* (DC) requiring only each decision to be evaluated once true and once false; and *multiple condition coverage* (MCC) which is an exhaustive testing of all possible input combinations of conditions to a decision. CC and DC are considered inadequate due to ignorance of the independence effect of conditions on the decision outcome. MCC requires 2^n tests for a decision with n inputs. This results in exponential growth in the number of test cases, and is therefore time-consuming and impractical for many test cases.

To address the limitations of the coverage criteria discussed above, *modified condition/decision coverage* (MC/DC) is considered and is required for safety critical systems such as in the avionics industry. MC/DC has been chosen as the coverage criterion for the highest safety level software because it is sensitive to the complexity of the decision structure [11] and requires only $n + 1$ test cases for a decision with n conditions [6,10]. In addition, MC/DC coverage criterion is suggested as a good candidate for model-based development (MBD) using tools such as Simulink and SCADE [7]. Thus, our model coverage analysis is based on MC/DC criterion. The following MC/DC definition is based on DO-178C [16]:

Definition 2 (Modified condition/decision coverage). *A program is MC/ DC covered and satisfies the MC/DC criterion if the following holds:*

- *every point of entry and exit in the program has been invoked at least once,*
- *every condition in a decision in the program has taken all possible outcomes at least once,*
- *every decision in the program has taken all possible outcomes at least once,*
- *each condition in a decision has shown to independently affect that decision's outcome by: (1) varying just that condition while holding fixed all other possible conditions, or (2) varying just that condition while holding fixed all other possible conditions that could affect the outcome.*

Table 1. MCC and selected MC/DC test cases for decision $D = (a \wedge b) \vee c$

TC	a	b	c	D	MC/DC pairs
1	0	0	0	0	
2	0	0	1	1	c(1,2)
3	0	1	0	0	
4	0	1	1	1	c(3,4)
5	1	0	0	0	
6	1	0	1	1	c(5,6)
7	1	1	0	1	a(3,7), b(5,7)
8	1	1	1	1	

(a) MCC test cases

TC	a	b	c	D	MC/DC pairs
1	0	?	0	0	
2	1	1	?	1	a(1,2)
3	1	0	0	0	b(2,3)
4	0	?	1	1	c(1,4)

(b) Selected MC/DC test cases

The coverage of program entry and exit in the Definition 2 is added to all control-flow criteria and is not directly connected with the main point of MC/DC [22]. The most challenging and discussed part is showing the independent effect, which demonstrates that each condition of the decision has a defined purpose. The item (1) in the definition defines the unique cause MC/DC and item(2) has been introduced in the DO-178C to clarify that so-called *Masked MC/DC* is allowed [3,15]. Masked MC/DC means that it is sufficient to show the independent effect of a condition by holding fixed only those conditions that could actually influence the outcome. Thus, in our analysis, we are interested in evaluation of expressions by checking the independence effect of each condition.

Example 1. Consider the decision $D = (a \wedge b) \vee c$. Table 1a presents all eight possible test cases (combinations) for MCC. The MC/DC pairs column for example, $c(1,2)$ specifies that from test case 1 and 2 we can observe that changing the truth value of c while keeping the values of a and b, we can affect the outcome of the decision. Comparing MCC to MC/DC in terms of the number of test cases, there are seven possible MC/DC test cases (1 through 7) that are part of an MC/DC pair, where condition c is represented by three MC/DC pairs of test cases. However, for a decision with three conditions, only four (i.e., $n + 1$) test cases are required to achieve MC/DC coverage as shown in Table 1b, where '?' represents the condition that was not evaluated due to short-circuiting.

3   Instrumentation of CPN Models

In this section, we describe our instrumentation approach on an example CPN model, and highlight the salient features of our coverage analysis library. Our overall goal is that through simulation or SSE, we (partially) fill a truth-table for each decision in the net inscriptions of the CPN model. Then, for each of these tables, and hence the decisions they are attached to, we determine whether the model executions that we have seen so far satisfy the MC/DC coverage criteria. If MC/DC is not satisfied, either further simulations are necessary, or if the state space is exhausted, developers need to consider the reason for this shortcoming, which may be related to insufficient exploration as per a limited set of initial markings, or a conceptual problem in that certain conditions indeed cannot contribute to the overall outcome of the decision.

3.1   MC/DC Coverage for CPN Models

MC/DC coverage (or any other type of coverage) is commonly used with executable programs: which decisions and conditions were evaluated by the test cases, and with which result. Specifically, these are decisions *from the source code* of the system (application) under test. Of course, a compiler may introduce additional conditionals into the code during code generation, but these are not of concern. CPN Tools already reports a primitive type of coverage as part of simulation (the transition and transition bindings that have been executed) and the state space exploration (transitions that have never occurred). These can be interpreted as variants of state- and branch coverage.

Hence, we first need to address what we want MC/DC coverage to mean in the context of CPN models. If we first consider guard expressions on transitions, then we have two interesting questions related to coverage: if there is a guard, we know from the state space (or simulation) report whether the transition has occurred, and hence whether the guard expression has evaluated to true. However, we do not know if during the calculation of enabling by CPN Tools it ever has been false. If the guard had never evaluated to false, this may indicate a problem in the model or the requirements it came from, since apparently that guard was not actually necessary. Furthermore, if a decision in a guard is a complex expression, then as per MC/DC, we would like to see evidence that each condition contributed to the outcome. Neither case can be deduced from the state space report or via the CTL model checker of CPN Tools as the executions only contain transition bindings that have occurred, and hence cases where the guard has evaluated to true.

3.2   Instrumentation of Net Inscriptions

In the following, we describe how we instrument the guards on transitions such that coverage data can be obtained. Arc expressions are handled analogously. Guards in a CPN model are written following the general form of a comma-separated list of Boolean expressions (decisions): $[bExp_0, \ldots, bExp_n]$. A special case is the expression: $var = exp$ which may have two effects: if the variable var is bound already via a pattern in another expression (arc or guard) of the transition, then this is indeed a Boolean equality test (decision). If, however, var is not bound via other expressions, then this assigns the value of exp to the variable var and does not contribute to any guarding effect.

We consider general Boolean expressions which may make use of the full feature set of the SML language for expressions, most importantly Boolean binary operations, negation, conditional expressions with if-then-else and function calls. Simplified, we handle:

$$\langle bExp \rangle ::= \text{not } \langle bExp \rangle \mid \langle var \rangle \mid \text{f } \langle exp \rangle_0 \ldots \langle exp \rangle_n$$
$$\mid \langle bExp \rangle \text{ andalso } \langle bExp \rangle \mid \langle bExp \rangle \text{ orelse } \langle bExp \rangle$$
$$\mid \text{if } \langle bExp \rangle \text{ then } \langle bExp \rangle \text{ else } \langle bExp \rangle$$
$$\mid \text{let } \ldots \text{ in } \langle bExp \rangle \text{ end}$$

Function symbols f cover user-defined functions as well as (built-in) relational operators such as $<, =$; we do not detail the overall nature of arbitrary SML expressions, but refer the reader to [21] for a comprehensive discussion.

SSE or simulation of the model is not in itself sufficient to determine the outcome of the overall expression and its subexpressions: guards are not explicitly represented, and we only have the event of taking the transition in the state space, but no value of the guard expressions. Hence, we need to rely on side-effects during model execution to record the intermediate results. Our key idea is to transform every subexpression and the overall decision into a form which will use SML's file input/output to emit a log-entry that we can collect and analyse.

Listing 1.1. Expressions

```
datatype condition =
    AND of condition * condition
  | OR of condition * condition
  | NOT of condition
  | ITE of condition * condition
                     * condition
  | AP of string * bool;
```

Listing 1.2. Evaluation function

```
fun eval (AP (cond,v))=([(cond, SOME v)],v)
  | eval (OR (a,b)) = let
        val (ares,a') = eval a;
        val (bres,b') = eval b;
    in
        (ares^^bres, a' orelse b')
    end
    ...
fun EXPR (name,expr) : bool = [ ... ]
```

For the necessary instrumentation which can be viewed as a transformation of guard and arc expressions, we essentially create an interpreter for Boolean expressions: when guards are checked (in a deterministic order due to SML's semantics from left to right), we traverse a term representation of the Boolean expression and output the intermediate results. Correspondingly, we design a data type (see Listing 1.1) that can capture the above constructs, and define an evaluation function (see Listing 1.2) on it. As we later need to map coverage reports back to code, for overall expressions EXPR and atomic proposition AP we introduce a component of type string that allows this identification. The evaluation function eval collects the result of intermediate evaluations in a list data structure, and the EXPR function (implementation not shown) turns this result into a single Boolean value that is used in the guard, and as a side-effect outputs the truth outcome for individual conditions. As an example, if we consider a guard: a>0 **andalso** (b **orelse** (c = 42)); then we can transform this guard in a straight forward manner into EXPR("Gid", AND(AP("1", a >0), OR(AP("2",b), AP("3", c = 42)))). It is important to notice that this does not give us the (symbolic) Boolean expressions, as we still leave it to the standard SML semantics to evaluate the a>0, while abstractly we refer to the AP as a condition named "1". We also elide expression- and proposition names for clarity in the text when not needed.

Any subexpression must be *total* and not crash and abort the model execution. A short-circuiting evaluation needs to explicitly incorporate the andalso or orelse operator and becomes more verbose, hence e.g. x = 0 **orelse** (y/x >0) becomes OR (AP("O1", x = 0)) (AP("O2", x = 0 **orelse** (y/x > 0))).

Figure 1 shows a module from one of our example models, the MQTT protocol [17] before and after instrumentation. The guard expressions for the transitions named Send CONNECT and Receive CONNACK have been instrumented. As the EXPR function contains input/output statements, we now as a side-effect observe intermediate results of conditions every time it evaluates a guard during execution of the CPN model. Note that the arc and guard expressions in the other modules were transformed in a similar manner. Even though we have done the instrumentation by hand, it could be automated based on the .cpn XML file of CPN Tools in combination with an SML parser.

4   Post Processing of Coverage Data

We now discuss the coverage analysis which is performed via post-processing of the coverage data recorded through the instrumentation. We did not implement the MC/DC coverage analysis in SML directly. Rather, we feed individual observations about decision outcomes and their constituent conditions into a Python tool that computes the coverage results. This allows us to reuse the backend in other situations, without being SML or CPN specific.

Coverage Analysis. The general format from the instrumentation step is a sequence of colon-delimited rows, where each triple in a row captures a single decision with the truth values of all conditions in a fixed order and the outcome.

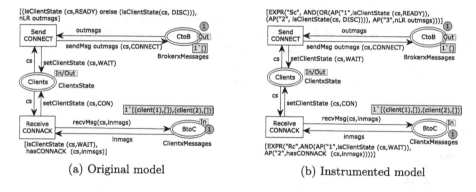

(a) Original model   (b) Instrumented model

Fig. 1. MQTT [17]: Guard expression before and after instrumentation

More details on the log format are given in our technical report [2]. The name (stemming from the first argument to an EXPR above) is configurable and should be unique in the model; and derived from the name of the element (guard or arc) the expression is attached to. This makes it easy to later trace coverage results for this name back to the element in the model.

Our Python tool parses the log file and calculates coverage information. It prints the percentage of decisions that are MC/DC and branch covered in textual mode and in GNU Plot syntax (see Fig. 3). The output contains individual reports in the form of the truth tables for each decision, which summarizes the conditions that are fired during the execution of the CPN model, and sets of pairs of test cases per condition that show the independence effect of that condition.

In the case that the decision is not MC/DC covered, the information provided by the Python script helps to infer the remaining valuations of the truth tables that should be evaluated in order to fulfill this criteria. This information can then be used by developers to drill down into parts of their model, e.g. through simulation, that have not been covered adequately yet.

Combining Coverage Data from Multiple Runs. Coverage- or testing frameworks rely on their correct use by the operator, only a sub-class of tools such as fuzzers are completely automated. Our central mcdcgen() function only explores the state space for the current configuration as determined by the initial markings. Compared to regular testing of software, this corresponds to providing a single input to the system under test.

It is straightforward to capture executions of multiple runs: our API supports passing initialisation functions that reconfigure the net between runs. However, as there is no standardised way of configuring alternative initial markings or configurations in CPN Tools, the user has to actively make use of this API. In the default configuration, only the immediate net given in the model is evaluated, and no further alternative configurations are explored.

As an example, we show in Listing 1.3 how we make use of this feature in the MQTT-model, where alternative configurations were easily discoverable for us: the signature of MC/DC-generation with a simple test-driver

is mcdcgenConfig = **fn** : int∗('a→'b)∗'a list∗string→unit, where the
first argument is a timeout for the SSE, the second is a function with side-effects
that manipulates the global configurations that are commonly used in CPN Tools
to parameterise models, the next argument is a list of different configurations,
followed by the filename for writing results to.

Listing 1.3. MC/DC tool invocation

```
use (cpnmcdclibpath^"config/simrun.sml");
(* Invocation with default settings (no timeout) *)
mcdcgen("path/to/mqtt.log");
(* Invocation without timeout; base model + 2 configurations *)
mcdcgenConfig(0, applyConfig,[co1,co2],"path/to/mqtt3.log");
```

This function will always first evaluate the initial model configuration, and
then have additional runs for every configuration. Internally, it calls into CPN
Tools' CalculateOccGraph() function for the actual SSE. Hence the first
mcdcgen-invocation in Listing 1.3 will execute a full SSE without timeout,
whereas the second mcdcgenConfig-invocation would produce three subsequent runs logged into the same file, again without a default timeout. The test-driver can easily be adapted to different scenarios or ways of reconfiguring a
model. Alternatively, traces can also be produced in separate files that are then
concatenated for the coverage analysis.

5 Evaluation on Example Models

In this section, we provide experimental results from an evaluation of our approach to model coverage for CPNs. We present the results of examining four
non-trivial models from the literature that are freely available as part of scientific publications: a model of the Paxos distributed-consensus algorithm [24],
a model of the MQTT publish-subscribe protocol [17], a model for distributed
constraint satisfaction problem (DisCSP), weak-commitment search (WCS) algorithms [13], and a complex model of the runtime environment of an actor-based
model (CPNABS) [5]. All models come with initial markings that allow state
space generation, in the case of MQTT and DisCSP finite, and infinite in the
case of Paxos and CPNABS.

Experimental Setup. Figure 2 gives an overview of our experimental setup.
First, we instrument the original CPN model under test by transforming each
guard and arc expression into a form that prints how conditions were evaluated
and the overall outcome of the decision (cf. Sect. 3). Second, we run the SSE on
the instrumented model and then reconfigure the configuration (initial marking)
with any additional initial configurations if they are obvious from the model. As
the side effect of SSE, we run the MC/DC generation which gives as output a log
file containing the information of evaluations of conditions in arcs expressions
and guards and the decision outcome. Finally, we run the MC/DC analyser (cf.
Sect. 4) that determines whether each decision is MC/DC-covered or not. In
addition, it reports the branch coverage (BC).

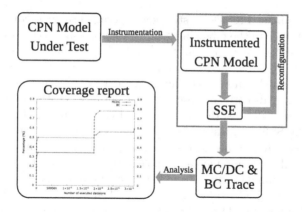

Fig. 2. Experimental setup

Experimental Results. Table 2 presents the experimental results for the four example models [5, 13, 17, 24]. For each model, we consider the number of executed decisions (second column) in arcs and guards. Column *Model decisions (m)* refers to the number of decisions that have been instrumented in the model. The number of decisions observed in the model and in the log-file may deviate in case some of decisions are never executed in which case they will not appear in the log file. We indicate them in brackets if during our exploration we did not visit, and hence log, each decision at least once. In the case of DisCSP, there are two guards decisions which never executed. The column *Non-trivial decisions* gives the number of the decisions (out of m) that have at least two conditions in the model, as they are the interesting ones while checking independence effect. If a decision has only one condition, it is not possible to differentiate MC/DC from DC. Columns $MC/DC(\%)$ and $BC(\%)$ present the coverage percentage for the CPN models under test. We record the ratio of covered decisions over the total number of decisions. For the CPNABS and Paxos models with an infinite state space, we aborted the SSE after two days at which point the number of arcs and guards expression executed no longer increased the coverage metrics.

Table 2. MC/DC coverage results for example CPN models

CPN Model	Executed decisions	Model decisions (m)	Non-trivial decisions	MC/DC (%)	BC (%)	State space
Paxos	2,281,466	27	11	37.03	40.74	Infinite
MQTT	3,870	18	14	11.11	22.22	Finite
CPNABS	3,716,896	32	13	59.37	84.37	Infinite
DisCSP	233,819	12 (10)	5	45.45	45.45	Finite

Discussion of Results. MC/DC is covered if all the conditions show the independence effect on the outcome. BC is covered if all the branches are taken at least once. This makes MC/DC a stronger coverage criterion compared to BC, which we will also see in the following graphs. Figure 3 shows the coverage results as the ratio of covered decisions and the number of executed decisions in guards and arcs for both MC/DC and BC. The plots show that the covered decisions increase as the model (and hence the decisions) is being executed. Note that the x-axis does not directly represent execution time of the model: the state space explorer prunes states that have been already visited (which takes time), and hence as the SSE progresses the number of expressions evaluated per time unit will decrease. In case an expression was executed with the same outcome, the coverage results do not increase, since those test cases have already been explored. Our instrumentation does not have a significant impact on the execution time of the model. Considering the time taken for the full SSE of the finite state models, for instance DisCSP model, both without and with instrumentation, it takes 212.346 s versus 214.922 s respectively. It is around 1% of overhead which is the cost for the instrumentation.

The CPNABS model has many single condition (trivial) decisions, and its coverage percentage is higher compared to other models. The Paxos model has less than a half covered decisions for both BC and MC/DC with a small percentage difference. MQTT has a high number of non-trivial decisions, which puts more weight on having a suitable test-suite to achieve good MC/DC coverage. It is interesting to observe the quality differences of the curves for the tested models. The later three models have less than half-covered decisions. This should attract the attention of developers and they should assess whether they have tested their models enough, as these results indicate that there is something that might be considered doubtful and require to revisit their test-suite.

Two factors affect the coverage percentage results presented for these models:

1. The tested models had no clear test suites; they might be lacking test cases to cover the remaining conditions. Depending on the purpose of each model, some of the test cases may not be relevant for the model or the model may not even have been intended for testing. This could be solved by using test case generation for uncovered decisions (see our future work).
2. The models might be erroneous in the sense that some parts (conditions) in the model are never or only partially executed due to a modelling issue, e.g. if the model evolved and a condition no longer serves any purpose or is subsumed by other conditions. For example in the DisCSP model, there are two decisions which were never executed, and we cannot tell if this was intentionally or not without knowing the goal of the developers.

A main result of our analysis of the example models is that none of the models (including those for which the state space could be fully explored) have full MC/DC or BC. This confirms our hypothesis that code coverage of net inscriptions of CPN models can be of interest to developers, such as revealing not taken branches of the if-then-else arc expressions, never executed guard decisions, conditions that do not independently affect the outcome and some model design errors. Our results show that even for full SSE, we may still find expressions that are not MC/DC covered. Assuming that the model is correct, improving

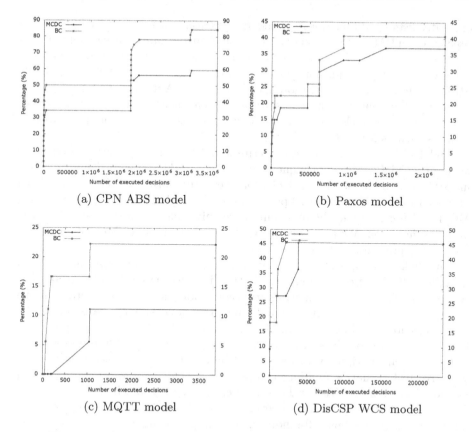

(a) CPN ABS model   (b) Paxos model

(c) MQTT model   (d) DisCSP WCS model

Fig. 3. MC/DC and BC coverage versus number of executed decisions

coverage then requires improving the test suite. This confirms the relevance and added value of performing coverage analysis of net inscriptions of CPN models over the dead places/transitions report as part of the state space generation. A natural next step in a model development process would be for the developers to revisit the decisions that are not MC/DC covered and understand the underlying reason. For the models that we have co-published, we can indeed confirm that the original models were not designed with a full test-suite in mind, neither from the initial configuration, nor through embedded configurations like for example the MQTT model.

6 Related Work

Coverage analysis has attracted attention in both academic and industrial research. Especially the MC/DC criterion is highly recommended and commonly used in safety critical systems, including avionic systems [15]. However, there is a limited number of research addressing model-based coverage analysis. Ghosh [20] expresses test adequacy criteria in terms of model coverage and explicitly lists *condition coverage* and *full predicate coverage criterion* for OCL predicates on UML interaction diagrams, which are semantically related to CPNs in that they express (possible) interactions. Test cases were not automatically generated. In [25], the authors present an automated test generation technique, MISTA (Model-based Integration and System Test Automation) for integrated functional and security testing of software systems using high-level Petri nets as finite state test models. None of the above works addressed structural coverage analysis such as MC/DC or BC on CPN models.

MC/DC is not a new coverage criterion. Chilenski [10] investigated three forms of MC/DC including Unique-Cause (UC) MC/DC, Unique-Cause + Masking MC/DC, and Masking MC/DC. Moreover, other forms of MC/DC have been discussed in [14]. More than 70 papers were reviewed and 54 of them discussed MC/DC definitions and the remaining were only focusing on the use of MC/DC in faults detection. We presented in [1], a tool that measures MC/DC based on traces of C programs without instrumentation.

Simulink [18] supports recording and visualising various coverage criteria including MC/DC from simulations via the Simulink Design Verifier. It also has two options for creating test cases to account for the missing coverage in the design. Test coverage criteria for autonomous mobile systems based on CPNs ware presented by Lill et al. in [12]. Their model-based testing approach is based on the use of CPNs to provide a compact and scalable representation of behavioural multiplicity to be covered by an appropriate selection of representative test scenarios fulfilling net-based coverage criteria. Simão et al. [19] provide definitions of structural coverage criteria family for CPNs, named CPN Coverage Criteria Family. These coverage criteria are based on checking if all-markings, all-transitions, all-bindings, and all-paths are tested at least once. Our work is different from the above presented work in that we are analysing the coverage of net inscriptions (conditionals in SML decisions) in CPN models based on structure coverage criteria defined by certification standards, such as DO-178C [16].

7   Summary and Outlook

We have presented a new approach and a supporting software tool to measure MC/DC and branch coverage (BC) of SML decisions in CPN models. There are three main contributions of this paper: (1) We provide a library and annotation mechanism that intercept evaluation of Boolean conditions in guards and arcs in SML decisions in CPN models, and record how they were evaluated; (2) we compute the conditions' truth assignment and check whether or not particular decisions are MC/DC-covered in the recordeded executions of the model; (3) we collect coverage data using our library from publicly available CPN models and report whether they are MC/DC and BC covered.

Our experimental results show that our library and post-processing tool can find how conditions were evaluated in all the net inscriptions in CPN models and measure MC/DC and BC. Results reveal that the MC/DC coverage percentage is quite low for all the four CPN models tested. This is interesting because it indicates that developers may have had different goals when they designed the model, and that the model only reflects a single starting configuration. We can compare this with the coverage of regular software: running a program will yield *some* coverage data, yet most programs will have to be run with many different inputs to achieve adequate coverage.

To the best of our knowledge, this is the first work on coverage analysis of CPN models based on BC and MC/DC criteria. This work highlighted that coverage analysis is interesting for CPN models, not only in the context of showing the covered guard and arcs SML decisions, but also the effect of conditionals in SML decisions on the model outcome and related potential problems.

Outlook. Our general approach to coverage analysis presents several directions forward which would help developers get a better understanding of their models: firstly, while generating the full state space is certainly the preferred approach, this is not feasible if the state space is inherently infinite or too large. Simulation of particular executions could then be guided by results from the coverage and try to achieve higher coverage in parts of the model that have not been explored yet. However, while selecting particular transitions to follow in a simulation is straight-forward, manipulating the data space for bindings used in guards is a much harder problem and closely related to test case generation (recall the CPNs also rely on suitable initial states, which are currently given by developers). Making use of feedback between the state of the simulation and the state of the coverage would, however, require much tighter integration of the tools. A related direction is to consider visualising coverage information in the graphical user interface: CPN Tools already supports a broad palette of visual options that could be used, e.g., to indicate successful coverage of guards through colour, or the frequency that transitions have been taken through their thickness [23].

As for the measured coverage results, it would be interesting to discuss with the original developers of the models if the coverage is within their expectations. While on the one hand low coverage could indicate design flaws, on the other hand our testing may not have exercised the same state space as the original developers did: they may have used their model in various configurations, whereof the state of the git repository just represents a snapshot, or we did not discover all possible configurations in the model. In the future, we may also try to generate test-cases specifically with the aim to increase coverage.

References

1. Ahishakiye, F., Jakšić, S., Stolz, V., Lange, F.D., Schmitz, M., Thoma, D.: Non-intrusive MC/DC measurement based on traces. In: Méry, D., Qin, S. (eds.) International Symposium on Theoretical Aspects of Software Engineering, pp. 86–92. IEEE (2019)
2. Ahishakiye, F., Jarabo, J.I.R., Kristensen, L.M., Stolz, V.: Coverage Analysis of Net Inscriptions in Coloured Petri Net Models (2020). https://arxiv.org/abs/2005. 09806v1

3. Certification Authorities Software Team (CAST): Rationale for Accepting Masking MC/DC in Certification Projects. Technical report, Position Paper CAST-6 (2001)
4. Cornett, S.: Code Coverage Analysis (1996–2014). https://www.bullseye.com/coverage.html. Accessed 6 June 2020
5. Gkolfi, A., Din, C.C., Johnsen, E.B., Kristensen, L.M., Steffen, M., Yu, I.C.: Translating active objects into Coloured Petri Nets for communication analysis. Sci. Comput. Program. **181**, 1–26 (2019)
6. Hayhurst, K.J., Veerhusen, D.S., Chilenski, J.J., Rierson, L.K.: A Practical Tutorial on Modified Condition/Decision Coverage. Technical report NASA/TM-2001-210876, NASA Langley Server (2001). https://dl.acm.org/doi/book/10.5555/886632
7. Heimdahl, M.P.E., Whalen, M.W., Rajan, A., Staats, M.: On MC/DC and implementation structure: an empirical study. In: Proceedings of IEEE/AIAA 27th Digital Avionics Systems Conference, pp. 5.B.3-1–5.B.3-13 (2008)
8. Jensen, K., Christensen, S., Kristensen, L.M., Michael, W.: CPN Tools (2010). http://cpntools.org/
9. Jensen, K., Kristensen, L.M.: Coloured Petri Nets: a graphical language for formal modeling and validation of concurrent systems. Commun. ACM **58**, 61–70 (2015)
10. John J., C.: An investigation of three forms of the Modified Condition Decision Coverage (MC/DC) criterion. Technical report, Office of Aviation Research (2001)
11. John, J., Steven, C.: Applicability of modified condition/decision coverage to software testing. Softw. Eng. J. **9**(5), 193–2001994 (1994)
12. Lill, R., Saglietti, F.: Model-based testing of cooperating robotic systems using Coloured Petri Nets. In: Proceedings of SAFECOMP 2013 - Workshop DECS (ERCIM/EWICS Workshop on Dependable Embedded and Cyber-physical Systems) of the 32nd International Conference on Computer Safety, Reliability and Security. Toulouse, France (Sep 2013). https://hal.archives-ouvertes.fr/hal-00848597
13. Pascal, C., Panescu, D.: A Coloured Petri Nets model for DisCSP algorithms. Concurr. Comput. Pract. Exp. **29**(18), 1–23 (2017)
14. Paul, T.K., Lau, M.F.: A systematic literature review on modified condition and decision coverage. In: Proceedings of the 29th Annual ACM Symposium on Applied Computing, pp. 1301–1308. SAC 2014, Association for Computing Machinery (2014)
15. Pothon, F.: DO-178C/ED-12C versus DO-178B/ED-12B: Changes and Improvements. Technical report, AdaCore (2012)
16. Rierson, L.: Developing Safety-Critical Software: A Practical Guide for Aviation Software and DO-178C Compliance. CRC Press, Boca Raton (2013)
17. Rodríguez, A., Kristensen, L.M., Rutle, A.: Formal modelling and incremental verification of the MQTT IoT protocol. In: Koutny, M., Pomello, L., Kristensen, L.M. (eds.) Transactions on Petri Nets and Other Models of Concurrency XIV. LNCS, vol. 11790, pp. 126–145. Springer, Heidelberg (2019). https://doi.org/10.1007/978-3-662-60651-3_5
18. Simulink: Types of Model Coverage. https://se.mathworks.com/help/slcoverage/ug/types-of-model-coverage.html. Accessed 06 March 2020
19. Simão, A., Do, S., Souza, S., Maldonado, J.: A family of coverage testing criteria for Coloured Petri Nets. In: Proceedings of 17th Brazilian Symposium on Software Engineering (SBES 2003), pp. 209–224 (2003)

20. Sudipto Ghosh, France, R., Braganza, C., Kawane, N., Andrews, A., Orest Pilskalns: Test adequacy assessment for UML design model testing. In: Proceedings of 14th International Symposium on Software Reliability Engineering, ISSRE 2003, pp. 332–343 (2003)
21. Tofte, M.: Standard ML language. Scholarpedia 4(2), 7515 (2009)
22. Vilkomir, S.A., Bowen, J.P.: Reinforced condition/decision coverage (RC/DC): a new criterion for software testing. In: Bert, D., Bowen, J.P., Henson, M.C., Robinson, K. (eds.) ZB 2002. LNCS, vol. 2272, pp. 291–308. Springer, Heidelberg (2002). https://doi.org/10.1007/3-540-45648-1_15
23. Wang, R., Artho, C., Kristensen, L.M., Stolz, V.: Visualization and abstractions for execution paths in model-based software testing. In: Ahrendt, W., Tapia Tarifa, S.L. (eds.) IFM 2019. LNCS, vol. 11918, pp. 474–492. Springer, Cham (2019). https://doi.org/10.1007/978-3-030-34968-4_26
24. Wang, R., Kristensen, L.M., Meling, H., Stolz, V.: Automated test case generation for the Paxos single-decree protocol using a Coloured Petri Net model. Logic. Algebraic Methods Program. **104**, 254–273 (2019)
25. Xu, D., Xu, W., Kent, M., Thomas, L., Wang, L.: An automated test generation technique for software quality assurance. IEEE Reliab. **64**(1), 247–268 (2015)

ON/OFF Control Trajectory Computation for Steady State Reaching in Batches Petri Nets

Ruotian Liu, Rabah Ammour$^{(\boxtimes)}$, Leonardo Brenner, and Isabel Demongodin

Aix-Marseille University, University of Toulon, CNRS, LIS, Marseille, France
{rabah.ammour,leonardo.brenner,isabel.demongodin}@lis-lab.fr

Abstract. This paper, dedicated to controlled generalized batches Petri nets without discrete nodes, presents a method for computing a control trajectory for reaching a steady state from a given initial marking. A steady state is characterized by a state in which the marking and the firing flow vector are constant. By controlling the firing flow vector of transitions, the proposed control strategy is an event-based one and relies on an algorithm solving a linear programming problem. This new control strategy, called in this paper as maximal flow based ON/OFF control, exploits the maximal firing flows and reduces the delay of the transient behavior for reaching the steady state. A practical communication system is provided to illustrate the relevance of such a control strategy for Cyber-Physical Systems.

Keywords: Petri nets · Hybrid systems · Event-driven control · Steady state.

1 Introduction

The prevalence of Cyber-Physical Systems (CPSs) in a wide range of critical applications requires the development of efficient control and monitoring methods for such systems. Since the communication between entities of a CPS is carried out through network and internet, this type of system is vulnerable to attacks and threats [6]. To deal with the modeling, control and analysis of CPSs, the use of Discrete Event Systems (DESs) formalisms, such as Automata and Petri Nets (PNs), has been widely considered [10].

Some important works have been devoted to the control problems of such models with continuous Petri nets under infinite server semantics, applying discrete-time control such as ON/OFF controllers [11] or Model Predictive Control [9]. As studied by [1], the event-driven control strategies provide an alternative to time-driven ones in hybrid systems where the control action is updated when an event occurs. In this spirit, [7] have proposed an event-driven control of timed hybrid Petri nets.

In this work, we consider controlled Generalized Batches Petri Nets (cGBPNs) [3] with the transition firing flows as control inputs. Such models

© Springer Nature Switzerland AG 2020
B. Ben Hedia et al. (Eds.): VECoS 2020, LNCS 12519, pp. 84–99, 2020.
https://doi.org/10.1007/978-3-030-65955-4_7

enrich the class of hybrid Petri nets [2] by introducing a new kind of nodes, called batch nodes. Batch transitions act as continuous ones with finite server semantics while batch places are hybrid ones defined by three continuous characteristics: a length, a transfer speed and a maximal density. Based on the concept of batches as marking, i.e., a group of entities moving through a batch place at a certain speed, cGBPNs allow variable delays on continuous flows to be represented. cGBPNs behavior is based on a switching continuous-time and event-driven dynamics.

A control strategy of cGBPNs was presented in our prior work [8] which is steady flow based ON/OFF control strategy. It assumes that the controlled firing flow of each transition cannot exceed its steady firing flow. Although the convergence to the steady state using this strategy was proved, the main drawback of using steady flows rather than the maximal values is the impact on the time performance. The main contribution of this work is to exploit the maximal firing flows when it is possible to improve the convergence delay. Since a certain configuration of the moving entities characterized by a steady state density should be reached, the use of the maximal firing flows is restricted and is only feasible when some conditions that are developed are satisfied.

The remainder of this paper is structured as follows. In Sect. 2, some preliminaries on cGBPNs and their steady states are presented. A running example is introduced and used throughout the paper to illustrate the different notions. Section 3 formalizes the control problem and the steady flow based ON/OFF control is recalled. In Sect. 4, the new proposed control strategy and the computation of the control trajectory are detailed. A communication system illustrates the presented approach in Sect. 5. Finally, some conclusions and future works are presented in Sect. 6.

2   Preliminaries

In this section, we introduce some basic definitions on controlled generalized batches Petri nets (cGBPNs) and identify the steady states.

2.1   Generalized Batches Petri Nets

We assume the reader to be familiar with hybrid Petri nets such as defined by [2]. For more details on batches Petri net formalisms, we refer readers to [3,5].

Definition 1. *A* Generalized Batches Petri net *(GBPN) is a 6-tuple* $N = (P, T, Pre, Post, \gamma, Time)$ *where:*

- $P = P^D \cup P^C \cup P^B$ *is finite set of places partitioned into the three classes of* discrete, continuous *and* batch *places.*
- $T = T^D \cup T^C \cup T^B$ *is finite set of transitions partitioned into the three classes of* discrete, continuous *and* batch *transitions.*

- *Pre, Post* : $(P^D \times T \rightarrow \mathbb{N}) \cup ((P^C \cup P^B) \times T \rightarrow \mathbb{R}_{\geq 0})$ [1] *are, respectively, the pre-incidence and post-incidence matrices, denoting the weight of the arcs from places to transitions and transitions to places.*
- $\gamma : P^B \rightarrow \mathbb{R}^3_{>0}$ *is the* batch place function. *It associates to each batch place $p_i \in P^B$ the triple $\gamma(p_i) = (V_i, d_i^{\max}, s_i)$ that represents, respectively,* maximal transfer speed, maximal density *and* length *of p_i.*
- *Time* $: T \rightarrow \mathbb{R}_{\geq 0}$ *associates a nonnegative number to every transition:*
  - *if $t_j \in T^D$, then $Time(t_j) = \psi_j$ denotes the* firing delay *associated with the discrete transition;*
  - *if $t_j \in T^C \cup T^B$, then $Time(t_j) = \Phi_j$ denotes the* maximal firing flow *associated with the continuous or batch transition.* ∎

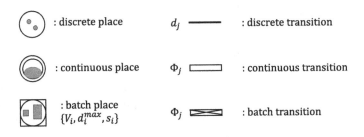

Fig. 1. Nodes of a GBPN.

The cardinality of the sets of places and transitions is denoted as $n^Y = |P^Y|$ and $h^Y = |T^Y|$, respectively, with $Y \in \{D, C, B\}$. Note that n and h are respectively the total number of places and transitions. Each node of a GBPN has a graphic representation, as shown in Fig. 1. The marking (i.e., net state) at time τ is a vector $\boldsymbol{m}(\tau) = [m_1(\tau) \ m_2(\tau) \ldots m_n(\tau)]^T$ that assigns to each discrete place a nonnegative integer, to each continuous place a nonnegative real number and assigns to each batch place p_i, a series of ordered batches, $m_i(\tau) = \{\beta_i^1(\tau), \ldots, \beta_i^r(\tau)\}$. A batch is a group of discrete entities, characterized by three continuous variables, a length (space unit), a density (number of entites/space unit) and a head position (space unit).

Definition 2. *A batch β_k at time τ is defined by a triple $\beta_k(\tau) = (l_k(\tau), d_k(\tau), x_k(\tau))$, where $l_k(\tau) \in \mathbb{R}_{\geq 0}$ is the length, $d_k(\tau) \in \mathbb{R}_{\geq 0}$ is the density and, $x_k(\tau) \in \mathbb{R}_{\geq 0}$ is the head position.* ∎

Note that the initial marking is denoted by $\boldsymbol{m}_0$ and time τ will be omitted in the rest of paper when there is no ambiguity.

Each batch place p_i is characterized by a maximal capacity given by $Q_i = s_i \cdot d_i^{\max}$. The output density of a batch place p_i, denoted d_i^{out}, is the density of

---

[1] We denote $\mathbb{R}_{\geq 0}$ (resp., $\mathbb{R}_{>0}$) the set of nonnegative (resp., positive) real numbers and $\mathbb{N}$ the set of natural numbers.

the batch $\beta_i^r(\tau)$ whose head position is equal to the length of the batch place (i.e., $x_i^r(\tau) = s_i$), then $d_i^{\text{out}}(\tau) = d_i^r(\tau)$, else $d_i^{\text{out}}(\tau) = 0$. The batch $\beta_i^r(\tau)$ is called *output batch* of p_i. A batch with a density that equals the maximal density of a batch place, i.e., $d_i^r(\tau) = d_i^{max}$, is called a *dense batch*.

Due to the existence of batches inside a batch place, we define the marking quantity as the total quantity of a place. The marking quantity of a continuous or discrete place is equal to its marking while it corresponds for a batch place to the sum of the quantities of all batches contained inside. The marking quantity vector is formally defined as follows:

Definition 3. *The* marking quantity vector $\boldsymbol{q} = \mu(\boldsymbol{m}) \in \mathbb{R}_{\geq 0}^n$ *associated with a marking* $\boldsymbol{m}$ *is defined as follows:*

$$q_i = \begin{cases} m_i & \text{if } p_i \in P^D \cup P^C, \\ \sum\limits_{\beta_i^k \in m_i} l_i^k \cdot d_i^k & \text{if } p_i \in P^B. \end{cases}$$

∎

Remark 1. $\mu(\boldsymbol{m})$ is an injective mapping which associates a given marking $\boldsymbol{m}$ to a single marking quantity vector $\boldsymbol{q}$. Its inverse is not, since more than one marking $\boldsymbol{m}$ may correspond to a given marking quantity vector $\boldsymbol{q}$.

The dynamics of a GBPN, ruled by the firing of enabled transitions and the hybrid dynamics inside batch places, uses the notion of *instantaneous firing flow vector* (IFF). More precisely, the IFF $\varphi_j(\tau) \leq \Phi_j$ of a transition $t_j \in T^C \cup T^B$ is used to represent the firing quantity by time unit. The IFF vector at time τ is denoted by $\boldsymbol{\varphi}(\tau) \in \mathbb{R}^{h^C + h^B}$. The *input* and *output flow* of a batch or continuous place p_i at time τ are the sum of all flows entering and leaving the place, respectively, denoted by: $\phi_i^{\text{in}}(\tau) = \text{Post}(p_i, \cdot) \cdot \boldsymbol{\varphi}(\tau)$ and $\phi_i^{\text{out}}(\tau) = \text{Pre}(p_i, \cdot) \cdot \boldsymbol{\varphi}(\tau)$.

The behavior of a GBPN is based on a timed discrete event dynamics. Between two timed events, the net state is characterized by an invariant behavior state (IB-state) [4] with linear or constant continuous evolution. Thus, the marking in discrete places, the firing flow of continuous and batch transitions and the output density of batch places are constant within an IB-state. Note that, as $\boldsymbol{\varphi}$ is constant in an IB-state, ϕ^{in} and ϕ^{out} are also constants between timed events.

The evolution of a GBPN is characterized by a fundamental equation expressed by:

$$\boldsymbol{q}(\tau) = \boldsymbol{q}(\tau_0) + \boldsymbol{C} \cdot \boldsymbol{z}(\tau) \tag{1}$$

where $\boldsymbol{C} = \text{Post} - \text{Pre}$ and $\boldsymbol{z}(\tau) \in \mathbb{R}_{\geq 0}^h$, called *firing quantity vector*, denotes the firing count of each transition during time interval $[\tau_0, \tau]$. Note that $\boldsymbol{z}(\tau_0) = \boldsymbol{0}$.

Example 1. Let us consider the GBPN represented in Fig. 2 composed of two continuous places and two batch places. The characteristics of each batch place (maximal transfer speed, maximal density and length) are depicted on the figure. The initial marking is $\boldsymbol{m}_0 = [8\ 8\ \emptyset\ \emptyset]^T$ which means that both batch places are empty. The maximal capacity of batch places are $Q_3 = 8$ and $Q_4 = 10$. Since the only enabled transition is t_1, the IFF vector is given by $\boldsymbol{\varphi}(\tau_0) = [2\ 0\ 0]^T$ at date $\tau_0 = 0$. Consequently, in the batch place p_3, a new batch is created as $\beta_3^1(\tau_0) = (0 \cdot V_3, \phi_3^{in}(\tau_0)/V_3, 0 \cdot V_3) = (0, 2, 0)$. At date $\tau_1 = 4$, this batch becomes an output batch as $\beta_3^1(\tau_1) = (4 \cdot V_3, 2, 4 \cdot V_3) = (4, 2, 4)$ and a new IFF vector $\boldsymbol{\varphi}(\tau_1) = [2\ 2\ 0]^T$ is applied since transition t_2 is also enabled. Thus the time interval $[\tau_0, \tau_1]$ corresponds to an IB-state. Note that, whatever the evolution, it holds : $q_1 + q_3 = 8$ and $q_2 + q_4 = 8$.

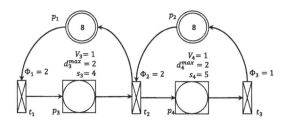

Fig. 2. A cGBPN system with initial marking $\boldsymbol{m}_0$.

2.2 Controlled GBPN and Steady State

In our work, we consider the instantaneous firing flow of continuous and batch transitions as control input. This leads to the following definition.

Definition 4. *A controlled Generalized Batches Petri net (cGBPN) is a GBPN N for which a control input $\boldsymbol{u}(\tau) = \boldsymbol{\varphi}(\tau)$ is defined with $0 \le \boldsymbol{u}(\tau) \le \boldsymbol{\Phi}$.* ∎

The following definition, taken from [3], gives the definition of a steady state of cGBPN.

Definition 5. (*Steady state*) *Let $\langle N, \boldsymbol{m}_0 \rangle$ be a cGBPN with $P^D = T^D = \emptyset$. The net is in a steady state at time τ_s if for $\tau \ge \tau_s$ the marking $\boldsymbol{m}^s$ and the instantaneous firing flow vector $\boldsymbol{\varphi}^s$ remain constant. Thus a steady state is defined by a pair $(\boldsymbol{m}^s, \boldsymbol{\varphi}^s)$. The marking quantity vector associated with a steady marking $\boldsymbol{m}^s$ is denoted as $\boldsymbol{q}^s = \mu(\boldsymbol{m}^s)$.* ∎

For a cGBPN with $P^D = T^D = \emptyset$, with an initial marking quantity vector $\boldsymbol{q}^0$ and a given steady marking quantity vector $\boldsymbol{q}^s$, we denote by $\boldsymbol{z}^s$ the *steady firing quantity vector*, i.e., the minimum firing quantity vector, that satisfies: $\boldsymbol{z}^s \ge \boldsymbol{0}$, and $\min \boldsymbol{z}^s$, s.t. $\boldsymbol{q}^s = \boldsymbol{q}^0 + \boldsymbol{C} \cdot \boldsymbol{z}^s$.

In a steady state $(\boldsymbol{m}^s, \boldsymbol{\varphi}^s)$ where $\boldsymbol{\varphi}^s > 0$, it has been proved in [3] that the marking of a batch place p_i has one of the following regular forms:

1. A single batch : $m_i^s = \{\beta_i^o\}$ with $\beta_i^o = (s_i, d_i^o, s_i)$ and density $d_i^o \leq d_i^{\max}$. In other terms, the steady state marking of p_i is composed by a single (dense or not) output batch whose length is equal to the length of p_i.
2. Two batches : $m_i^s = \{\beta_i^e, \beta_i^o\}$ with $\beta_i^e = (l_i^e, d_i^e, l_i^e)$, $\beta_i^o = (l_i^o, d_i^{\max}, s_i)$ and $l_i^e + l_i^o = s_i$. The steady state marking of p_i is composed by a dense output batch in contact with one input batch.

In [3], (q^s, φ^s) of a cGBPN without discrete nodes, has been computed by solving a linear programming problem that only considers the net structure and the initial marking (see Proposition 4.5 in [3]). From (q^s, φ^s), the steady state marking m^s could be characterized (see Proposition 4.4 in [3]). Note that, in a steady state, the input flow and output flow of place p_i are equal and denoted as ϕ_i^s.

Definition 6. *Let* $\langle N, m_0 \rangle$ *be a cGBPN with* $P^D = T^D = \emptyset$ *and* (m^s, φ^s) *a reachable steady state. The event-driven control problem for reaching* (m^s, φ^s) *deals with the computation of a timed control trajectory* $(u^0, \tau_0), (u^1, \tau_1), \cdots, (u^i, \tau_i), \cdots, (u^s, \tau_s)$ *feasible in the cGBPN from* m_0 *such that* $m(\tau_f) = m^s$ *and* $u(\tau_f) = u^s = \varphi^s, \forall \tau_f \geq \tau_s$. ∎

Note that the controlled firing flow vector u^i is applied at date τ_i and remains constant until τ_{i+1}, corresponding to a timed interval of an IB-state. When a cGBPN is without discrete nodes, the fundamental equation Eq. (1) for the control trajectory $(u^0, \tau_0), (u^1, \tau_1), \cdots, (u^i, \tau_i), \cdots, (u^s, \tau_s)$ becomes:

$$q(\tau_s) = q^0 + C \cdot \left(\int_{\tau_0}^{\tau_1} u^0 \cdot d\rho + \cdots + \int_{\tau_{s-1}}^{\tau_s} u^{s-1} \cdot d\rho \right). \tag{2}$$

where $\int_{\tau_{i-1}}^{\tau_i} u^{i-1} \cdot d\rho$ denotes the sum of firing quantity of continuous or batch transitions during time interval $[\tau_{i-1}, \tau_i]$.

Example 2. A reachable steady state for the net presented in Example 1 is given by : $m^s = [4 \ 0 \ \{(4,1,4)\} \ \{(2,1,2),(3,2,5)\}]^T$ (see Fig. 3) and $\varphi^s = [1 \ 1 \ 1]^T$. Batch place p_3 is in the first regular form with only one batch inside $\beta_3^o(\tau \geq \tau_s) = (4,1,4)$ while batch place p_4 has two batches: a dense output batch $\beta_4^o(\tau \geq \tau_s) = (3,2,5)$ in contact with an input batch $\beta_4^e(\tau \geq \tau_s) = (2,1,2)$. The marking remains constant since the input and output flow of each place are equal.

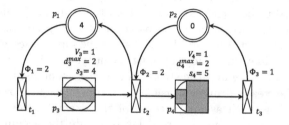

Fig. 3. The cGBPN of Example in a steady state.

3   Steady Flow Based ON/OFF Control

The problem addressed in this paper is the control of the transient behavior of a cGBPN. More specifically, the objective is to compute a control trajectory that drives the net from a given initial marking m_0 to a reachable steady state (m^s, φ^s). Such a control has been proposed in [8] denoted as steady flow based ON/OFF control, in this paper. This section aims to present this previous proposed control strategy. In the rest of the paper, we consider the following assumptions already done in [8].

A1 No discrete nodes $(P^D = T^D = \emptyset)$.
A2 The steady firing flow vector is positive $(\varphi^s > \mathbf{0})$.
A3 The net is conservative.

Assumption (A1) preserves the restricted class of cGBPNs for which the steady state analysis has been studied [3]. Assumption (A2) ensures that the net is consistent, i.e., there exits a positive T-semiflow. Assumption (A3) imposes that each place is contained in the support of a P-semiflow.

The basic idea of steady flow based ON/OFF control is, from the initial marking, to firstly reach the steady marking quantity vector and then to reach the steady state marking. More precisely, enabled transitions that at least one of their input places has less marking quantity than its steady marking quantity, are blocked. For the other enabled transitions, their controlled firing flows are maximized to their steady flows. For any place that reaches its steady marking quantity, an additional constraint imposes that its input flow equals its output flow. This control law is applied until the target steady state marking is reached. One can remark that the firing flow of a transition cannot exceed its steady flow. To illustrate this approach, the introduced example is considered again.

Example 3. Let us consider again the cGBPN depicted in Fig. 2 with initial marking $m_0 = [8\ 8\ \emptyset\ \emptyset\}]^T$. The control trajectory, obtained by the steady flow based ON/OFF approach, to reach the steady state $m^s = [4\ 0\ \{(4,1,4)\}\ \{(2,1,2),(3,2,5)\}]^T$ showed in Fig. 3 is given by $(u^0, 0), (u^1, 4), (u^2, 9), (u^3, 12)$, with $u^0 = [1\ 0\ 0]^T$, $u^1 = [1\ 1\ 0]^T$, $u^2 = [1\ 1\ 0]^T$, $u^3 = [1\ 1\ 1]^T$. At the initial marking, only the flow of transition t_1 is maximized to the steady flow since only t_1 is enabled and its input place has greater marking quantity than the steady one. At date $\tau_1 = 4$, batch place p_3 has reached its steady marking quantity and transition t_2 is enabled. The input and output controlled flows of p_3 are imposed to be equal as order to hold the marking quantity of p_3 and to feed place p_4. At date $\tau_2 = 9$, the batch created in p_4 becomes an output batch, transition t_3 is enabled and blocked since the input place p_4 has still less marking quantity than the steady marking quantity. At date $\tau_3 = 12$, all places reach the steady marking and the steady flow vector is imposed.

The use of the steady firing flow as maximal control value allows to construct the regular marking such as for place p_3 of the example. However, for place p_4, this limitation causes much more delay to create the accumulated part of the steady marking. Thus the use of the maximal firing flow could reduce the delay for reaching the steady state. From this motivation, the following work is dedicated to developing a new control strategy that exploits the maximal firing flow to improve the time performance.

4 Maximal Flow Based ON/OFF Control

In this section, we propose a new method that improves the time performance compared with the steady flow based ON/OFF control strategy. Before introducing the control strategy, some necessary notions are first presented hereafter.

According to the regular forms in steady state marking (see Sect. 2), the steady marking quantity of a batch place p_i could be divided into two components of marking quantities $q_i^{f,s}$ and $q_i^{a,s}$ defined as follows.

Definition 7. *The steady marking quantity of place p_i associated with a steady state marking $\boldsymbol{m}^s$ could be divided as:*

$$q_i^s = q_i^{f,s} + q_i^{a,s},$$

where $q_i^{f,s}$ and $q_i^{a,s}$ represent, respectively, the steady marking quantity of free part and the steady marking quantity of accumulated part. They are defined as follows:

1. *If p_i is a batch place :*
   (a) if $m_i^s = \{\beta_i^e, \beta_i^o\}$, $q_i^{f,s} = l_i^e \cdot d_i^e$ and $q_i^{a,s} = l_i^o \cdot d_i^{\max}$.
   (b) if $m_i^s = \{\beta_i^o\}$,
      $- \; d_i^o = \phi_i^s / V_i$, such that $q_i^{f,s} = s_i^o \cdot d_i^o$, and $q_i^{a,s} = 0$.
      $- \; d_i^o \neq \phi_i^s / V_i$, such that $q_i^{a,s} = s_i^o \cdot d_i^o$, and $q_i^{f,s} = 0$.
2. *If p_i is a continuous place : $q_i^{f,s} = 0$ and $q_i^{a,s} = q_i^s$.* ∎

Definition 8. *For a cGBPN at marking $\boldsymbol{m}$, given its steady state marking $\boldsymbol{m}^s$ and steady firing quantity vector $\boldsymbol{z}^s$. The minimum remaining quantity that enters into a place p_i from an input transition t_j for reaching the steady marking quantity vector $\boldsymbol{q}^s$ is denoted as $q_{i,j}^{rs}(\boldsymbol{m})$ and given by $q_{i,j}^{rs}(\boldsymbol{m}) = Post(p_i, t_j)(z_j^s - z_j(\boldsymbol{m}))$.* ∎

Let us remark that when $q_{i,j}^{rs}(\boldsymbol{m}) \leq 0$ holds, it is not necessary to fire transition t_j to reach the steady marking quantity of place p_i at current marking $\boldsymbol{m}$.

Example 4. Let us consider the cGBPN with the given steady state $\boldsymbol{m}^s = [4\ 0\ \{(4,1,4)\}\ \{(2,1,2),(3,2,5)\}]^T$ as shown in Fig. 3. The steady flow vector is $\boldsymbol{\varphi}^s = [1\ 1\ 1]^T$ and the steady marking quantity vector is $\boldsymbol{q}^s = [4\ 0\ 4\ 8]^T$. For continuous places, q_1^s corresponds to the steady marking quantity of accumulated part $q_1^{a,s}$, i.e., $q_1^s = q_1^{a,s} = 4$, and $q_2^s = q_2^{a,s} = 0$ while $q_1^{f,s} = q_2^{f,s} = 0$. For batch place p_3, $d_3^o = \phi_3^s / V_3 = 1\ /\ 1 = 1$, thus the steady marking quantity of free part is $q_3^{f,s} = l_3^o \cdot d_3^o = 4 \cdot 1 = 4$ and the steady marking quantity of accumulated part is $q_3^{a,s} = 0$. For batch place p_4, the steady marking quantity of free part is $q_4^{f,s} = l_4^e \cdot d_4^e = 2$, and $q_4^{a,s} = l_4^o \cdot d_4^o = 6$ for the accumulated part.

For the minimum remaining quantity, the steady firing quantity vector to reach $\boldsymbol{q}^s$ at marking $\boldsymbol{m}_0$ is $\boldsymbol{z}^s = [12\ 8\ 0]$, thus, for batch place p_3 and its input transition t_1, $q_{3,1}^{rs} = \text{Post}(p_3, t_1)(z_1^s - z_1(\boldsymbol{m}_0)) = 1 \cdot (12 - 0) = 12$ corresponds to the remaining quantity entering into p_3 for reaching steady marking quantity. Similarly, $q_{4,2}^{rs} = 8$, $q_{2,3}^{rs} = 0$ and $q_{2,1}^{rs} = 8$ which means that transition t_3 does not need to be fired to reach $\boldsymbol{q}^s$ as already specified by $\boldsymbol{z}^s$.

4.1 Proposed Control Strategy

Due to the existence of batches in a batch place, a steady state marking is reached by guaranteeing both the steady marking quantity and the regular forms of batches in a batch place. Hence, the use of the maximal flows of transitions as the threshold to feed places improves the time performance to reach $\boldsymbol{q}^s$ but more time will be needed to generate regular forms given by $\boldsymbol{m}^s$. Therefore, a condition on the minimum remaining quantity is introduced to switch the threshold from the maximal flow into the steady flow for getting the steady marking quantity and creating the regular forms as follows:

$$q_{i,j}^{rs}(\boldsymbol{m}) \leq q_i^{f,s}, \tag{3}$$

where $q_{i,j}^{rs}(\boldsymbol{m}) = \text{Post}(p_i, t_j)(z_j^s - z_j(\boldsymbol{m}))$ is the minimum remaining quantity given by Definition 8. This quantity is thus compared with the free part quantity of place p_i.

The basic idea of the proposed control strategy is to limit the firing flow of an enabled transition to its steady value when the condition given by Eq. (3) is satisfied. For any enabled transition that does not satisfy this condition, its flow is maximized to its maximum flow value. Finally, the flow of an enabled transition is set to zero when the condition given by Eq. (3) is satisfied but at least one of its input places has less quantity than steady marking quantity. At a given marking $\boldsymbol{m}(\tau)$, the proposed control strategy could be summarized as follows[2]:

- ON_{ss}: maximize $u_j(\tau)$ to φ_j^s if t_j is enabled with $\exists\, p_k \in t_j^\bullet : q_{k,j}^{rs}(\boldsymbol{m}) \leq q_k^{f,s}$.
- ON_{max}: maximize $u_j(\tau)$ to Φ_j if t_j is enabled with $\nexists\, p_k \in t_j^\bullet : q_{k,j}^{rs}(\boldsymbol{m}) \leq q_k^{f,s}$.
- OFF: $u_j(\tau) = 0$ if t_j is not enabled or if t_j is enabled with $\exists\, p_i \in {}^\bullet t_j : q_i(\tau) < q_i^s$ and $\exists\, p_k \in t_j^\bullet : q_{k,j}^{rs}(\boldsymbol{m}) \leq q_k^{f,s}$.

The controlled flow values are maintained during an IB-state and updated when a particular event (or possibly several events at the same time) occurs. Compared with our previous work in [8], three more events are considered. One is that the remaining marking quantity satisfies $q_{i,j}^{rs}(\boldsymbol{m}) = q_i^{f,s}$. The two other considered events are: a continuous place becomes empty and batch place becomes full where the control actions are updated to ensure the nonnegativity of marking

---

[2] For $t \in T$, ${}^\bullet t = \{p \in P \mid \text{Pre}(p,t) > 0\}$, and $t^\bullet = \{p \in P \mid \text{Post}(p,t) > 0\}$..

quantity and avoid increasing the marking quantity, respectively. All the events considered by the controller are the following:

- The marking quantity of place p_i reaches its steady quantity q_i^s;
- The remaining marking quantity $q_{i,j}^{rs}(\boldsymbol{m})$ becomes equal to $q_i^{f,s}$;
- A continuous place becomes empty;
- A batch place becomes full;
- The value of batch place output density changes.

4.2   Computation of the Timed Control Trajectory

Before we present the algorithm that allows one to compute a control trajectory, several sets must be defined at any marking $\boldsymbol{m}$.

- $P_\emptyset(\boldsymbol{m}) = \{p_i \in P^C \mid m_i = 0\}$ be the subset of empty continuous places.
- $P_F(\boldsymbol{m}) = \{p_i \in P^B \mid q_i = Q_i\}$ be the subset of full batch places.
- $S_L(\boldsymbol{m}) = \{p_i \in P \mid q_i < q_i^s\}$ be the subset of places whose marking quantities are lower than their steady state values.
- $S_E(\boldsymbol{m}) = \{p_i \in P \mid q_i = q_i^s\}$ be the subset of places whose marking quantities are equal to their steady state values.
- $T_L(\boldsymbol{m}) = \{t_j \in T \mid \exists p_i \in \{S_L(\boldsymbol{m}) \cap {}^\bullet t_j\}\}$ be the subset of transitions with at least one of its input places belonging to $S_L(\boldsymbol{m})$.
- $T_N(\boldsymbol{m})$ be the subset of transitions that are not enabled at $\boldsymbol{m}$.
- $T_Z(\boldsymbol{m}) = \{t_j \in T \mid \exists \ p_i \in t_j^\bullet : q_{i,j}^{rs}(\boldsymbol{m}) \leq q_i^{f,s}\}$ be the subset of transitions with a remaining firing quantity greater or equal than the free part steady marking quantity of one of their output place.
- $T_{ZL}(\boldsymbol{m}) = T_Z(\boldsymbol{m}) \cap T_L(\boldsymbol{m})$ be the subset of transitions that belongs to $T_Z(\boldsymbol{m})$ and at least one of their input places belongs to $S_L(\boldsymbol{m})$.
- $S_{ZE}(\boldsymbol{m}) = \{p_i \in S_E(\boldsymbol{m}) \mid \forall \ t_j \in \{{}^\bullet p_i \cup p_i^\bullet\}$ such that $t_j \in T_Z(\boldsymbol{m})\}$ be the subset of places whose marking quantities are equal to their steady state values and the firing quantities of all their input and output transitions satisfy $T_Z(\boldsymbol{m})$.

Algorithm 1 computes a timed control trajectory from an initial state to a given reachable steady state.   It starts at initial marking $\boldsymbol{m}_0$ with a control firing flow vector $\boldsymbol{u}^0$ initialized to $\boldsymbol{0}$ at line 2. Each timed event that refers to an execution of the while-end structure (lines $3 - 8$), a new marking $\boldsymbol{m}^i$ is built and compared with the steady marking $\boldsymbol{m}^s$. Line 4 determines the necessary constraint sets to solve the LPP which maximizes the controlled firing flow vector at each timed event. Constraints (a) are used to maximize the flow approaching the maximal firing value. Constraints (b) show that the maximal flow of each transition t_j cannot exceed the steady firing flow φ_j^s when the remaining marking quantity associated with one of its output places p_k is less than or equal to the free part steady marking quantity i.e., $\exists \ p_k \in t_j^\bullet : q_{k,j}^{rs}(\boldsymbol{m}) \leq q_k^{f,s}$. Constraints (c) block transition t_j when it is not enabled or when belongs to $T_{ZL}(\boldsymbol{m})$. Constraints (d) require that the marking quantity cannot increase when the batch place is full. Constraints (e) claim that the marking cannot decrease when the

---

Algorithm 1: Computation of control trajectory

Input: A cGBPN $\langle N, m_0 \rangle$, a reachable steady state (m^s, φ^s) and the steady firing quantity vector z^s

1 . **Output:** Control trajectory $(u^0, \tau_0), (u^1, \tau_1), \cdots$
2 Initialize : $q^0 = \mu(m_0)$, $m^0 = m_0$, $u^0 = \mathbf{0}$, $z^0 = \mathbf{0}$, $\tau_0 = 0$, $i = 0$;
3 **while** $m^i \neq m^s$ **do**
4 $\quad$ Determine $T_\mathcal{N}(m^i)$, $T_L(m^i)$, $T_Z(m^i)$, $T_{ZL}(m^i)$, $P_\emptyset(m^i)$, $P_F(m^i)$ $S_L(m^i)$, $S_E(m^i)$, $S_{ZE}(m^i)$;
5 $\quad$ Solve the following LPP: max $\mathbf{1}^T \cdot u^i$ s.t.

$$
\begin{cases}
\text{(a) } 0 \le u_j^i \le \Phi_j & \forall t_j \in T \\
\text{(b) } 0 \le u_j^i \le \varphi_j^s & \forall t_j \in T_Z(m^i) \\
\text{(c) } u_j^i = 0 & \forall t_j \in T_\mathcal{N}(m^i) \cup T_{ZL}(m^i) \\
\text{(d) } C\,(p_k, \cdot) \cdot u^i \le 0 & \forall p_k \in P_F(m^i) \\
\text{(e) } C\,(p_k, \cdot) \cdot u^i \ge 0 & \forall p_k \in P_\emptyset(m^i) \\
\text{(f) } C\,(p_k, \cdot) \cdot u^i = 0 & \forall p_k \in S_{ZE}(m^i) \\
\text{(g) } \text{Post}\,(p_k, \cdot) \cdot u^i \le V_k \cdot d_k^{\max} & \forall p_k \in P^B \\
\text{(h) } \text{Pre}\,(p_k, \cdot) \cdot u^i \le V_k \cdot d_k^{\text{out}} & \forall p_k \in P^B
\end{cases}
$$

6 $\quad$ Determine all the next timed events, select the nearest in time and deduce time τ_{i+1} ;
7 $\quad$ Determine the new marking m^{i+1}, the marking quantity vector q^{i+1} and the current firing quantity vector z^{i+1};
8 $\quad$ i = i + 1;
9 $u^i = \varphi^s$;
10 Return $(u^0, \tau_0), \cdots, (u^i, \tau_i)$

---

continuous place is empty. Constraints (f) hold the marking quantity to the steady marking quantity when place p_i belongs to $S_{ZE}(m)$. Constraints (g) ensure that the input flow of a batch place should be less than or equal to the maximal flow that can be accepted by its transfer element. Constraints (h) require that the output flow of a batch place should be lower than or equal to the flow out-coming from the place. Lines 6–7 determine the nearest event and update the new marking. The procedure is repeated until the steady marking is reached. When the marking m^i reaches the steady state value m^s, the controlled firing flow vector is set to its steady firing flow vector φ^s as shown in line 9. The complexity of this algorithm is polynomial since it is obtained by solving a LPP at each timed step.

Remark 2. When transitions are in structural conflicts that are effective ones, the proportional policy is used to solve these conflicts by adding the following constraints : $u_a^i \cdot \varphi_b^s = u_b^i \cdot \varphi_a^s$ if $\forall t_a, t_b \in T, {}^\bullet t_a \cap {}^\bullet t_b \neq \emptyset$ and $t_a, t_b \notin T_\mathcal{N} \cup T_{ZL}$.

4.3   Transient Behavior of the Controlled Net

In this part, we focus on the convergence of the proposed algorithm for reaching a give steady state from an initial one. Different from the previous work that all the places converge and maintain the steady marking quantities, here we consider all the transitions should be included in the set of T_Z, then this set will hold on until the steady state marking is reached.

Lemma 1. *Consider a cGBPN $\langle N, \boldsymbol{m}_0 \rangle$ that satisfies assumptions (A1) – (A3). By Algorithm 1, a transition included in T_Z remains in T_Z whatever the net evolution.*

Proof. A transition t_j is included in T_Z if $\exists\ p_i \in t_j^\bullet : q_{i,j}^{rs}(\boldsymbol{m}) \leq q_i^{f,s}$, which is equivalent to $\exists\ p_i \in t_j^\bullet : z_j(\boldsymbol{m}) \geq z_j^s - q_i^{f,s}/\mathrm{Post}(p_i, t_j)$. Since $z_j^s - q_i^{f,s}/\mathrm{Post}(p_i, t_j)$ is a constant value and the firing quantity z_j is monotonically increasing, any new reachable marking $\boldsymbol{m}'$ will satisfy $z_j(\boldsymbol{m}') \geq z_j(\boldsymbol{m}) \geq z_j^s - q_i^{f,s}/\mathrm{Post}(p_i, t_j)$ for place p_i which means that t_j will remain in T_Z.   ◇   □

Lemma 2. *Consider a cGBPN $\langle N, \boldsymbol{m}_0 \rangle$ that satisfies assumptions (A1) – (A3). By Algorithm 1, all the transitions in the net N will be included in T_Z in finite time.*

Proof. Based on Lemma 1, the transitions that are already included in T_Z at $\boldsymbol{m}_0$ remain in T_Z. In case that transition t_j is not initially included in T_Z at $\boldsymbol{m}_0$, which means $\forall p_i \in t_j^\bullet : q_{i,j}^{rs}(\boldsymbol{m}_0) > q_i^{f,s}$, let us assume that this transition will never be included in T_Z and show that this leads to a contradiction.

The transition t_j cannot be included in T_Z in finite time means that it is blocked before satisfying the condition $\exists p_i \in t_j^\bullet : q_{i,j}^{rs}(\boldsymbol{m}_0) \leq q_i^{f,s}$ and remains blocked whatever the net evolution. According to Algorithm 1, a transition that does not belong to T_Z is blocked only when it is not enabled. More precisely, one of input places p_k of transition t_j should be empty. Due to assumption (A3), the empty place p_k must belong to a conservative component. This means that there must exist at least one upstream place p_l that has a marking quantity which is greater than its steady one. By Algorithm 1, the output transition of place p_l cannot be blocked for infinite time. Thus, place p_k cannot remain empty which contradicts the initial assumption. Consequently, transition t_j can be included in T_Z in finite time.   □

Proposition 1. *Given a cGBPN $\langle N, \boldsymbol{m}_0 \rangle$ that satisfies assumptions (A1) – (A3) and a given reachable steady state $(\boldsymbol{m}^s, \boldsymbol{\varphi}^s)$. The control trajectory computed by Algorithm 1 drives the system from its initial state to the given steady state in finite time.*

Proof. According to Lemmas 1 and 2, all the transitions of the net will be included in T_Z in finite time. When $T_Z = T$, the controlled firing flows are maximized to the steady ones by constraint (b) in Algorithm 1 and the sets T_{ZL} and S_{ZE} will satisfy $T_{ZL} = T_L$ and $S_{ZE} = S_E$. In this case, the algorithm is equivalent to the one presented in our previous work in [8] where the convergence of the marking to the steady state marking has been proved (see Proposition 12 in [8]). $\diamond$ $\square$

Example 5. Let us consider again the cGBPN with an initial marking $m_0 = [8\ 8\ \emptyset\ \emptyset\}]^T$ in Fig. 2 and the reachable steady state (m^s, φ^s) with $m^s = [4\ 0\ \{(4,1,4)\}\ \{(2,1,2),(3,2,5)\}]^T$ as shown in Fig. 3, and $\varphi^s = [1\ 1\ 1]^T$. The steady marking quantity vector is $q^s = [4\ 0\ 4\ 8]^T$, and the steady firing quantity vector is $z^s = [12\ 8\ 0]^T$. Under the proposed strategy, a control trajectory that drives the net from its initial state to the steady state is obtained as follows.

- At $\tau_0 = 0$, $T_N(m_0) = \{t_2, t_3\}$. As $q^0 = [8\ 8\ 0\ 0]^T$, the subsets of places are: $S_E(m_0) = \emptyset$, $S_L(m_0) = \{p_3, p_4\}$, $P_F(m_0) = \emptyset$ and, $T_L(m_0) = \{t_2, t_3\}$. The firing quantity vector is $z(m_0) = [0\ 0\ 0\ 0]$, by checking the condition $q_{i,j}^{rs}(m_0) \leq q_i^{f,s}$, $T_Z(m_0) = \{t_3\}$, $T_{ZL}(m_0) = \{t_3\}$ and, $S_{ZE}(m_0) = \emptyset$. Consequently, transition t_1 is enabled and ON_{max} since it is not in $T_Z(m_0)$ while transitions t_2 and t_3 are not enabled, thus $u_2^0 = u_3^0 = 0$. Algorithm 1 returns the controlled firing flow vector $u^0 = [2\ 0\ 0]^T$.
- At $\tau_1 = 2$, places p_1 and p_3 reach steady marking quantities so $S_E(m^1) = \{p_1, p_3\}$ where $m^1 = [4\ 8\ \{(2,2,2)\}\ \emptyset]^T$ and $q(\tau_1) = [4\ 8\ 4\ 0]^T$. This implies $T_N(m^1) = \{t_2, t_3\}$, $P_F(m^1) = \emptyset$, $S_L(m^1) = \{p_4\}$ and $T_L(m^1) = \{t_3\}$. The firing quantity vector is $z(m^1) = [4\ 0\ 0]$, $T_Z(m^1) = \{t_3\}$, $T_{ZL}(m^1) = \{t_3\}$, and $S_{ZE}(m^1) = \emptyset$. Consequently, transition t_1 is enabled and ON_{max} since it is not in $T_Z(m^1)$. Transitions t_2 and t_3 are not enabled, thus $u_2^1 = u_3^1 = 0$. The obtained controlled firing flow vector is $u^1 = [2\ 0\ 0]^T$.

By continuing the Algorithm execution, the resulting control trajectory is given by $(u^0, 0), (u^1, 2), (u^2, 4), (u^3, 6), (u^4, 7), (u^5, 8), (u^6, 9), (u^7, 10)$ with $u^0 = [2\ 0\ 0]^T$, $u^1 = [2\ 0\ 0]^T$, $u^2 = [0\ 2\ 0]^T$, $u^3 = [1\ 2\ 0]^T$, $u^4 = [1\ 0\ 0]^T$, $u^5 = [1\ 1\ 0]^T$, $u^6 = [1\ 1\ 0]^T$, $u^7 = [1\ 1\ 1]^T$.

One can remark that the delay $\tau_s = 10$ for reaching the steady state is improved compared with the steady flow based ON/OFF control $\tau_s = 12$ obtained in Example 3 although the number of steps is greater (7 generated events with the proposed strategy and 3 events using the steady flow strategy). The time improvement is due to the use of the maximal firing flow (vectors u^0, u^1, u^2, and, u^3).

5   Study Case

The considered system depicted in Fig. 4 is a communication system composed of three buffers and four routers. Its corresponding cGBPN model is represented in Fig. 5(a) where three buffers are respectively modeled by three batch places p_4, p_5, p_6 with $\gamma(p_4) = (V_4, d_4^{max}, s_4) = (8, 4, 4)$, $\gamma(p_5) = (2, 3, 5)$, $\gamma(p_6) = (2, 4, 4)$, and the routers are modeled with five controlled transitions such that the processing rates of the routers are controllable variables. The maximal flow of the transitions are given by: $\Phi_1 = \Phi_2 = \Phi_4 = \Phi_5 = 4$, $\Phi_3 = 8$. Data arrived in buffer B_1 from router R_1 are processed by the router R_2 which allows to direct the data flows throughout two different channels represented by (B_2, R_3) and (B_3, R_4). The desired (nominal) behavior of the system is represented by the steady state showed in Fig. 5(a) and characterized by $(\boldsymbol{m}^s, \boldsymbol{\varphi}^s)$ with $\boldsymbol{m}^s = [10\ 0\ 0\ \{(4, 0.5, 4)\}\ \{(5, 1, 5)\}\ \{(2, 1, 2), (2, 4, 4)\}]^T$ and $\boldsymbol{\varphi}^s = [4\ 2\ 2\ 2\ 2]^T$. It means that only buffer B_3 has an accumulation while data in buffers B_1 and B_2 are processed as soon as they are received by the routers. The steady marking quantity vector is $\boldsymbol{q}^s = [10\ 0\ 0\ 2\ 5\ 10]^T$.

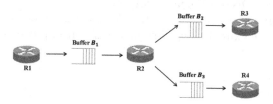

Fig. 4. The considered communication system.

Let us consider that some faults or cyber-attack causes a deviation from the normal behavior which leads the system to the state represented in Fig. 5 with $\boldsymbol{m}_0 = [0\ 0\ 10\ \{(3, 4, 4)\}\ \{(2.5, 2, 5)\}\ \emptyset]^T$. The steady firing quantity vector for reaching the steady marking quantity vector is $\boldsymbol{z}^s = [0\ 0\ 10\ 0\ 0]^T$.

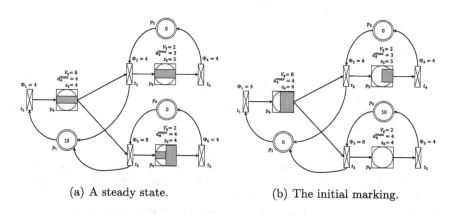

(a) A steady state.   (b) The initial marking.

Fig. 5. A cGBPN of study case.

The proposed control strategy is used to compute a control trajectory that drives the net from its abnormal state given by $\boldsymbol{m}_0$ to the steady state $\boldsymbol{m}^s$. By applying Algorithm 1, the following control trajectory is obtained $(\boldsymbol{u}^0, 0), (\boldsymbol{u}^1, 0.625), (\boldsymbol{u}^2, 1), (\boldsymbol{u}^3, 1.5), (\boldsymbol{u}^4, 2), (\boldsymbol{u}^5, 2.5)$ with $\boldsymbol{u}^0 = [0\ 2\ 8\ 2\ 0]^T$, $\boldsymbol{u}^1 = [0\ 2\ 8\ 2\ 0]^T$, $\boldsymbol{u}^2 = [4\ 2\ 2\ 2\ 0]^T$, $\boldsymbol{u}^3 = [4\ 2\ 2\ 2\ 0]^T$, $\boldsymbol{u}^4 = [4\ 2\ 2\ 2\ 2]^T$, $\boldsymbol{u}^5 = [4\ 2\ 2\ 2\ 2]^T$.

Consequently, 5 events are generated and the delay for reaching the steady state with the maximal flow based ON/OFF strategy is $\tau_s = 2.5$.

Under the steady flow based ON/OFF control strategy, a control trajectory that drives the net from its initial state to the steady state is $(\boldsymbol{u}^0, 0), (\boldsymbol{u}^1, 0.625), (\boldsymbol{u}^2, 2), (\boldsymbol{u}^3, 2.5), (\boldsymbol{u}^4, 3), (\boldsymbol{u}^5, 5)$ with $\boldsymbol{u}^0 = [0\ 2\ 2\ 2\ 0]^T$, $\boldsymbol{u}^1 = [0\ 2\ 2\ 2\ 0]^T$, $\boldsymbol{u}^2 = [0\ 2\ 2\ 2\ 0]^T$, $\boldsymbol{u}^3 = [4\ 2\ 2\ 2\ 0]^T$, $\boldsymbol{u}^4 = [4\ 2\ 2\ 2\ 0]^T$, $\boldsymbol{u}^5 = [4\ 2\ 2\ 2\ 2]^T$.

Although the number of generated events is equal (5 events), the delay for reaching the steady state is $\tau_s = 5$ and is much higher compared with the maximal flow based ON/OFF control strategy.

From the results above, one can conclude that the proposed control method that considers the maximal firing flow of transition improves the time performance.

6   Conclusions

An algorithm that incrementally computes a control trajectory for reaching the steady state from an initial one has been proposed. The main advantage of this algorithm is to exploit the maximal firing flow of transitions compared with the previous one developed in [8]. The convergence of the algorithm has been proved and the proposed method is applied to recover a communication systems from its abnormal state to its nominal steady state. In future work, we plan to compare the two methods to quantify the improvement. We will focus on the optimality to address the minimum-time control problem.

References

1. Cassandras, C.G.: The event-driven paradigm for control, communication and optimization. J. Control Decis. **1**(1), 3–17 (2014)
2. David, R., Alla, H.: Discrete, Continuous, and Hybrid Petri Nets, vol. 1. Springer, Heidelberg (2005)
3. Demongodin, I., Giua, A.: Dynamics and steady state analysis of controlled generalized batches Petri nets. Nonlinear Anal. Hybrid Syst. **12**, 33–49 (2014)
4. Demongodin, I., Giua, A.: Linear programming techniques for analysis and control of batches Petri nets. In: 10th International Workshop on Discrete Event Systems, WODES 2010, Berlin, Germany (2010)
5. Demongodin, I.: Generalised batches Petri net: hybrid model for high speed systems with variable delays. Discrete Event Dyn. Syst. **11**(1–2), 137–162 (2001)
6. Humayed, A., Lin, J., Li, F., Luo, B.: Cyber-physical systems security–a survey. IEEE Internet Things J. **4**(6), 1802–1831 (2017)
7. Júlvez, J., Di Cairano, S., Bemporad, A., Mahulea, C.: Event-driven model predictive control of timed hybrid Petri nets. Int. J. Robust Nonlinear Control **24**(12), 1724–1742 (2014)
8. Liu, R., Ammour, R., Brenner, L., Demongodin, I.: Event-driven control for reaching a steady state in controlled generalized batches Petri nets. In: 15th International Workshop on Discrete Event Systems, WODES 2020, Rio de Janeiro, Brazil (2020)

9. Mahulea, C., Giua, A., Recalde, L., Seatzu, C., Silva, M.: Optimal model predictive control of timed continuous Petri nets. IEEE Trans. Autom. Control **53**(7), 1731–1735 (2008)

10. Rashidinejad, A., Wetzels, B., Reniers, M., Lin, L., Zhu, Y., Su, R.: Supervisory control of discrete-event systems under attacks: an overview and outlook. In: 18th European Control Conference (ECC 2019), pp. 1732–1739 (2019)

11. Wang, L., Mahulea, C., Júlvez, J., Silva, M.: ON/OFF strategy based minimum-time control of continuous Petri nets. Nonlinear Anal. Hybrid Syst. **12**, 50–65 (2014)

Towards Efficient Partial Order Techniques for Time Petri Nets

Kuangze Wang[1,2], Hanifa Boucheneb[2(✉)], Kamel Barkaoui[3], and Zhiwu Li[1]

[1] School of Electro-Mechanical Engineering, Xidian University, Xi'an 710071, China
kuangzewang@stu.xidian.edu.cn, zhwli@xidian.edu.cn
[2] Laboratoire VeriForm, Department of Computer and Software Engineering,
Polytechnique de Montréal, P.O. Box 6079, Station Centre-ville, Montréal,
Québec H3C 3A7, Canada
hanifa.boucheneb@polymtl.ca
[3] Laboratoire CEDRIC, Conservatoire National des Arts et Métiers,
192 rue Saint Martin, Paris Cedex 03, France
kamel.barkaoui@cnam.fr

Abstract. Time Petri nets (TPN for short) are established as a powerful formalism for modeling and verifying systems with explicit soft and hard time constraints. However, their verification techniques run against the state explosion problem that is accentuated by the fact that the diamond property is difficult to meet, even for conflict-free transitions.

To deal with this limitation, the partial order reduction (POR) techniques of Petri nets (PN for short) are used in combination with the partially ordered sets (POSETs). Nevertheless, POSETs introduce extra selection conditions that may offset the benefits of the POR techniques.

This paper establishes a subclass of TPN for which the POR techniques of PN can be used without resorting to POSETs.

Keywords: Time petri nets · Reachability analysis · State explosion problem · Stubborn sets · Partial order techniques.

1 Introduction

Time Petri nets (TPN for short) are established as a powerful formalism to describe and verify the behaviour of real time systems while taking into account their hard and soft time constraints. The verification of such systems is mainly based on the state space abstractions combined with an interleaving semantics [4,5,9,11,20].

Several TPN state space abstractions are proposed in the literature such as the *State Class Graph (SCG)* [4], the *Contracted State Class Graph (CSCG)* [11], the *Geometric Region Graph (GRG)* [20], the *Strong State Class Graph (SSCG)* [5] and the *Zone Based Graph (ZBG)* [9]. In these state space abstractions, all states reachable by the same sequence of transitions, independently of their firing times are grouped together in the same node and considered modulo some

© Springer Nature Switzerland AG 2020
B. Ben Hedia et al. (Eds.): VECoS 2020, LNCS 12519, pp. 100–115, 2020.
https://doi.org/10.1007/978-3-030-65955-4_8

relation of equivalence. Each node, called an abstract state (state class or state zone), is represented by the common marking of its states and a set of atomic time constraints characterising the union of time domains of its states. These abstractions are finite for all bounded TPN and are exact w.r.t. the markings and the firing sequences of the TPN[1]. They however run against the state explosion problem that is accentuated by the fact that in the TPN state space abstractions different interleavings of conflict-free transitions lead, in general, to different abstract states. This claim is valid for all the TPN state space abstractions proposed in the literature [7].

Partial order reduction (POR for short) techniques, proposed for Petri nets (PN for short) and automata, such as stubborn sets [17–19], persistent sets [13] and ample sets [12, 16], are well-accepted to tackle the state explosion problem. They aim at reducing the state space to be explored, by selecting, at each state encountered during the exploration, as few as possible the transitions to be fired while preserving the properties of interest. They ensure, at least, the preservation of the deadlock states (i.e., states with no enabled transitions). Thus, the set of the selected transitions is only empty for the deadlock states. Furthermore, for each maximal firing sequence of an encountered state[2], there is an equivalent sequence[3] that starts with a selected transition.

For the stubborn sets, the selection criteria are applied on the enabled and non enabled transitions and guarantee the preservation of the deadlock states. The persistent sets are a particular case of the stubborn sets, where the selection criteria are only applied on the enabled transitions. The selection criteria of the ample sets at a state consist of four conditions that ensure preservation of a given $LTL_{\{-X\}}$ property[4] [12, 16]. Two of these conditions are sufficient to preserve deadlock states. The first one, common to all the POR techniques, stipulates that the set of transitions is empty for deadlock states only. The second one guarantees independency between the selected transitions and the non selected ones. Two transitions are independent (i.e., satisfy the diamond property), if whenever they are enabled, they are firable in both orders, the firing of one of them does not inhibit the occurrence of the other, and their firings in both orders lead to the same state.

From a practical point of view, each POR technique provides a useful selection procedure that mainly relies on the diamond property. In the reachability graphs of PN, the diamond property holds for any pair of conflict-free transitions. By contrast, in the TPN state space abstractions, the last condition of the diamond property does not necessarily hold even for conflict-free transitions. Moreover, the abstract states reachable by different interleavings of transitions

---

[1] All and only all markings and firing sequences of the TPN are represented in the state space abstraction.

[2] A maximal firing sequence is either an infinite firable sequence or a finite firable sequence leading to a deadlock state.

[3] Two sequences of transitions ω and ω' are equivalent, denoted by $\omega \equiv \omega'$ iff they are identical or each one can be obtained from the other by a series of permutations of transitions.

[4] $LTL_{\{-X\}}$ is the classical linear temporal logic without the next-state operator X.

may have different behaviours, To overcome these issues, the POR techniques proposed in the literature for the TPN model rely on the partially ordered sets (POSETs) [1,3,14,15,21].

The idea of the POR techniques based on POSETs is to compute the successor of each selected transition t, without fixing any firing order between t and the non selected transitions [1,3,14,15,20,21]. The POSETs allow then to handle simultaneously several equivalent firing sequences, by fixing partially the firing order of transitions. Each partially ordered transitions stands then for some equivalent sequences of transitions and the abstract states reachable by these firing sequences are represented by the convex-hull of their union (i.e., the smallest enclosing abstract state). The correctness of a POSET is ensured if the union of the abstract states reachable by its firing sequences is identical to their convex-hull. In the context of the TPN state space abstractions, this union is not necessarily identical to the convex-hull, even for conflict-free transitions [6].

In [6,7], the authors have established some practical sufficient conditions for the correctness of POSETs. Intuitively, these sufficient conditions ensure that the effects[5] of the partially ordered transitions are independent of their firing orders. In a POR technique based on POSETs, the selection criteria of the POR technique (stubborn sets, persistent sets, ample sets, etc.) are completed with extra conditions that guarantee the correctness of POSETs. However, these extra conditions may offset the benefits of the POR techniques.

This paper identifies an interesting subclass of TPN for which the POR techniques of PN can be used without POSETs. Among the POR techniques of PN, this paper focuses on the stubborn sets [17–19]. Note that however the results established here are still valid for other POR techniques of PN.

The rest of the paper is organised as follows. Section 2 is devoted to the TPN model, its semantics and its state space abstraction considered here. Section 3 starts with the classical definition of the stubborn sets [17–19] followed by the revisited stubborn sets used in combination with POSETs. Finally, it proves that for a subclass of TPN the revisited stubborn sets can be used without POSETs while preserving the deadlock markings. Conclusions are presented in Sect. 4.

2   Time Petri Nets

2.1   Definition and Semantics

Let $\mathbb{N}$, $\mathbb{Q}^+$ and $\mathbb{R}^+$ be the sets of non-negative integer, rational and real numbers, respectively, and $INT_{\mathbb{X}}$, for $\mathbb{X} \in \{\mathbb{N}, \mathbb{Q}^+, \mathbb{R}^+\}$, the set of intervals whose lower and upper bounds are in $\mathbb{X}$ and $\mathbb{X} \cup \{\infty\}$, respectively.

A multi-set over a non-empty set P is a function $M : P \longrightarrow \mathbb{N}$ defined also by $\sum_{p \in P} M(p) \times p$. We denote by P_{MS} the set of all multi-sets over P. Operations on multi-sets are defined as usual.

---

[5] The effect of a transition is the set of transitions disabled plus those newly enabled by its firing.

A PN is a tuple $\mathcal{N}_u = (P, T, pre, post, M_0)$ where P and T are finite and non-empty sets of places and transitions such that $P \cap T = \emptyset$; pre and $post$ are the backward and forward incidence functions over the set of transitions T ($pre, post : T \longrightarrow P_{MS}$); $M_0 \in P_{MS}$ is the initial marking of $\mathcal{N}_u$.

A TPN is a pair $\mathcal{N} = (\mathcal{N}_u, Is)$, where $\mathcal{N}_u$ is a PN and Is is the static firing function ($Is : T \rightarrow INT_{\mathbb{Q}+}$). For every transition t, $\downarrow Is(t)$ and $\uparrow Is(t)$ denote the lower and the upper bounds of the static firing interval of t, respectively.

For $t \in T$, ${}^{\bullet}t$ and $t^{\bullet}$ denote the sets of input and output places of t, respectively. They are defined by ${}^{\bullet}t = \{p \in P \mid pre(t)(p) > 0\}$ and $t^{\bullet} = \{p \in P \mid post(t)(p) > 0\}$. Similarly, the sets of input and output transitions of a place $p \in P$ are defined by ${}^{\bullet}p = \{t \in T \mid post(t)(p) > 0\}$ and $p^{\bullet} = \{t \in T \mid pre(t)(p) > 0\}$, respectively. Two transitions t and t' are in structural conflict iff ${}^{\bullet}t \cap {}^{\bullet}t' \neq \emptyset$. We denote by $CFS(t) = ({}^{\bullet}t)^{\bullet}$ the set of transitions in structural conflict with t.

A marking of a TPN indicates the distribution of tokens over its places. It is defined as a multi-set over places. For the following, we fix a TPN $\mathcal{N}$, a marking $M \in P_{MS}$ and a transition $t \in T$ of $\mathcal{N}$. The transition t is enabled at M, denoted by $M[t\rangle$ iff all the required tokens for firing t are present in M, i.e., $M \geq pre(t)$.

In case t is enabled at M, its firing leads to the marking $M' = M - pre(t) + post(t)$. The notation $M[t\rangle M'$ means that t is enabled at M and M' is the marking reached from M by t. We denote by $En(M)$ the set of transitions enabled at M, i.e., $En(M) = \{t \in T \mid M \geq pre(t)\}$. The marking M is a deadlock iff $En(M) = \emptyset$.

For any subset $X \subseteq T$ of transitions, we denote by $\Omega(X)$ the sets of all sequences of transitions of X, including the empty sequence ϵ. For any non-empty sequence of transitions $\omega = t_1 t_2 ... t_n \in \Omega(T) - \{\epsilon\}$, the usual notation $M[t_1 t_2 ... t_n\rangle$ means that there exist markings $M_1, ..., M_n$ such that $M_1 = M$ and $M_i[t_i\rangle M_{i+1}$, for $i \in [1, n-1]$ and $M_n[t_n\rangle$. The sequence ω is said to be a firing sequence of M. The notation $M[t_1 t_2 ... t_n\rangle M'$ gives, in addition, the marking reached by the sequence. The marking M' is said to be reachable from M by ω. By convention, we have $M[\epsilon\rangle M$.

Among the semantics proposed in the literature for the TPN model [10], we focus here on the classical intermediate semantics. Let $M \in P_{MS}$ be a marking, $t \in En(M)$ a transition enabled in M, and M' the successor marking of M by t. We denote by $Nw(M, t)$ the set of transitions newly enabled in M'. Formally, $Nw(M, t)$ contains t, if t is enabled at M', and also all transitions enabled at M' but not enabled at the intermediate marking $M - pre(t)$, i.e., $Nw(M, t) = \{t' \in En(M') \mid t' = t \vee M - pre(t) \not\geq pre(t')\}$.

Starting from the initial marking M_0, the marking of $\mathcal{N}$ evolves by firing transitions at irregular intervals of time. When a transition t is newly enabled, its firing interval is set to its static firing interval. The bounds of this interval decreases synchronously with time. The transition t is firable, if the lower bound of its firing interval reaches 0. It must fire immediately, without any additional delay, when the upper bound of its firing interval reaches 0, unless it is disabled by another firing. The firing of a transition takes no time but leads to a new marking.

Syntactically, in the context of $\mathcal{N}$, a state is defined as a pair $s = (M, I)$, where M is a marking and I is a firing interval function (I: $En(M) \to INT_{\mathbb{R}+}$), which associates a firing interval with each enabled transition at M. The initial state of $\mathcal{N}$ is $s_0 = (M_0, I_0)$, where $I_0(t) = Is(t)$ and for all $t \in En(M_0)$. Let $\mathcal{S} = \{s = (M, I) \mid M \in P_{MS} \wedge I\colon En(M) \to INT_{\mathbb{R}+}\}$, $s = (M, I) \in \mathcal{S}$, $s' = (M', I') \in \mathcal{S}$, $dh \in \mathbb{R}^+$, $t \in T$ and $\to$ the transition relation defined by:

- $s \xrightarrow{dh} s'$ (s' is also denoted by $s + dh$) iff s' is reachable from s by dh time units, i.e., $\forall t \in En(M), dh \leq \uparrow I(t), M' = M$, and $\forall t' \in En(M'), I'(t') = [Max(0, \downarrow I(t') - dh), \uparrow I(t') - dh]$. • $s \xrightarrow{t} s'$ iff t is immediately firable from s and its firing leads to s', i.e., $t \in En(M), \downarrow I(t) = 0, M' = M - pre(t) + post(t)$, and

$$\forall t' \in En(M'), I'(t') = \begin{cases} Is(t') & \text{if } t' \in Nw(M, t) \\ I(t') & \text{otherwise.} \end{cases}$$

The semantics of $\mathcal{N}$ is defined by the transition system $(S, \to, s_0)$, where $S \subseteq \mathcal{S}$ is the set of all states reachable from s_0 by $\xrightarrow{*}$ (the reflexive and transitive closure of $\to$). A *run* in $(S, \to, s_0)$, starting from a state s_1 of S, is a maximal sequence $\eta = s_1 \xrightarrow{dh_1} s_1 + dh_1 \xrightarrow{t_1} s_2 \xrightarrow{dh_2} s_2 + dh_2 \xrightarrow{t_2} s_3 \ldots$. By convention, for any state s_i, the relation $s_i \xrightarrow{0} s_i$ holds. A marking M is reachable in $\mathcal{N}$ iff $\exists s \in S$ s.t. the marking of s is M. Runs, timed traces and firing sequences of $\mathcal{N}$ are all runs, timed traces and firing sequences of its initial state s_0. A PN (TPN) is k-bounded (i.e., k-safe) iff for every reachable marking in $\mathcal{N}$, the number of tokens in each place does not exceed k. It is bounded iff it is k-bounded for some natural number k.

Let $M \in P_{MS}$ be a marking, $t \in T$ a transition and $k \in \mathbb{N}$ a non-negative number. The transition t is k-enabled in a marking M iff $M \geq k \times pre(t) \wedge M \not\geq (k + 1) \times pre(t)$. It is said to be multi-enabled in M if $k > 1$. In this paper, we suppose that all transitions are not multi-enabled in each reachable marking (i.e., single-server semantics). However, the results established here could be extended to the multiple-server semantics, as the different enabling instances of the same transition are considered as totally independent. Indeed, each of the enabling instances of the same transition t has its own firing delay and is not in conflict with all other enabling instances of t (i.e., firing an enabling instance of t does not disable its other enabling instances). In [10], the authors have first shown that multiple-server semantics is strictly more expressive than its single-server counterpart. Then, they have established that the First Enabled First Fired (FEFF) policy, where the different firable instances of the same transition are fired starting with the oldest one, generates the same timed traces than the non deterministic firing order choice[6]. Therefore, in the context of the multiple-server semantics, the FEFF policy leads to a more compact state-space and is more appropriate to detect the deadlocks of the TPN model.

---

[6] All possible firing orders of the firable instances of the same transition are considered.

2.2   Contracted State Class Graph

The contracted state class graph CSCG of $\mathcal{N}$ is a TPN state space abstraction, where the abstract state, called state class, is defined by the pair $\alpha = (M, F)$ composed with a marking M and a conjunction of triangular atomic constraints F^7 over the firing delays of the enabled transitions in M. By convention, $F = true$ if the number of the enabled transitions in M is less than 2 (i.e., there is no triangular atomic constraint in F).

The initial state class of $\mathcal{N}$ is $\alpha_0 = (M_0, F_0)$, where M_0 is the initial marking of $\mathcal{N}$ and $F_0 = \bigwedge\limits_{t,t' \in En(M_0) \text{ s.t. } t \neq t'} \underline{t} - \underline{t}' \leq \uparrow Is(t) - \downarrow Is(t')$, $\underline{t}$ and $\underline{t}'$ being real-valued variables representing the firing delays of transitions t and t', respectively.

From a practical point of view, the firing domain F of a state class $\alpha = (M, F)$ is encoded by a difference bound matrix (DBM in short) [2]. The DBM of F is a square matrix B, indexed by variables of F. Each entry b_{ij} represents the triangular atomic constraint $\underline{t}_i - \underline{t}_j \leq b_{ij}$. Although the same nonempty domain may be encoded by different DBM, they have a canonical form. The canonical form of a DBM is the representation with tightest bounds on all differences between variables, computed by propagating the effect of each entry through the DBM. The canonical form can be computed in $O(n^3)$, n being the number of variables in the DBM, using a shortest path algorithm, like Floyd-Warshall's all-pairs shortest path algorithm. Canonical forms make operations over DBM much simpler [2].

Let $\mathcal{C}$ be the set of all state classes, $\alpha = (M, F) \in \mathcal{C}$ a state class, B the DBM of F in canonical form and $t_f \in T$ a transition of $\mathcal{N}$. The transition t_f is firable from α, denoted by $\alpha \xrightarrow{t_f}$, iff $t_f \in En(M) \wedge \forall t_i \in En(M), b_{if} \geq 0$. Intuitively, this formula, called the firing time condition of t_f from α, means that there is, at least one state in α such that the firing delay of t_f does not exceed those of all transitions enabled in M. If $\alpha \xrightarrow{t_f}$ then $\alpha \xrightarrow{t_f} \alpha'$, for some $\alpha' = (M', F') \in \mathcal{C}$ such that $M' = M - pre(t_f) + post(t_f)$ and the canonical form of the DBM B' of F' is computed as follows: $\forall t_i, t_j \in En(M')$,

$$
b'_{ij} = \begin{cases}
0 & \text{if } i = j \\
\uparrow Is(t_i) - \downarrow Is(t_j) & \text{if } t_i, t_j \in Nw(M, t_f), \\
b_{if} - \downarrow Is(t_j) & \text{if } t_i \notin Nw(M, t_f) \wedge t_j \in Nw(M, t_f), \\
\uparrow Is(t_i) + \underset{t_u \in En(M)}{Min}\, b_{uj} & \text{if } t_i \in Nw(M, t_f) \wedge t_j \notin Nw(M, t_f), \\
Min(b_{ij}, b_{if} + \underset{t_u \in En(M)}{Min}\, b_{uj}) & \text{otherwise.}
\end{cases}
$$

We denote by $Fr(\alpha)$ the set of transitions firable from α.

The CSCG of $\mathcal{N}$ is the structure $(C, \longrightarrow, \alpha_0)$, where $C \subseteq \mathcal{C}$ is the set of all state classes reachable from the initial state class α_0 by $\xrightarrow{*}$ (the reflexive and transitive closure of $\longrightarrow$). A sequence $\omega \in \Omega(T)$ is a firing sequence of $\mathcal{N}$

---

[7] A triangular atomic constraint is of the form $x - y \prec c$, where x, y are real valued variables, c is a rational constant and $\prec \in \{<, >, \leq, \geq, =\}$.

iff $\alpha_0 \xrightarrow{\omega}$. The CSCG of $\mathcal{N}$ is finite iff $\mathcal{N}$ is bounded. Moreover, the CSCG preserves the markings and the firing sequences of $\mathcal{N}$. The boundedness of $\mathcal{N}$ is not decidable but there are some useful sufficient conditions, which ensure its boundedness (and then the finiteness of its CSCG) [4].

Let $\alpha = (M, F)$ and $\alpha' = (M', F')$ be two state classes such that $M = M'$, B and B' the DBM of F and F' in canonical form, respectively. The state classes α and α' are equal iff $M = M'$ and $B = B'$. The state class α is included in α', denoted by $\alpha \subseteq \alpha'$ iff [2]: $\forall t_i, t_j \in En(M), b_{ij} \leq b'_{ij}$. State classes are not closed under union (i.e., the union of state classes is not necessarily a state class). The smallest enclosing state class of the union of α and α', denoted by $\alpha \sqcup \alpha'$, is the state class $\alpha'' = (M, F'')$, where the DBM B'' of F'' is computed as follows: $\forall t_i, t_j \in En(M), b''_{ij} = Max(b_{ij}, b'_{ij})$. The union $\alpha \cup \alpha'$ is a state class iff $\alpha \sqcup \alpha' = \alpha \cup \alpha'$. The state class $\alpha \sqcup \alpha'$ is also called the convex-hull of $\alpha \cup \alpha'$.

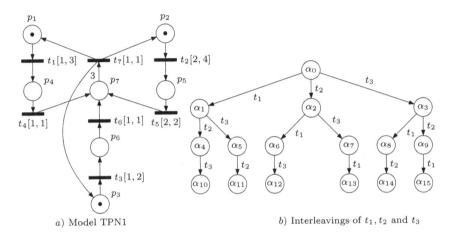

a) Model TPN1   b) Interleavings of t_1, t_2 and t_3

Fig. 1. Model TPN1 and some of its state classes

2.3   Interleavings in the CSCG

Like in other TPN state space abstractions, in the CSCG, different interleavings of conflict-free transitions lead generally to different, even incomparable, abstract states. As an example, consider the model TPN1 in Fig. 1.a. From its initial state, the transitions t_1, t_2 and t_3 are firable in 6 orders leading to 6 different state classes ($\alpha_{10}, \alpha_{11}, \alpha_{12}, \alpha_{13}, \alpha_{14}$ and α_{15}), reported in Fig. 1.b and Table 1. None of these state classes is larger than all the others. Due to this drawback, the number of state classes sharing the same marking may explode.

To attenuate this explosion, we can group state classes linked by an inclusion relation (i.e., inclusion based abstractions) while preserving the set of reachable states. State classes α_6 and α_7 can be then replaced by α_4 and α_9, respectively. State classes α_5 and α_8, reachable from α_0 by sequences $t_1 t_3$ and $t_3 t_1$, are not linked by any inclusion relation but their union is a state class. Therefore, these

Table 1. State classes of paths in Fig. 1.b

α_0	α_1	α_2	α_3
$p_1 + p_2 + p_3$	$p_2 + p_3 + p_4$	$p_1 + p_3 + p_5$	$p_1 + p_2 + p_6$
$-3 \le \underline{t_1} - \underline{t_2} \le 1 \wedge$	$0 \le \underline{t_2} - \underline{t_3} \le 3 \wedge$	$0 \le \underline{t_1} - \underline{t_3} \le 1 \wedge$	$-3 \le \underline{t_1} - \underline{t_2} \le 1 \wedge$
$-1 \le \underline{t_1} - \underline{t_3} \le 2 \wedge$	$-1 \le \underline{t_2} - \underline{t_4} \le 2 \wedge$	$-2 \le \underline{t_1} - \underline{t_5} \le -1 \wedge$	$-1 \le \underline{t_1} - \underline{t_6} \le 1 \wedge$
$0 \le \underline{t_2} - \underline{t_3} \le 3$	$-1 \le \underline{t_3} - \underline{t_4} \le 0$	$\underline{t_3} - \underline{t_5} = -2$	$-1 \le \underline{t_2} - \underline{t_6} \le 2$
α_4	α_5	α_6	α_7
$p_3 + p_4 + p_5$	$p_2 + p_4 + p_6$	$p_3 + p_4 + p_5$	$p_1 + p_5 + p_6$
$-1 \le \underline{t_3} - \underline{t_4} \le 0 \wedge$	$-1 \le \underline{t_2} - \underline{t_4} \le 2 \wedge$	$\underline{t_3} - \underline{t_4} = -1 \wedge$	$-2 \le \underline{t_1} - \underline{t_5} \le -1 \wedge$
$\underline{t_3} - \underline{t_5} = -2 \wedge$	$-1 \le \underline{t_2} - \underline{t_6} \le 2 \wedge$	$\underline{t_3} - \underline{t_5} = -2 \wedge$	$-1 \le \underline{t_1} - \underline{t_6} \le 0 \wedge$
$-2 \le \underline{t_4} - \underline{t_5} \le -1$	$-1 \le \underline{t_4} - \underline{t_6} \le 0$	$\underline{t_4} - \underline{t_5} = -1$	$1 \le \underline{t_5} - \underline{t_6} \le 1$
α_8	α_9	α_{10}	α_{11}
$p_2 + p_4 + p_6$	$p_1 + p_5 + p_6$	$p_4 + p_5 + p_6$	$p_4 + p_5 + p_6$
$-1 \le \underline{t_2} - \underline{t_4} \le 2 \wedge$	$-2 \le \underline{t_1} - \underline{t_5} \le -1 \wedge$	$-2 \le \underline{t_4} - \underline{t_5} \le -1 \wedge$	$-1 \le \underline{t_4} - \underline{t_5} \le -1 \wedge$
$-1 \le \underline{t_2} - \underline{t_6} \le 2 \wedge$	$-1 \le \underline{t_1} - \underline{t_6} \le 1 \wedge$	$-1 \le \underline{t_4} - \underline{t_6} \le 0 \wedge$	$-1 \le \underline{t_4} - \underline{t_6} \le 0 \wedge$
$0 \le \underline{t_4} - \underline{t_6} \le 1$	$1 \le \underline{t_5} - \underline{t_6} \le 2$	$1 \le \underline{t_5} - \underline{t_6} \le 1$	$1 \le \underline{t_5} - \underline{t_6} \le 2$
α_{12}	α_{13}	α_{14}	α_{15}
$p_4 + p_5 + p_6$	$p_4 + p_5 + p_6$	$p_4 + p_5 + p_6$	$p_4 + p_5 + p_6$
$\underline{t_4} - \underline{t_5} = -1$	$-1 \le \underline{t_4} - \underline{t_5} \le 0$	$-2 \le \underline{t_4} - \underline{t_5} \le -1$	$-1 \le \underline{t_4} - \underline{t_5} \le 0$
$\underline{t_4} - \underline{t_6} = 0 \wedge$	$0 \le \underline{t_4} - \underline{t_6} \le 1 \wedge$	$0 \le \underline{t_4} - \underline{t_6} \le 1 \wedge$	$0 \le \underline{t_4} - \underline{t_6} \le 1 \wedge$
$\underline{t_5} - \underline{t_6} = 1$	$1 \le \underline{t_5} - \underline{t_6} \le 1$	$1 \le \underline{t_5} - \underline{t_6} \le 2$	$1 \le \underline{t_5} - \underline{t_6} \le 2$

state classes can be replaced by their union while preserving the set of reachable states of TPN1 (i.e., union based abstractions). However, it is not always true that the union of state classes is a state class. Moreover, over-approximating this union by its convex-hull does not necessarily preserve the reachable states of the model or even its markings. Indeed, consider the model TPN2 in Fig. 2.a and its contracted state class graph in Fig. 2.b. The union of state classes α_3 and α_5 reachable from α_0 by sequences $t_1 t_2$ and $t_2 t_1$, respectively, is not a state class. Their convex-hull α_{35} (reported in Table 2) generates an infinite number of unreachable markings. The properties of the model are not preserved by this over-approximation. Figure 2.c shows the firing sequence $t_4 t_3 t_6 t_7 t_7 \dots$ that is feasible neither from α_3 nor from α_5 but is feasible from α_{35}. This firing sequence produces an infinite number of markings that are not reachable in the bounded model TPN2. Thus, the convex-hull based abstraction does not preserve the boundedness property of the TPN model. The convex-hull based abstraction does not also preserve the deadlocks of the TPN model. The counterexample is the model TPN3 in Fig. 3. This model is deadlock free but the convex-hull of the union of state classes reachable from its initial state class by $t_1 t_2$ and $t_2 t_1$ introduces a deadlock marking.

Let α be a state class and $\nu \subseteq Fr(\alpha)$ a subset of transitions firable from α. In [6,7], the authors have established some sufficient conditions that ensure for the transitions of ν to be firable in different orders from α and, in addition,

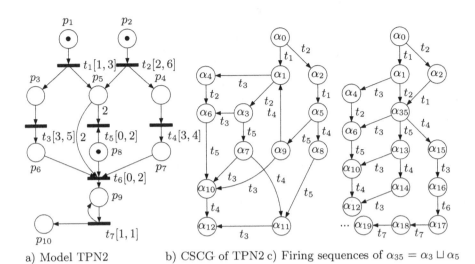

a) Model TPN2   b) CSCG of TPN2   c) Firing sequences of $\alpha_{35} = \alpha_3 \sqcup \alpha_5$

Fig. 2. Model TPN2, its CSCG and convex-hull based abstraction

Table 2. Some CSCG state classes of TPN2

α_0	α_3	α_5	$\alpha_{35} = \alpha_3 \sqcup \alpha_5$
$p_1 + p_2 + p_8$	$p_3 + p_4 + 2p_5 + p_8$	$p_3 + p_4 + 2p_5 + p_8$	$p_3 + p_4 + 2p_5 + p_8$
$-5 \le \underline{t_1} - \underline{t_2} \le 1 \wedge$	$-4 \le \underline{t_3} - \underline{t_4} \le 2 \wedge$	$-1 \le \underline{t_3} - \underline{t_4} \le 3 \wedge$	$-4 \le \underline{t_3} - \underline{t_4} \le 3 \wedge$
	$-2 \le \underline{t_3} - \underline{t_5} \le 5 \wedge$	$1 \le \underline{t_3} - \underline{t_5} \le 5 \wedge$	$-2 \le \underline{t_3} - \underline{t_5} \le 5 \wedge$
	$1 \le \underline{t_4} - \underline{t_5} \le 4$	$0 \le \underline{t_4} - \underline{t_5} \le 4$	$0 \le \underline{t_4} - \underline{t_5} \le 4$

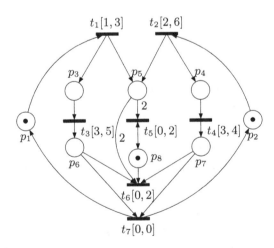

Fig. 3. Model TPN3

their different interleavings lead to state classes such that their convex-hull is identical to their union. We report in the following the structural one, used in [7] (denoted by sc):

$$\forall t, t' \in \nu, \ t \neq t' \Rightarrow \ ((^\bullet t)^\bullet \cup (t^\bullet)^\bullet) \ \cap \ ((^\bullet t')^\bullet \cup (t'^\bullet)^\bullet) = \emptyset.$$

Intuitively, this sufficient condition guarantees that the effects[8] of transitions of ν are independent of their firing orders, no matter the state class from which these transitions are fired.

For example, consider the model TPN1 and its initial state class α_0. The set $\nu = \{t_1, t_2, t_3\} = Fr(\alpha_0)$ satisfies the condition sc. Then, its transitions are firable in all orders from α_0 and the state classes reachable by these different firing orders can be merged while preserving the set of reachable states of TPN1. This union can be computed from α_0 without fixing any firing order between t_1, t_2 and t_3 [6].

For the model TPN2 and its initial state class α_0, the set $\nu = \{t_1, t_2\} = Fr(\alpha_0)$ does not satisfy Condition sc because $((^\bullet t_1)^\bullet \cup (t_1^\bullet)^\bullet) = \{t_1, t_3, t_5\}$, $((^\bullet t_2)^\bullet \cup (t_2^\bullet)^\bullet) = \{t_2, t_4, t_5\}$ and then $((^\bullet t_1)^\bullet \cup (t_1^\bullet)^\bullet) \cap ((^\bullet t_2)^\bullet \cup (t_2^\bullet)^\bullet) = \{t_5\} \neq \emptyset$. Since sc is a sufficient condition, it does not allow us to conclude whether or not the union of state classes reachable from α_0 by $t_1 t_2$ and $t_2 t_1$ is a state class. To achieve further state space reduction, POR techniques are established as a good alternative.

The next section is devoted to the stubborn sets in the context of TPN.

3   Stubborn Sets Method Without POSETs

3.1   Stubborn Sets of State Classes and Their Limitations

The stubborn sets of state classes can be defined in a similar way as the stubborn sets of markings [17–19]. Let α be a state class of a TPN $\mathcal{N}$ and $\mu \subseteq T$. The subset μ is a stubborn set of α, if it satisfies all the following conditions:

$D0$: $Fr(\alpha) \neq \emptyset \Leftrightarrow \mu \neq \emptyset$.
$D1w$: $\exists t \in \mu, \forall \omega \in \Omega(T - \mu), \alpha \xrightarrow{\omega} \Rightarrow \alpha \xrightarrow{\omega t}$.
$D2$: $\forall t \in \mu, \forall \omega \in \Omega(T - \mu), \forall \alpha' \in \mathcal{C}, \alpha \xrightarrow{\omega t} \alpha' \Rightarrow \alpha \xrightarrow{t\omega} \alpha'$.

However, the part of $D2$, imposing that sequences ωt and $t\omega$ lead to the same state class α', is difficult to meet in the context of the CSCG. Can we do without this part? To answer this question, consider the model TPN4 in Fig. 4 and its initial state class α_0. The set $\mu = \{t_1\}$ satisfies $D0$, $D1w$ and partially $D2$. Indeed, for $D2$, it holds that $\forall \omega \in \Omega(T - \mu), \alpha_0 \xrightarrow{\omega t_1} \Rightarrow \alpha_0 \xrightarrow{t_1 \omega}$ but for $\omega = t_2$, the sequences $t_1 t_2$ and $t_2 t_1$ lead to different state classes α_3 and α_4. If t_1 is the only transition selected in α_0 then only the cycle $\alpha_0 \alpha_1 \alpha_3 \alpha_5 \alpha_0$ will

---

[8] The effect of a transition t is the set of transitions disabled plus those newly enabled by firing t.

be explored and the deadlock marking p_3 will not be detected. Moreover, firing t_1 followed by t_2 from α_0, without imposing any firing order between t_1 and t_2, leads to the state class α_3. This POSET is then not sufficient to detect the deadlock marking p_3.

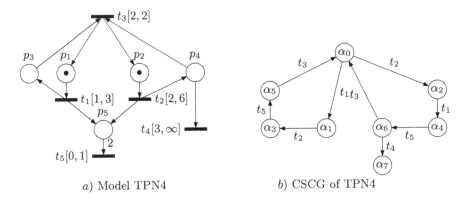

a) Model TPN4

b) CSCG of TPN4

Fig. 4. Model TPN4 and its CSCG

Table 3. State classes of the CSCG in Fig. 4.b

α_0	α_1	α_2	α_5
$p_1 + p_2$	$p_2 + p_3 + p_5$	$p_1 + p_3 + p_5$	$p_3 + p_4$
$-5 \leq \underline{t_1} - \underline{t_2} \leq 1$	$true$	$\underline{t_1} - \underline{t_4} \leq -2$	$\underline{t_3} - \underline{t_4} \leq -1$

α_3	α_4	α_6	α_7
$p_3 + p_4 + 2p_5$	$p_3 + p_4 + 2p_5$	$p_3 + p_4$	p_3
$\underline{t_3} - \underline{t_4} \leq -1 \wedge$	$\underline{t_3} - \underline{t_4} \leq 0 \wedge$	$\underline{t_3} - \underline{t_4} \leq 0$	$true$
$1 \leq \underline{t_3} - \underline{t_5} \leq 2 \wedge$	$1 \leq \underline{t_3} - \underline{t_5} \leq 2 \wedge$		
$2 \leq \underline{t_4} - \underline{t_5}$	$1 \leq \underline{t_4} - \underline{t_5}$		

We have shown, by means of an example, that combining naively the stubborn sets with POSETs may fail to detect the deadlocks of $\mathcal{N}$. We have to be sure, that the used POSETs cover all deadlocks of $\mathcal{N}$ and, at the same time, they do not introduce unreachable deadlocks. To avoid these issues, extra selection conditions are needed to ensure the preservation of deadlocks of $\mathcal{N}$. One solution consists in ensuring the correctness of the POSETs induced by the selected transitions, i.e., the union of the state classes reachable by the partially ordered sequences is identical to their convex-hull.

In this aim, in [7], the set of the selected transitions μ from each state class α satisfies $D0$, $D1$, $D2'$: $\forall t \in \mu, \forall \omega \in \Omega(T - \mu), \alpha \xrightarrow{\omega t} \Rightarrow \alpha \xrightarrow{t\omega}$ and, in addition, if a firable transition t is selected then all transitions of $((^\bullet t)^\bullet \cup (t^\bullet)^\bullet \cup {}^\bullet(^\bullet t))$ are selected too. This extra condition ensures that no transition outside the set μ can affect directly the effect of t. In [8], some time dependent weaker selection conditions are established, where the selection is limited to the transitions that may affect the effects of t and, at the same time, may occur before transition t.

For example, consider the model TPN2 in Fig. 2.a and its initial state class α_0. The set $\mu = \{t_1\}$ satisfies conditions $D0$, $D1$ and $D2'$ for α_0 but the resulting POSET $\nu = \{t_1, t_2\}$ does not satisfy Condition sc. If sc is used to ensure the correctness of the POSET, we need to add the transition t_2 to μ and then all firable transitions are selected.

For the model TPN1 in Fig. 1.a and its initial state class α_0, transitions t_1, t_2 and t_3 are enabled in M_0 and firable from α_0 in different orders. The set $\mu = \{t_2\}$ satisfies conditions $D0$, $D1w$ and $D2'$ for α_0. To ensure the correctness of the resulting POSETs, the approaches proposed in [7,8] will select the output transition t_5 of t_2 and the input transition t_7 of t_2. Since the newly selected transitions t_5 and t_7 are not enabled, their input transitions t_4 and t_5 will be selected too. Finally, the input transitions of t_4 and t_5 will be selected (i.e., t_1 and t_3). At the end, all the enabled transitions will be selected.

3.2   Revisited Stubborn Sets Without POSETs

As shown above, for the TPN model, the POR techniques based on POSETs need extra conditions to ensure the correctness of POSETs. These extra conditions may offset the benefits of the POR techniques. The following revisited definition of stubborn sets, called lightweight stubborn sets, can be used without POSETs while preserving the deadlocks of the TPN. The proof of this claim is similar to the one provided in [19] for the stubborn sets of PN.

Let α be a state class of a TPN $\mathcal{N} = (P, T, pre, post, M_0, Is)$, $\mu \subseteq T$ and $\mathcal{D}$ the set of deadlocks of $\mathcal{N}$. The subset μ is a lightweight stubborn set of α, if it satisfies all the following conditions:

$D0$: $Fr(\alpha) \neq \emptyset \Leftrightarrow \mu \neq \emptyset$.
$D1w$: $\exists t \in \mu, \forall \omega \in \Omega(T - \mu), \alpha \xrightarrow{\omega} \Rightarrow \alpha \xrightarrow{\omega t}$.
$D2w$: $\forall t \in \mu, \forall \omega \in \Omega(T - \mu), \forall \omega' \in \Omega(T), \forall D \in \mathcal{D},$

$$\alpha \xrightarrow{\omega t \omega'} D \Rightarrow \exists \omega'' \in \Omega(T), \omega'' \equiv \omega' \wedge \alpha \xrightarrow{t \omega \omega''} D.$$

In the following, we define a subclass of TPN and show that for this subclass, conditions $D0$, $D1$ and $D2'$ (i.e., $\forall t \in \mu, \forall \omega \in \Omega(T - \mu), \alpha \xrightarrow{\omega t} \Rightarrow \alpha \xrightarrow{t\omega}$) are sufficient to preserve deadlocks. This subclass, denoted by $\mathcal{TPN}$, contains, at least, any TPN $\mathcal{N} = (P, T, pre, post, M_0, Is)$ such that the following condition holds: $\forall \alpha = (M, F) \in C, \forall t_i \in Fr(\alpha),$

$\uparrow Is(t_i) = \infty \vee \forall t_j \in CFS(t_i), t_j \in En(M) \wedge (F \wedge \underline{t_j} \leq \underline{t_i}$ is consistent).

Intuitively, this condition means that if a transition t_i is firable from a state class $\alpha = (M, F)$ then all transitions in conflict with t_i are enabled in M and firable before t_i. Using the canonical form B of the DBM of F, the constraint $(F \wedge \underline{t_j} \leq \underline{t_i})$ is consistent if and only if $b_{ij} \geq 0$.

Theorem 1. *Let $\mathcal{N} = (P, T, pre, post, M_0, Is)$ be TPN of $\mathcal{TPN}$. The selective search w.r.t. D0, D1w and D2' from the initial state class of $\mathcal{N}^9$ preserves the deadlock markings of $\mathcal{N}$.*

Proof. Let $\alpha = (M, F)$ be a state class reachable by the selective search from α_0. Obviously, α is reachable in $\mathcal{N}$ and all deadlocks reachable by the selective search from α are also reachable in $\mathcal{N}$. Let D be a deadlock marking reachable from α in $\mathcal{N}$. Let us show that D is also reachable by the selective search from α. Let ω be a firing sequence leading to the marking D from α in $\mathcal{N}$ and μ a set of transitions satisfying conditions D0, D1w and D2' for α. The rest of the proof is by induction on the length of ω.

1) If $\omega = \epsilon$ then $M = D$.
2) If $\omega = t_1$ and $t_1 \in \mu$ then D is reachable by the selective search from α. If $\omega = t_1$ and $t_1 \notin \mu$ then, by D1w, $\exists t \in \mu, \alpha \xrightarrow{t_1 t}$. Then, D is not a deadlock marking, which contradicts the assumption (i.e., D is a deadlock).
3) Suppose that if D is reachable from α by a sequence ω' such that $|\omega'| < |\omega|$ then it is also reachable by the selective search. Let us show that it is also the case for ω.
   3.1) If there is no transition of μ in ω then, by D1w, $\exists t \in \mu, \alpha \xrightarrow{\omega t}$, which is in contradiction with the fact that firing ω from α leads to a deadlock.
   3.2) If there is, at least, a transition t of μ in ω then $\omega = \omega_1 t \omega_2$, for some $\omega_1 \in \Omega(T - \mu)$ and $\omega_2 \in \Omega(T)$. By D2', $\alpha \xrightarrow{\omega_1 t} \Rightarrow \alpha \xrightarrow{t\omega_1}$. Therefore, for some state class α_t of $\mathcal{N}$, it holds that $\alpha \xrightarrow{t} \alpha_t, \alpha_t \xrightarrow{\omega_1}$ and $\alpha \xrightarrow{\omega_1 t}$. Thus, α_t is reachable by the selective search. Let us show that D is reachable from α_t by the selective search.
      3.2.1) If $\omega_1 \omega_2$ is firable from α_t then D is reachable from α_t by the sequence $\omega_1 \omega_2$ and $|\omega_1 \omega_2| < |\omega|$. By induction hypothesis, D is also reachable by the selective search from α_t.
      3.2.2) Otherwise, suppose that for every sequence $\omega_2' \in \Omega(T)$ such that $\omega_2' \equiv \omega_2$, it holds that $\alpha_t \xnrightarrow{\omega_1 \omega_2'}$. Let ω_{21} be the longest common prefix of ω_2 and ω_2', i.e., $\omega_2 = \omega_{21} \omega_{22}, \omega_2' = \omega_{21} \omega_{22}'$, sequences ω_{22} and ω_{22}' are equivalent but their first transitions are different. Therefore, it holds that $\alpha \xrightarrow{\omega_1 t \omega_{21}} \alpha_1 = (M_1, F_1), \alpha \xrightarrow{t\omega_1 \omega_{21}} \alpha_2 = (M_2, F_2)$ with $M_1 = M_2$, $\alpha_1 \xrightarrow{\omega_{22}}$, and $\alpha_2 \xnrightarrow{\omega_{22}'}$. In addition, α_2 has no firable sequence that is equivalent to ω_{22}'. It follows that there is a transition t_2 that must fire from α_2 before all transitions of ω_{22}. Therefore, it holds that $t_2 \in Fr(\alpha_2) \wedge \uparrow Is(t_2) < \infty$. Moreover, t_2 is disabled by some transition t_1 when ω_{22} is fired from α_1, which implies that $t_1 \neq t_2 \wedge t_1 \in CFS(t_2)$. Let us now consider the two possible cases: $t_1 \in En(M_2)$ and $t_1 \notin En(M_2)$.

---

9 A selective search w.r.t. $D0, D1w$ and $D2'$, from the initial state class of $\mathcal{N}$, is a partial state space exploration, where the set of transitions selected to be fired, from the initial state class and each computed state class, satisfies $D0, D1w$ and $D2'$.

- If $t_1 \in En(M_2)$ then since, from α_2, t_2 must fire before all transitions of ω_{22}, it follows that $(F_2 \wedge t_1 \leq t_2)$ is not consistent. Such a situation is not possible for the subclass $\mathcal{TPN}$ of TPN. Indeed, for α_2, it holds that $t_2 \in Fr(\alpha_2) \wedge \uparrow Is(t_2) < \infty \wedge t_1 \in En(M) \cap CFS(t_2) \wedge (F_2 \wedge t_1 \leq t_2)$ is not consistent.
- If $t_1 \notin En(M_2)$ then for α_2, it holds that $t_2 \in Fr(\alpha_2) \wedge CFS(t_2) \not\subseteq En(M_2)$. Such a situation is not possible too for the subclass $\mathcal{TPN}$ of TPN.

The deadlock marking D is then reachable from α_t by a sequence smaller than ω. By induction hypothesis, D is reachable by the selective search from α_t. Note that, consequently, Condition $D2w$ holds for μ.   □

Thanks to Theorem 1, for any $\mathcal{N} \in \mathcal{TPN}$, a selective search w.r.t. $D0$, $D1w$ and $D2'$ allows to detect the deadlocks of $\mathcal{N}$, without resorting to POSETs.

The subclass $\mathcal{TPN}$ obviously includes the conflict-free TPN. The free-choice TPN[10] are also part of this subclass, since if a transition t is enabled then all transitions in conflict with t are enabled too and share the same maximal firing delay. Moreover, if t is firable and, at the same time, another transition in conflict is not firable then the firing of t can be postponed until all transitions in conflict with t become firable.

The set $\mathcal{TPN}$ includes also the weighted comparable preset TPN[11] such that $\forall t_i \in T, \forall t_j \in CFS(t_i), pre(t_i) \leq pre(t_j) \Rightarrow \downarrow Is(t_i) \leq \uparrow Is(t_j) \wedge \uparrow Is(t_i) = \infty$. Let t_i and t_j be two transitions such that $t_j \in CFS(t_i) \wedge pre(t_i) \leq pre(t_j)$. Then, $\downarrow Is(t_i) \leq \uparrow Is(t_j) \wedge \uparrow Is(t_i) = \infty$. If t_j is firable from a state class $\alpha = (M, F)$ then t_i is enabled in M and $F \wedge t_i \leq t_j$ is consistent, because t_i is enabled before or at the same time as t_j and $\downarrow Is(t_i) \leq \uparrow Is(t_j)$. With the fact that $\uparrow Is(t_i) = \infty$, it follows that this subclass of TPN belongs to $\mathcal{TPN}$.

Finally, the subclass $\mathcal{TPN}$ includes any TPN $\mathcal{N}$ such that:
$\forall t_i \in T, |CFS(t_i)| > 1 \Rightarrow \uparrow Is(t_i) = \infty$.

For example, the model TPN1 is a conflict-free TPN (i.e., TPN1 $\in \mathcal{TPN}$). Its CSCG consists of 39 state classes and 69 arcs. Using the stubborn sets with POSETs, this size is reduced to 16 state classes and 15 arcs. It is reduced to 7 state classes and 7 arcs when the condition $D2'$ is used without POSETs.

4   Conclusion

This paper discusses the limitations of using the POR techniques of PN in combination with POSETs, in the context of TPN. Then, it provides a subclass of TPN that takes advantage of the POR techniques of PN, without resorting to POSETs. This subclass includes, at least, the conflict-free TPN and the free-choice TPN.

As future work, we will investigate the expansion of this subclass of TPN as well as sufficient structural conditions to belong to this subclass.

---

[10] A free-choice TPN is a safe TPN such that $\forall t \in T, \forall t' \in CFS(t), pre(t) = pre(t') \wedge \uparrow Is(t) = \uparrow Is(t')$.

[11] A weighted comparable preset TPN is a safe TPN such that $\forall t \in T, \forall t_j \in CFS(t_i), pre(t_i) \leq pre(t_j) \vee pre(t_j) \leq pre(t_i))$.

References

1. Belluomini, W., Myers, C.J.: Timed state space exploration using POSETs. IEEE Trans. Comput. Aided Des. Integr. Circuits **19**(5), 501–520 (2000)
2. Bengtsson, J.: Clocks, DBMs and States in Timed Systems. PhD thesis, Dept. of Information Technology, Uppsala University (2002)
3. Bengtsson, J., Jonsson, B., Lilius, J., Yi, W.: Partial order reductions for timed systems. In: Sangiorgi, D., de Simone, R. (eds.) CONCUR 1998. LNCS, vol. 1466, pp. 485–500. Springer, Heidelberg (1998). https://doi.org/10.1007/BFb0055643
4. Berthomieu, B., Diaz, M.: Modeling and verification of time dependent systems using time Petri nets. IEEE Trans. Software Eng. **17**(3), 259–273 (1991)
5. Berthomieu, B., Vernadat, F.: State class constructions for branching analysis of time Petri nets. In: Garavel, H., Hatcliff, J. (eds.) TACAS 2003. LNCS, vol. 2619, pp. 442–457. Springer, Heidelberg (2003). https://doi.org/10.1007/3-540-36577-X_33
6. Boucheneb, H., Barkaoui, K.: Reducing interleaving semantics redundancy in reachability analysis of time Petri nets. ACM Trans. Embedded Comput. Syst. (TECS) **12**(1), 259–273 (2013)
7. Boucheneb, H., Barkaoui, K.: Stubborn sets for time Petri nets. ACM Trans. Embedded Comput. Syst. (TECS) **14**(1), 11:1–11:25 (2015)
8. Boucheneb, H., Barkaoui, K.: Delay-dependent partial order reduction technique for real time systems. Real-Time Syst. **54**(2), 278–306 (2017). https://doi.org/10.1007/s11241-017-9297-0
9. Boucheneb, H., Gardey, G., Roux, O.H.: TCTL model checking of time Petri nets. Logic Comput. **19**(6), 1509–1540 (2009)
10. Boucheneb, H., Lime, D., Roux, O.H.: On multi-enabledness in time Petri nets. In: Colom, J.-M., Desel, J. (eds.) PETRI NETS 2013. LNCS, vol. 7927, pp. 130–149. Springer, Heidelberg (2013). https://doi.org/10.1007/978-3-642-38697-8_8
11. Boucheneb, H., Rakkay, H.: A more efficient time Petri net state space abstraction useful to model checking timed linear properties. Fundamenta Informaticae **88**(4), 469–495 (2008)
12. Gerth, R., Kuip, R., Peled, D., Penczek, W.: A partial order approach to branching time logic model checking. Inf. Comput. **150**, 132–152 (1999)
13. Godefroid, P. (ed.): Introduction. Partial-Order Methods for the Verification of Concurrent Systems. LNCS, vol. 1032, pp. 11–18. Springer, Heidelberg (1996). https://doi.org/10.1007/3-540-60761-7_28
14. Lugiez, D., Niebert, P., Zennou, S.: A partial order semantics approach to the clock explosion problem of timed automata. Theoret. Comput. Sci. TCS **345**(1), 27–59 (2005)
15. Mercer, E.G., Myers, C.J., Yoneda, T.: Improved POSET timing analysis in timed Petri nets. In: International Workshop on Synthesis and System Integration of Mixed Technologies, pp. 1–6 (2001)
16. Peled, D.: All from one, one for all: on model checking using representatives. In: Courcoubetis, C. (ed.) CAV 1993. LNCS, vol. 697, pp. 409–423. Springer, Heidelberg (1993). https://doi.org/10.1007/3-540-56922-7_34
17. Valmari, A.: A stubborn attack on state explosion. Formal Methods Syst. Des. **1**(4), 297–322 (1992)
18. Valmari, A.: The state explosion problem. In: Reisig, W., Rozenberg, G. (eds.) ACPN 1996. LNCS, vol. 1491, pp. 429–528. Springer, Heidelberg (1998). https://doi.org/10.1007/3-540-65306-6_21

19. Valmari, A., Hansen, H.: Can stubborn sets be optimal? Fundam. Inform. **113**(3–4), 377–397 (2011)
20. Yoneda, T., Schlingloff, B.H.: Efficient verification of parallel real-time systems. Formal Methods Syst. Des. **11**(2), 187–215 (1997)
21. Yoneda, T., Shibayama, A., Schlingloff, B.-H., Clarke, E.M.: Efficient verification of parallel real-time systems. In: Courcoubetis, C. (ed.) CAV 1993. LNCS, vol. 697, pp. 321–332. Springer, Heidelberg (1993). https://doi.org/10.1007/3-540-56922-7_27

Towards a Generic Framework for Formal Verification and Performance Analysis of Real-Time Scheduling Algorithms

Salwa Habbachi[1]([✉]), Zhiwu Li[1,2], and Mohamed Khalgui[3]

[1] Institute of Systems Engineering (MISE), Macau University of Science
and Technology (MUST), Taipa, Macau
salwahabbachi29@gmail.com
[2] School of Electro-Mechanical Engineering, Xidian University, Xi'an 710071, China
[3] School of Electrical and Information Engineering, Jinan University, Zhuhai, China

Abstract. In real-time systems, several tasks must be executed under strict time constraints. The time requirements satisfaction depends considerably on task scheduling algorithms. Thus, performance analysis, formal verification, and comparison between a set of existing scheduling algorithms or newly proposed ones in such systems could be of great interest to designers. However, when systems are complex, models are typically used to represent their various aspects and evaluate their performance. In this paper, we propose a generic scalable framework for the description, analysis, and verification of various types of scheduling protocols based on periodic task models with the integration of their task splitting versions. The proposed framework is hierarchical based on a graphical formal model defined as a network of timed automata. This model is generic as it can be easily instantiated to describe widely used scheduling protocols or any newly proposed ones. Such modeled protocols are then formally analyzed and a set of their crucial properties is formally verified. We evaluate the feasibility of the proposed approach by implementing well-known load-balancing heuristic-based protocols.

Keywords: Real-time system · Formal verification · Scheduling algorithm · Task-splitting · Model-checking

1 Introduction

Currently, real-time systems are increasingly used in a variety of daily life applications of which some are relate to human safety [5]. Real-time systems very often consist of multiple components, such as processors. These components perform multi-tasking operations concurrently, which mainly depend on scheduling algorithms. Consequently, the choice of the scheduling algorithm affects directly the design of the systems as the interference between the tasks are critical. Thus, it is hard to imagine that without formal analysis, real-time requirements can be guaranteed.

© Springer Nature Switzerland AG 2020
B. Ben Hedia et al. (Eds.): VECoS 2020, LNCS 12519, pp. 116–130, 2020.
https://doi.org/10.1007/978-3-030-65955-4_9

Monitoring real-time requirements and evaluating the performance of real-time scheduling algorithms have become a new research issue [7,13]. Ensuring efficient scheduling protocols is then a major challenge for designers in order to guarantee appropriate results [3,15]. The use of formal methods draws a great interest to deal with any type of scheduling algorithm. The main principle behind formal methods is to build a computer-based formal model of a system under construction and formally verify that this model meets rigorous specifications of intended behavior. Thus, verification of task scheduling systems safety has become a serious challenge for engineers and raises grand defies [16].

Formal methods have become a useful way, which use allows the proving or disproving of the correctness of intended algorithms with respect to a pre-defined formal specification (set of properties). Recently, designers frequently use model-based approaches in order to increase the flexibility to predict the behavior of such kind of systems which proprieties can be easily checked using automated verification techniques. Two of the most commonly used formal verification methods are model-checking [4] and theorem proving. Model-checking is an automatic verification approach to systems that can be expressed as a finite state machine. Recently, performance analysis of real-time scheduling systems is carried out through model-based approaches based on model-checking using timed automata, which is one of the most under use formalism to check the performance of a system.

In this paper, we propose a generic model-based framework for formal verification and performance analysis of any real-time scheduling algorithms based on periodic task models. We focus on such kind of task models because existing and newly proposed scheduling algorithms are broadly based on periodic task models. These algorithms are usually applied to hierarchical scheduling systems, which are the most widely used in the industry [11,17]. Our model is based on formal semantics as used by UPPAAL framework [20] to provide a formal description of hierarchical scheduling systems modeled as a network of timed automata. The proposed model is a high-level formal model that is presented as a layer-based architecture in which the upper layer defines the scheduling layer, where the studied scheduling algorithm is given as an input to our model. Such a framework allows designers to verify the correctness and evaluate the performance of existing or newly proposed scheduling strategies at a high abstraction level without any need to implement it in the real world. Formal verification of a set of properties (deadlock freedom, invariance, CTL formulas, etc.) could be automatically checked using the UPPAAL model-checker. In addition, performance analysis could be performed using an integrated simulator. To illustrate the applicability and the feasibility of the proposed generic framework, we consider the verification, analysis, and performance comparison of different well-known scheduling protocols: Scheduling Points-based [18], Round-Robin [19], and a second version of each protocol using a task-splitting strategy as well, where the task-splitting is a technique that could be invoked when a task cannot be accommodated as a whole by any of the clusters.

2   Contribution and Related Work

2.1   Related Work

Timing aspects affect real-time systems performance and correctness. Therefore, task scheduling algorithms play a major role in such systems efficiency. However, the formal analysis of the various problems related to scheduling protocols is generally carried out using informal or semi-formal techniques and is based on simulations. For example, in [6], authors present a solution for modeling real-time systems based on simulations of software and hardware parts (co-simulation) with SystemC. In [2], a simulator, called SimSo, is designed for the comparison and the understanding of real-time scheduling policies. Over the years, significant progress has been made in the development of software verification methods and tools [13]. Therefore, rigorous formal frameworks for real-time scheduling algorithms verification and evaluation have been proposed in the context of safety-critical systems [7–9,15] and different others contexts [10]. Currently, many researchers attach great importance to the proposal of formal verification frameworks and approaches in different application areas [12]. For instance, authors in [11] proposed a compositional framework for real-time scheduling analysis that supports multiple levels of hierarchy and well-known scheduling policies of periodic task models for real-time systems. Their framework is only limited to RM and EDF protocols. Similarly, in [8,9], a framework for schedulability analysis using UPPAAL tool is proposed, which also supports limited types of scheduling algorithms. However, in our work, we seek to propose a flexible framework able to embed various scheduling algorithms based on periodic task models. Authors in [14] proposed a compositional framework for the specification and analysis of hierarchical scheduling systems, in which a generic formal model that consists of templates for specifying sporadic tasks, schedulers, and processing units, is presented. Their approach focuses, particularly, on the analysis of the impact of scheduling strategies on different resource models which are general enough to capture multi-core architectures, preemptiveness and nondeterminism. Probabilistic analysis frameworks have been recently proposed [21], based on dynamic stochastic updates of real-time attributes. Their approaches focus more on architectural and low level aspects (energy consumption and budget) of real-time systems rather than an exhaustive approach offering more flexibility and complexity in the system design, and expanding the scope of properties that can be analyzed.

2.2   Contribution

In this paper, we propose a formal methodology for the verification and analysis of scheduling algorithms based on periodic task models in real-time systems. This methodology is based on a generic formal framework that can be easily instantiated to model diverse scheduling strategies by offering a way to formally check/analyze them with minimal manual efforts. Using such a framework, the designers can implement any scheduling strategy based on periodic task models

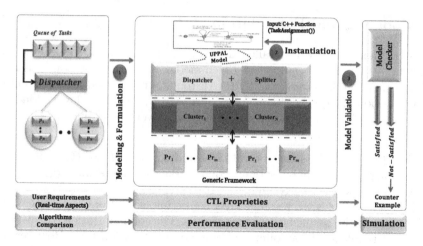

Fig. 1. A verification methodology for scheduling algorithms in Real-time Systems.

on our formal architecture, since it has already defined in a correct, by construction manner, the most critical operations that are technically sophisticated and error-prone for software implementation. Thus, it is arguably safe to implement, verify, and analyze scheduling algorithms without any risk to introduce crucial bugs. Figure 1 presents the essential workflow of our approach which can be summarized in three main steps:

1. *Step 1: Modeling and Formulation of the Framework Architecture:*
   The architecture of our framework is defined as the superposition of three layers, where the upper one represents the scheduling layer. Using UPPAAL, we have modeled the set of three layers as a set of timed automata with a set of synchronization channels. Our model is heterogeneous as it considers various resources types. It also supports scalability, as it gives the possibility to increase automatically the number of resources. In addition, a splitting module is integrated in our framework allowing to study the impact of a task-splitting strategy on the modeled scheduling algorithm.
2. *Step 2: Instantiation of the framework to specify scheduling strategies:*
   The proposed framework allows to easily implement various scheduling algorithms based on the same studied architecture. Thus, modeling any scheduling algorithm could be simply performed by implementing it into a predefined function called *TaskAssignment()*. Such a function is written in **C++** code and given as an input to our framework.
3. *Step 3: Performance analysis and formal verification:*
   Once a given scheduling strategy is implemented, our framework allows designers to verify the correctness and evaluate the performance of this strategy at a high abstraction level without any need to implement it in the real world. In particular, we propose to verify a set of timing properties for different real-time configurations. In addition to formal verification, performance analysis could be also performed using the integrated simulator of UPPAAL.

3   Modeling and Formulation

In this section, we give a detailed description of our formal framework. Our framework is based on a component-based model, in which a component is modeled as a timed automaton using UPPAAL tool. Our proposed model is built as a superposition of three layers, namely, Lower-layer (Processors layer), Medium-layer (Clusters layer) and Top-layer (Scheduling layer). The different layers are defined as follows:

- *Scheduling layer*: It is the control layer and consists of two types of components, namely, a Dispatcher that assigns tasks to clusters and a Splitter that ensures a task splitting technique.
- *Clusters layer*: It is the resource layer consisting of a set of Clusters $\{Cl_i\}_{i=1}^{N}$. Each Cluster is assigned a set of processors. The communication cost between clusters is taken into account and assumed to be a constant. However, the communication cost, task migration cost and context switch cost in the same cluster are ignored.
- *Processors layer*: It consists of a set of processors $\{Pr_i\}_{i=1}^{m}$. Each processor ensures the execution of its assigned tasks.

In our model, a set of tasks is given as an input at the level of the scheduling layer. The number and the characteristics of these tasks are parameters of the model, i.e., the user could initialize these variables as needed (see Sect. 5). Only periodical tasks are modeled in this paper. We assume that there is no dependency between tasks. Each task $T = (ID, c, r, p, d)$ is defined by five parameters belonging to $\mathbb{Q}$ [1], namely, ID is the task ID, c is the worst-case execution time, p is its period, r is its release time, and d is its relative deadline. The number of processors requested by a task T_i to be executed is calculated as follows: $np = c_i/p_i$. To describe our model, we proceed in a bottom-up manner through the different layers. The idea is to first present the two lower layers that are independent of the scheduling strategy. Then in the next section, the upper layer is detailed and an instantiation to two well-known scheduling strategies is given.

3.1   Formal Model of the Processors Layer

The lower layer of the proposed model is defined by a set of m processors. For each cluster, we define a set of components $\{Pr_1, Pr_2, \ldots, Pr_m\}$ modeling the resources that execute the assigned tasks of each cluster. Figure 2(a) describes the behavior of a processor Pr as follows:

- Initially, in state *Idle*, Pr can fire the transition labeled by $Run[Tid][cid]$? as a receiver of the synchronization triggered by its cluster. Such a transition allows Pr to move to the state *Start* and to get the different parameters of the task to be executed (ID, c, r, p, d). The transition from *Start* state has a guard based on the variable r (release time) of the task and then two cases are possible:

---
[1] $\mathbb{Q}$ denotes the set of rational numbers.

- If the release time r is equal to 0, then, the processor component fires immediately a local transition and goes to state *Executing*. In this state we define an invariant corresponding to the worst-case execution time of the task: $x \leq L[cid].T[Tid].c$, where x is a local clock. Once the execution of the task is finished, the processor triggers the synchronization *Release[cid]*! as a sender and goes back to the initial state. Such a synchronization informs the corresponding cluster about the task termination, such that it can update the number of its free processors.
- If the release time r is not equal to 0, Pr waits until the global clock reaches the release time, then, Pr fires a transition allowing to go to the state *Executing*. Note that, this is in general the case of sub-tasks dispatched by the *Splitter*.
- In *Executing* state, two transitions are possible:
  - *PR* finishes the execution of its task before or at deadline, in such case it fires the transition *Release[cid]*!. This is guaranteed by the guard $x = L[cid].T[Tid].d$ defined on this transition.
  - The deadline of the task under execution is missed, in such case the processor moves to *Error* state.
- In the *Error* state, *PR* triggers the transition *Release[cid]*! to reset while waiting for a new executable task.

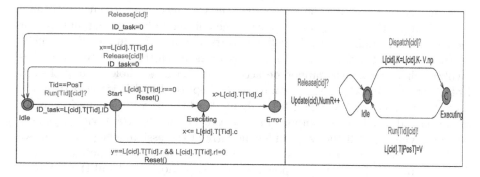

Fig. 2. (a) The Behavior of a Processor Pr and (b) the behavior of a cluster Cl.

3.2   Formal Model of the Clusters Layer

The medium layer of the proposed model consists of a set of N clusters $\{Cl_1, \ldots, Cl_N\}$. The behavior of each cluster is given in Fig. 2(b) and it is defined as follows:

- Initially in state *Idle*, each cluster can fire the first transition to synchronize with the *Dispatcher* through the synchronization *Dispatch[cid]*? as a receiver, which means that it accepts to execute the dispatched task. The execution of this transition is triggered by the *Dispatcher* or the *Splitter*, which are responsible of ensuring whether this cluster has enough available processors or not.

- Once a task is accepted to be executed, the cluster goes to state *Executing*, where it effectively runs the task through a synchronization $Run[Tid][cid]!$ triggered by the cluster and received by a component modeling the processor.
- In the state *Idle*, the cluster can also perform the synchronization $Release[cid]?$ as a receiver at any time, which corresponds to a synchronization with a processor that finishes the execution of a given task. The firing of this transition updates a set of variables, e.g.. it increasing the number of available processors of this cluster.

Note that at the level of these two lower layers, only the execution of the tasks is ensured. However, the assignment of tasks is decided at the level of the upper layer, namely, the scheduling layer. This layer allows designers to easily implement any scheduling policy as an input function. The detailed description of this layer and how it can be instantiated to model scheduling protocols is given in the next Section.

4   Instantiation at the Scheduling Layer

The scheduling layer defines the assignment policy adopted to assign tasks to clusters, taking into account different factors and trying to ensure different goals. Indeed, scheduling strategies are defined to improve the availability and the performance by balancing the load among the various clusters to ensure optimal resource utilization, maximum response time and maximum throughput. In our model, the scheduling layer is defined by two types of components, namely, *Dispatcher* and *Splitter*. The first assigns tasks to clusters and the second is invoked whenever a task cannot be accommodated as a whole by any cluster. Note that invoking the *Splitter* component is optional and it is decided by the user of our framework. We have formally modeled this layer in a generic manner as we have defined both components (*Dispatcher* and *Splitter*) with a generic behavior which is still the same independently of the assignment policy and we have encapsulated the task assignment policy and the splitting policy as input functions provided to both components. In this section, we detail the behavior of each component and illustrate how two well-known protocols could be easily modeled at this level. The behavior of the *Dispatcher* is given in Fig. 3(a) and defined as follows:

- Initially, the *Dispatcher* is in an *Idle* state where a local transition makes it go automatically to *Ready* state. This local transition allows to initialize the different variables of our model, in particular, the number of clusters and their respective capacities by the means of the function *initialize()*.
- In *Ready* state, the *Dispatcher* is ready to handle new tasks. Thus, whenever a new task is added in the queue, the transition, labeled by the function *ReadT()*, allows the *Dispatcher* to read the set of task characteristics. Then, a transition involving a function *Check()* computes whether there is a cluster that can accommodate the new task or not. Once such a transition is fired, the *Dispatcher* goes to state *Checking*. Note that to guarantee that each call

for the *Check()* function is preceded by a unique call for *ReadT()*, we use a set of guards and actions defined on a Boolean variable *R*.

- In *Checking* state, the function *Check()* is already performed, and thus depending on its result, the *Dispatcher* can go back to the state *Ready* or go to the state *Dispatching*. More precisely, if there is no cluster with enough resources (*Test = false*), the *dispatcher* comes back to *Ready* state and fires as much as needed the *Check()* function until one or more clusters become available to execute the task. However, if there is at least one cluster with enough processors to run the task (*Test = True*), then the *Dispatcher* goes to state *Dispatching*, where it computes the most suitable cluster to which the task will be assigned. This computation is performed through the function *TaskAssignment()* which defines the task assignment policy. Indeed, in our framework the only behavior depending on the scheduling strategy is this *TaskAssignment()*, which is given as a **C++** code. In other words, whenever a user of our framework wants to model, analyze, or study a given scheduling protocol, it has only to provide the code of such a strategy in this function.
- Once in state *Dispatching*, the *dispatcher* knows the *id* of the chosen cluster to accommodate the task through the function *TaskAssignment()*. To dispatch the task to its corresponding cluster, the *dispatcher* fires the transition labeled by the synchronization *Dispatch[cid]!* allowing to synchronize with the corresponding cluster. Once this synchronization is performed, an *update()* function updates a set of variables such as the number of busy processors and the variable *R* is set to *true* such that the *Dispatcher* can read a new task from the queue.

To illustrate the feasibility of our methodology, we have implemented two task assignment algorithms:

1) **Scheduling Points-based** [18]: It is an algorithm based on the computation of scheduling points of the set of tasks assigned to each cluster. The scheduling points mechanism is defined by the computation of the least common multiple of a set of periods. The objective of this algorithm is to choose an appropriate cluster for a new arriving task such that the total increase of scheduling points is minimal.
2) **Round-Robin** [19]: It is a well-known classic scheduling algorithm for managing clustered tasks. It divides the traffic equally, which means that tasks are distributed equivalently among clusters. More clearly, Round-Robin algorithm sends a request to the cluster with the fewest assigned tasks, which can also accommodate the arriving task.

After task assignment, in order to maximize the processors' utilization, a task splitting version could be invoked when a task cannot be accommodated as a whole by any cluster. This means that the task is divided into subsets of sub-tasks. Then, these subsets could be executed separately on different clusters. Contrary to classic distributed computing, in which split tasks are executed in parallel, task splitting in this case has to ensure that split tasks have to be executed serially even though they are assigned to different clusters. In such

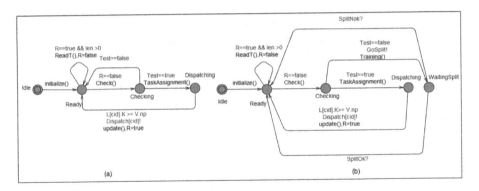

Fig. 3. (a) The behavior of the *Dispatcher* without Task-Splitting and (b) the behavior of the *Dispatcher* with Task-Splitting.

a configuration, different issues have to be taken into account, in particular, communication cost between clusters involved in the execution of the same task.

As task-splitting is a mechanism invoked only when the task-assignment fails to find a cluster to accommodate a task, in this case the *Dispatcher* can invoke the *Splitter* using a particular synchronization called *GoSplit* (see Fig. 3(b)). This synchronization allows the *Dispatcher* to go to a new state called *WaitingSplit*. In this state, the *Dispatcher* is waiting for an answer from the *Splitter*. A first possible synchronization is *SplitNok*, which means that for the task under consideration there is no possible splitting configuration. A second synchronization that may be triggered by the *Splitter* is *SplitOk*, which means that the task can be handled by a set of clusters.

The splitting mechanism and sub-tasks dispatching are ensured by the behavior of the *Splitter* described in Fig. 4, defined as follows:

- When the *Splitter* receives *GoSplit?* message from the Dispatcher, it goes to state *Splitting* by executing the function *Compute()*. This function computes whether there is a possible splitting configuration of the task under consideration or not. In other words, it checks whether the sum of all available processors of different cluster is higher or equal to the requested number of processors and whether the splitting still guarantees the schedulability of the task. If no splitting configuration is possible, the *Splitter* sends back a *SplitNok* message to the Dispatcher. However, if any splitting configuration is possible, then the *Splitter* goes to state *Ready* by executing the function *Split()*.
- *Split()* function implements the main functionality of the *Splitter* as it computes a table of the sets of sub-tasks, the *id* of the associated cluster to each set of sub-tasks, and also the release time of each set. Indeed, as the different sets of sub-tasks have to be executed serially, the computation of the release time for each set is important.

- In *Ready* state, the *Splitter* reads each set of sub-tasks and dispatches it to its associated cluster. Once all sub-tasks are dispatched, the *Splitter* informs the *Dispatcher* through the synchronization *SplitOk*.

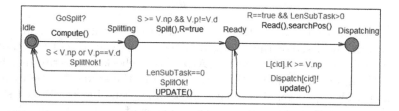

Fig. 4. The Behavior of the *Splitter* component.

5   Formal Verification and Performance Analysis

Formal verification is an integral part of the design phase of hardware and software components. It involves creating a formal design in a specification language to prove the satisfaction of a set of relevant properties required to guarantee its correctness. In the previous section, we have described our generic model allowing to specify different scheduling level and which we have instantiated for two existing algorithms. In this section, we focus on the formal verification and the performance analysis of the scheduling protocols implemented in our framework: scheduling points (SP), Round Robin (RR) and their versions with task-splitting (SP+Split and RR+Split). In particular, we first propose a formal verification of these protocols based on the automatic verification of different properties. Then, we analyze and compare their performance based on a set of simulation results.

5.1   Formal Verification

We perform the formal analysis of our model and the already specified scheduling algorithms so as to compare them and prove their satisfaction of a set of properties. Our verification is based on the method of *Model-Checking* which allows to formally prove the correctness of a system by checking if a set of desired properties are satisfied in every reachable state of that system. To this end, we use UPPAAL model-checker as a verification tool. Note that any desired property expressed as a CTL formula could be automatically checked over the proposed model.

Deadlock-Freedom Verification Deadlock-freedom is a crucial property to be checked in our model. Indeed, deadlock freedon ensures that our formal model has been correctly designed, such that it cannot reach a deadlock state, that is to say a state in which time is blocked and no action is available. Such property cannot be proven using simulation. However, deadlock detection at a high-level facilitate for us to study the correctness of the algorithms instantiated in our model, without going further into implementation. Such a property is formulated as a CTL formula as follows:

$$A[] \; not \; deadlock$$

We have automatically proven the satisfaction of the deadlock-freedom for the studied algorithms based on the proposed model at a high abstract level. Tables 1 and 2 give the time taken by the UPPAAL model-checker to verify deadlock-freedom. Using our framework, the architecture of the studied system can be easily extendable to any number of clusters or processors. Indeed, Table 1 shows the verification time when increasing the number of clusters for the different studied protocols. Extending our proposed model to a large number of clusters offers a way to study the scheduling protocols response in complex architectures. Table 2 shows the time taken for the deadlock-freedom verification when increasing the number of processors with a configuration of 10 clusters. Note that for a small number of tasks, the studied algorithms take almost the same verification time. Indeed, as already explained in Sect. 3, task-splitting is only applied when there is no cluster available to accommodate a given task. Such a situation is not relatively recurrent when the number of tasks is small.

Table 1. Verification-time (second) of deadlock-freedom when increasing the number of clusters.

Nbr of components	RR	RR+Split	SP	SP+Split
40	1.003	1.136	1.085	1.195
80	3.62	4.15	4.275	4.615
120	13.75	14.935	14.05	15.39
150	24.46	24.85	25.025	26.511
170	29.35	30.85	31.79	34.475

Table 2. Verification-time (second) of deadlock-freedom when increasing the number of processors.

Nbr of components	RR	RR+Split	SP	SP+Split
800	12.973	13.78	13.345	14.6
1200	33.62	35.68	35.89	38.5
1400	42.94	43.65	43.29	46.576

Verification of invariance and timing constraints One of the most important issues, when studying task-scheduling protocols, is related to the set of timing constraints associated to each task. In particular, in the case of real-time systems, a task is defined by a set of parameters such as execution time and deadlines that the system has to ensure when executing these tasks (Sect. 3). To check the respect of such parameters, we have defined first an invariant to guarantee the execution time and a we have added a particular state called *error* to detect deadline violation (see Sect. 3). More precisely, in our model, we define an invariant in the state *Executing* of each processor, which guarantees that a given task cannot be executed more than its predefined worst-case execution time c. This invariant is formally described using Temporal-Logic as follows :

$$Invariance: A[] \, Pr(Tid, cid).Executing \implies (Pr(Tid, cid).x \leq L[cid].T[Tid].c)$$

The verification time of the invariance property is given in Fig. 5 when the number of both clusters and processors increases. Note that for formal verification, the measurement of the verification time is very interesting as it gives an idea about the cost of the verification process.

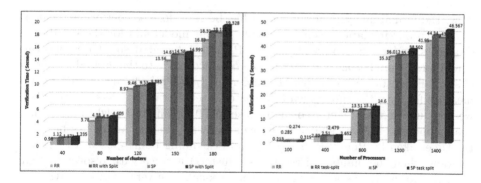

Fig. 5. Invariance verification time (second) for task-scheduling protocols.

Ensuring the respect of deadlines is very important as it represents the performance and efficiency of scheduling algorithms. In our model, this is related to detect the deadline violation through the reachability of a state *error*. Such a state is reached whenever the deadline of the task is missed. The reachability of this state is verified using the following CTL property: :

$$Reachability: A[] \, not \, Pr(Tid, cid).Error$$

The different results obtained when studying the reachability property for different topologies are given in Tables 3 and 4.

Table 3. Verification-time (second) of the Reachability property when increasing the number of clusters.

Nbr of components	RR	RR+Split	SP	SP+Split
7	22.945	35.597	23.865	33.791
10	35.78	41.937	37.02	43.238
20	53.8	67.523	51.703	68.123

Table 4. Verification-time (second) of the Reachability property when increasing the number of processors.

Nbr of components	RR	RR+Split	SP	SP+Split
10	0.031	0.45	0.125	0.153
14	0.859	1.839	32.8885	33.45
20	42.02	48.36	55.32	58.238

5.2   Performance Analysis

Now, the formal analysis of our model has been performed, performance analysis based on simulations is also possible for our model and for the scheduling strategies under study. Indeed, using UPPAAL integrated simulator, one can run different simulations and measure metrics according to the desired aspect to observe. For instance, to compare the different studied scheduling algorithms, we measure for each the time taken to dispatch a set of waiting tasks. In Fig. 6(a), we measure the time taken to dispatch a set of 50 waiting tasks, when increasing the number of clusters. Therefore, we can easily see that the difference between the time taken by Round-Robin (RR) and Scheduling Points-based (SP) is not relevant when the number of resources (clusters) is large with respect to that of tasks. However, the performance of the task-splitting version of both RR and

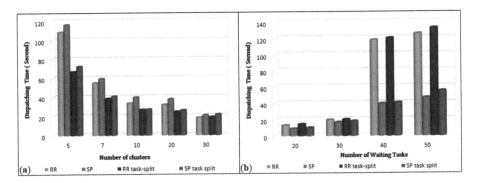

Fig. 6. The impact of Splitting on the time taken to dispatch tasks when: (a) increasing the number of clusters and (b) increasing the number of waiting tasks.

SP is noticeable when the resources number is small. Hence, we can conclude that the performance comparison between the scheduling protocols depends on the number of tasks to be executed and that of resources provided, which means that the choice of the algorithm can be differentiated according to the needs. In Fig. 6(b), we can observe that the task-splitting version of both algorithms decreases considerably the time taken to dispatch tasks, in particular when the number of waiting tasks is large. In other words, the performance aspect of the task-splitting algorithms over load balancing is remarkable when the number of waiting tasks increases.

6 Conclusion

In this work we have proposed a generic framework for formal verification of task-scheduling protocols and their performance analysis in real-time systems based on timed automata. In particular, we have focused on the hierarchical scheduling systems, where we model such systems as the superposition of three layers. We have instantiated our framework to formally support a set of scheduling protocols, studied their performance, and proved a set of properties on these protocols. In future work, we intend to extend our framework in order to handle different types of tasks, the notion of preemption, and the task migration between clusters.

References

1. Saifullah, A., et al.: Parallel real-time scheduling of DAGs. IEEE Trans. Parall. Distrib. Syst. **25**(12), 3242–3252 (2014)
2. Chéramy, M., Hladik, P.-E., Déplanche, A.-M.: Simso: a simulation tool to evaluate real-time multiprocessor scheduling algorithms (2014)
3. Cho, S., et al.: Efficient real-time scheduling algorithms for multiprocessor systems. IEICE Trans. Commun. **85**(12), 2859–2867 (2002)
4. Alur, R., Courcoubetis, C., Dill, D.: Model-checking for real-time systems. In: 1990 Proceedings Fifth Annual IEEE Symposium on Logic in Computer Science, IEEE (1990)
5. Krishna, C.M.: Real-Time Systems. Wiley Encyclopedia of Electrical and Electronics Engineering (2001)
6. Le Moigne, R., Pasquier, O., Calvez, J.: A generic RTOS model for real-time systems simulation with systemC. In: Proceedings Design, Automation and Test in Europe Conference and Exhibition, vol. 3, Paris, France, pp. 82–87 (2004)
7. Wang, M., Tian, C., Zhang, N., Duan, Z., Du, H.: Verifying a scheduling protocol of safety-critical systems. J. Comb. Optim. **37**(4), 1191–1215 (2018). https://doi.org/10.1007/s10878-018-0343-1
8. D. Alexandre, I. Jacob, K. G. LARSEN, and al, "Model-based framework for schedulability analysis using UPPAAL 4.1. Model-based design for embedded systems, vol. 1, no. 1, pp. 93–119 (2009)
9. Boudjadar, A., et al.: Hierarchical scheduling framework based on compositional analysis using Uppaal. In: Fiadeiro, J.L., Liu, Z., Xue, J. (eds.) FACS 2013. LNCS, vol. 8348, pp. 61–78. Springer, Cham (2014). https://doi.org/10.1007/978-3-319-07602-7_6

10. Ahmad, S., Malik, S., Ullah, I., Park, D.-H., Kim, K., Kim, D.: Towards the design of a formal verification and evaluation tool of real-time tasks scheduling of IoT applications. Sustainability **11**(1), 204 (2019)
11. Shin, I., Lee, I.: Compositional real-time scheduling framework with periodic model. ACM Trans. Embed. Comput. Syst. (TECS) **7**(3), 1–39 (2008)
12. Beek, M.H., Gnesi, S., Knapp, A.: Formal methods an automated verification of critical systems. Int. J. Softw. Tools Technol. Transfer (STTT) **20**(4), 355–358 (2018)
13. Shan, L., Graf, S., Quinton, S., Fejoz, L.: A framework for evaluating schedulability analysis tools. In: Aceto, L., Bacci, G., Bacci, G., Ingólfsdóttir, A., Legay, A., Mardare, R. (eds.) Models, Algorithms, Logics and Tools. LNCS, vol. 10460, pp. 539–559. Springer, Cham (2017). https://doi.org/10.1007/978-3-319-63121-9_27
14. Boudjadar, J., Kim, J.H., Phan, L.T.X., Insup, L., Kim, G.L., Ulrik, N.: Generic formal framework for compositional analysis of hierarchical scheduling systems. In: Proceedings IEEE 21st International Symposium on Real-Time Distributed Computing (ISORC), pp. 51–58 (2018)
15. Quinton, S.: Evaluation and comparison of real-time systems analysis methods and tools. In: Howar, F., Barnat, J. (eds.) FMICS 2018. LNCS, vol. 11119, pp. 284–290. Springer, Cham (2018). https://doi.org/10.1007/978-3-030-00244-2_19
16. Liu, J., Zhou, M., Song, X., Ming, G., Jiaguang, S.: Formal modeling and verification of a rate-monotonic scheduling implementation with real-time maude. IEEE Trans. Ind. Electron. **64**(4), 3239–3249 (2016)
17. Alsheikhy, A., Elfouly, R., Alharthi, M., Reda, A., Abdulrahma, A.: Hybrid scheduling algorithm for periodic tasks in real-time systems. J. King Abdulaziz Univ. **27**(2), 3–13 (2016)
18. Chen, Y., Chen, Y.: Real-time scheduling in cyber-physical systems. J. Control Eng. Appl. Inform. **13**(3), 41–50 (2011)
19. Sharma, S., Singh, S., Sharma, M.: Performance analysis of load balancing algorithms. World Acad. Sci. Eng. Technol. **38**(3), 269–272 (2008)
20. David, A., Larsen, K.G., Legay, A., Mikucionis, M., Bøgsted Poulsen, D.: Uppaal SMC tutorial. Softw. Tools Technol. Transfer **14**, 397–415 (2015)
21. Chadli, M., Kim, J.H., Larsen, K.G., Legay, A., Naujokat, S., Steffen, B., Traonouez, L.-M.: High-level frameworks for the specification and verification of scheduling problems. Int. J. Softw. Tools Technol. Transfer **20**(4), 397–422 (2017). https://doi.org/10.1007/s10009-017-0466-1

Accurate Strategy for Mixed Criticality Scheduling

Yasmina Abdeddaïm[(✉)]

LIGM, Univ Gustave Eiffel, CNRS, ESIEE Paris, 77454 Marne-la-Vallée, France
`yasmina.abdeddaim@esiee.fr`

Abstract. In mixed criticality systems, functionalities with different
criticalities share the same execution platform. The goal of a mixed criticality scheduling algorithm is to ensure a safe execution of the highest
criticality tasks while using efficiently the execution platform. Classical
real-time scheduling algorithms for mixed criticality systems propose to
stop the execution of low criticality tasks to ensure that the highest
criticality tasks meet their deadlines. In this work, we propose the Accurate Fault Mode (AFM) strategy for the scheduling problem of mixed
criticality real-time systems. The advantages of this strategy are that
firstly it can reduce the number of stopped low criticality tasks and secondly, that it allows the designer to define the low criticality tasks to be
stopped depending on the criticality configuration of the system. Using
model checking for timed game automata, we propose an exact feasibility
test and exact schedulability tests for fixed priority and earliest deadline
first scheduling algorithms for AFM strategy.

Keywords: Mixed criticality · Real-time scheduling · Timed game
automata

1 Introduction

A mixed criticality system is a system that incorporates the functionalities of
different criticalities on the same platform. The highest criticality functionalities are usually related to safety critical applications and need to fulfill strict
certification requirements, while lower criticality functionalities are non-safety
critical functionalities with less stringent certification requirements. To avoid
interference between applications of different criticalities, traditional methods
use temporal or spatial isolation between applications of different criticalities
[17], however, this isolation may lead to an inefficient use of the execution platform.

With the increasing complexity of embedded systems in the industry of critical systems, standards in automotive and avionic, are now mentioning that
"mixed criticality" must be supported on their platform [10].

The first work dealing with real-time scheduling for mixed criticality systems
is that of Vestal [19]. In Vestal's model, a real-time task has several worst-case

B. Ben Hedia et al. (Eds.): VECoS 2020, LNCS 12519, pp. 131–146, 2020.
https://doi.org/10.1007/978-3-030-65955-4_10

execution time (WCET) estimates, one per possible "mode of execution" of the system, called the criticality mode of the system. For example, a WCET in the case of a "regular mode" and a WCET in the case of a "fault mode" where the "fault mode" is more critical than a "regular mode" and the higher the criticality mode, the larger the WCET estimate. The intuition is that larger execution time values are less likely to occur, but if they occur, they may be indicative of an erroneous event and the system moves in this case to a high criticality mode.

A system designed to be executed in a high criticality mode is safe but does not use the platform efficiently as high criticality WCETs are pessimistic, however, considering lowest criticality WCETs is unsafe even if the probability of occurrence of high criticality WCETs is small. The challenging issue is to ensure that the system is safe while using the execution platform efficiently.

Since Vestal's model, a standard approach for the scheduling problem of mixed criticality real-time systems has emerged. In this approach, the system starts its execution in the lowest criticality mode and moves to the highest criticality mode if the execution time of a job exceeds the WCET of the low criticality mode. In the lowest mode, all the tasks have to respect their timing constraints and when the criticality mode of the system increases, lowest criticality tasks are no more activated so that higher criticality tasks fulfill their requirements.

As stated in the frequently updated review paper [10], criticisms of the standard approach are addressed concerning the following hypothesis:

1. The criticality mode of the system can only increase: it should be possible for the system to recover and move from a high to a low criticality mode.
2. When the system is in a high criticality mode, the execution time of all the high tasks is supposed to be equal to the WCET of the high criticality mode: this hypothesis is too abusive, a high criticality mode, could be the consequence of the occurrence of an error in only one critical task.
3. When the criticality mode of the system increases, less critical tasks are definitely no more activated: even if lower criticality functionalities are non-safety critical functionalities they are relevant for the good functioning of the system [9]. Abandoning all low criticality tasks may have an impact on the execution of the system and degrade the quality of service.

In this paper, we present a dual criticality model where:

1. The criticality mode of the system increases if an erroneous behavior occurs (a job exceeds its low criticality WCET estimates), however, if the erroneous behavior disappears the criticality of the system decreases.
2. We introduce a more accurate measure of the criticality mode of the system called the criticality configuration of the system. The criticality configuration gives the set of tasks that are exhibiting an erroneous behavior.
3. The designer can specify the subset of lowest criticality tasks to stop when the system is in high criticality mode depending on the subset of high criticality tasks exceeding their low criticality WCET.

For this model, we propose an exact feasibility test and exact schedulability tests for fixed priority and earliest deadline first scheduling algorithms. These tests are derived from CTL model checking for timed game automata.

Section 2 reviews related research. Section 3 introduces the model and the mixed criticality scheduling problem. Section 4 presents the Accurate Fault Mode (AFM) strategy. Section 5 presents the feasibility and schedulability tests for the AFM strategy. Section 6 is dedicated to an example used to illustrate the advantages of our approach. We conclude and give future research directions in Sect. 7.

2   Related Work

For a complete review of mixed criticality real-time scheduling see [10]. We focus in this section on uniprocessor scheduling addressing one of the three criticisms cited in the introduction. Since Vestal's model, many scheduling policies have been proposed to handle the problem of stopping all low criticality tasks when the criticality of the system is high. In [12], the AMC Weakly Hard (AMC-WH) policy is proposed, in this policy, only a number of consecutive jobs of low critically tasks is allowed to be stopped when the system is in high criticality mode, the authors proposed a sufficient schedulability test for this policy. In [11], the authors introduce the notion of importance. When the system is in high criticality mode, low criticality tasks are stopped in the inverse order of their importance. In these two works, the set of tasks that are no more activated do not depend on the configuration of the system as proposed in our work.

In [13], the authors introduce the notion of interference constraint graph, where the designer can specify the subset of tasks to be stopped when a particular high criticality task exceeds a certain budget of time. This approach has some similarities with our work, however, in this approach, when a task is stopped, it is no more activated, and the strategy is defined for a task and not for a particular configuration.

Concerning the possibility of a system to recover and move from a high to a low criticality mode, a simple protocol presented in [18] is to switch to a low criticality mode at the first instant where no job is active. A more complex protocol, called the bailout protocol, is presented in [6,7]. In this protocol, when a high criticality task exceeds its WCET at low criticality level, it is assigned a bailout fund, this bailout is funded by the execution times of low tasks or by the execution times of tasks that terminate before their estimated WCET.

None of the mentioned work proposes a model to handle the three criticisms cited in Sect. 1 at the same time and does provide exact feasibility and schedulability tests.

3   The Problem Statement

We model the mixed criticality real-time system as a set of n sporadic real-time tasks $\Gamma = \{\tau_i : i = 1 \ldots n\}$ scheduled on a single processor. We restrict this work to dual criticality systems, i.e. systems with only two possible criticalities LO and HI, LO is the lowest criticality and HI is the highest criticality. Every task is assigned a criticality level defined by the system designer, nLO is the number of LO criticality tasks, and nHI is the number of HI criticality tasks. Without loss of generality, we suppose that the set Γ is sorted in a decreasing order of criticality. We denote Γ^{LO} the set of low criticality tasks, and Γ^{HI} the set of HI criticality tasks. Each task $\tau_i \in \Gamma$ generates an infinite number of jobs $\tau_{i,k}$.

A task $\tau_i \in \Gamma$ is defined as a tuple $(L_i, pr_i, T_i, D_i, C_i)$ where:

- $L_i \in \{LO, HI\}$ is the criticality of the task.
- $pr_i \in \{1 \ldots n\}$ is the priority of the task with 1 as the highest priority and n the lowest priority. We use $lp(i)$ to be the indexes of tasks of lower priority than τ_i.
- T_i is the minimum time separation between jobs releases. Two tasks can have the same priority.
- $D_i \leq T_i$ is its relative deadline, i.e. the maximal time length between the arrival time and the completion time of a job.
- C_i is a tuple $(C_i(LO), C_i(HI))$, where $C_i(l) \in \mathbb{N}$ is the WCET budget of task τ_i at criticality level $l \in \{LO, HI\}$ s.t. if $L_i = LO$, $C_i(HI) = C_i(LO)$ and if $L_i = HI$, $C_i(LO) \leq C_i(HI)$.

For every task τ_i, the $C_i(l)$ WCET budget is a constant interval of time allocated to the execution of every job of the task when the criticality of the system is $l \in \{LO, HI\}$. We use the notion of run-time monitoring of execution times introduced in [5]. If a job of a task does not signal its completion after the execution of its allocated budget at its own criticality it is stopped, as a consequence, a LO criticality task cannot exceed its LO WCET budget.

A job is *active* at instant t if it is triggered by a task at time $t' \leq t$ and the job has not yet notified its completion at instant t. We note S_t the set of active jobs at time t. The response time $R_{i,k}$ of a job $\tau_{i,k}$ is the duration between the activation date of the job and its completion date. The current response time of an active job is the duration between the activation date of the job and the current time. A job $\tau_{i,k}$ respects its deadline if and only if $R_{i,k} \leq D_i$ and a task τ_i respects its deadline if and only if all the jobs activated by the task respect their deadlines.

An active job of a task τ_i is *critical*, if and only if τ_i is a high criticality task and the job does not notify its completion after the execution of $C_i(LO)$ time unit. An execution scenario of a task set in an interval of time $[t_1, t_2]$, gives for every instant t, $t_1 \leq t \leq t_2$ $Exec(t)$ the sets of active jobs, preempted and executed jobs at time t and for every active task its current response time and the duration since its activation. We note $\mathbb{E}$ the set of possible execution scenarios.

The notion of criticality mode of the system is used to characterize a particular execution scenario of a task set. The criticality mode of the system is HI at instant t if there exists an active critical job, otherwise the criticality mode of the system is LO. A formal definition is given in Definition 1.

Definition 1. *Criticality mode at time t. The criticality mode of the system at time t is a function $Cr(t) : \mathbb{R}^+ \rightarrow \{LO, HI\}$ with*

$$Cr(t) = \begin{cases} LO \text{ if for all } \tau_{i,k} \in S_t, \tau_{i,k} \text{ is not a critical job} \\ HI \text{ if there exists } \tau_{i,k} \in S_t \text{ a critical job at time } t \end{cases}$$

Given a task set, a scheduling algorithm gives at each instant the job to be executed among the set of active jobs for every execution scenario.

Definition 2. *Scheduling algorithm.* *A scheduling algorithm is a function* $Sched : (\mathbb{E}, \mathbb{R}^+) \rightarrow \{\tau_{i,k} \mid i = 1 \dots n, k = 1 \dots \infty\} \bigcup \{\perp\}$ *with*

- *if* $Sched(Exec(t), t) = \tau_{i,k}$ *then execute the active job* $\tau_{i,k} \in S_t$ *at time t*
- *if* $Sched(Exec(t), t) = \perp$ *then the processor is idle at time t.*

A scheduling algorithm is preemptive, if the execution of a job of a task can be preempted by a job of an other task. A scheduling algorithm is a job level fixed priority algorithm if a fixed priority is assigned to every active job of a task and at each time the job with the highest priority among the set of active jobs is executed. A scheduling algorithm is a fixed priority algorithm if priorities are assigned to tasks, i.e. all the jobs of a task have the same priority.

In definition 3, we define the mixed criticality scheduling problem. In this problem, when the criticality of the system is LO, no job is allowed to miss its deadline, but when the criticality of the system is HI, only jobs of HI criticality must respect their deadlines. The idea is to relax the timing constraints of low criticality functionalities of the system to focus on the good functioning of the high criticality functionalities.

Definition 3. *Mixed Criticality Schedulability.* *A task set* Γ *is mixed critical- ity schedulable according to a scheduling algorithm iff (1) all the tasks respect their deadlines when the criticality mode of the system is LO and (2) all the HI criticality tasks respect their deadlines when the criticality mode of the system is HI.*

A task set Γ is *mixed criticality feasible* if and only if there exists a scheduling algorithm such that Γ is schedulable according to this scheduling algorithm.

Note that the schedulability problem of Definition 3 is equivalent to the stan- dard scheduling problem of mixed criticality real-time systems. The difference is in the way that we compute the criticality mode of the system, see Definition 1, as we take into account the fact that a system can recover and the critical- ity mode of the system can return to low if no job exceeds its WCET at LO criticality.

4   Accurate Fault Mode Strategy

Classical scheduling strategies for mixed criticality scheduling propose to stop the activation of low criticality tasks when the criticality mode of the system is HI as low criticality tasks are not constraint to respect their deadlines in a high criticality mode, see Definition 3. As we mentioned in the introduction, stopping the activation of all low criticality tasks has been criticized.

In this section, we introduce the "Accurate Fault Mode" (AFM) strategy. In this strategy, the designer can specify, by setting up a "fault mode policy", the set of LO criticality tasks that are no more activated when the system is in HI criticality mode. The idea is that even if they are not critical, some LO criticality tasks may have an impact on the quality of service of the system or have to be executed during the HI criticality period to ensure the degraded mode. This subset of LO criticality tasks can change depending on the execution configuration of the task set. For example, if only one job is critical, the designer may decide to ensure a certain quality of service, otherwise, if more jobs are critical the quality of service is only ensured at its minimum.

To evaluate the criticality mode of the system in a more accurate way, we introduce the notion of criticality configuration. If the criticality mode of the system is HI, the criticality configuration returns the set of tasks that are exhibiting an erroneous behavior.

Definition 4. *Criticality configuration. The criticality configuration of the system $\overrightarrow{HI} = (HI(1), \dots, HI(nHI))$ at time t is a boolean vector of size nHI with*

$$HI(i) = \begin{cases} 1 \ \textit{if there exits } \tau_{i,k} \ \textit{an active critical job of } \tau_i \ \textit{at } t \\ 0 \ \textit{if there is no active critical job of } \tau_i \ \textit{at } t \end{cases}$$

We note $\mathbb{H}$ the set of possible criticality configurations.

To define the policy for dealing with LO criticality tasks, the designer gives for every criticality configuration, the set of LO criticality tasks for which the activation can be stopped. The fault mode policy is formalized in Definition 5.

Definition 5. *Fault mode policy. Given a task set Γ, a fault mode policy of a criticality configuration $\overrightarrow{HI} \in \mathbb{H}$ is a function $policy^{\overrightarrow{HI}}$ from Γ^{LO} to $\{0,1\}$ with,*

$$policy^{\overrightarrow{HI}}(\tau_i) = \begin{cases} 1 \ \textit{denotes that jobs of } \tau_i \textit{ are not activated when} \\ \quad \textit{the criticality configuration is } \overrightarrow{HI} \\ 0 \ \textit{denotes that jobs of } \tau_i \textit{ are activated when the} \\ \quad \textit{criticality configuration is } \overrightarrow{HI} \end{cases}$$

A fault mode policy $policy_i$ is defined by $policy_i = \bigcup\limits_{\overrightarrow{HI} \in \mathbb{H}} policy^{\overrightarrow{HI}}$.

In Definition 6, we define the Accurate Fault Mode (AFM) strategy for mixed criticality scheduling.

Definition 6. *Accurate Fault Mode (AFM) strategy (w.r.t. a fault mode policy). Given a task set Γ and a fault mode policy $policy_i$, a scheduling algorithm Sched respects an Accurate Fault Mode (AFM) strategy if and only if when the criticality of the system is HI the schedule computed using Sched respects the fault mode policy $policy_i$.*

In Definition 7, we define the AFM schedulability problem for mixed criticality scheduling. In this problem, a subset of LO criticality tasks may have to respect their deadlines even in a HI criticality mode of the system. This subset of tasks is defined using the fault mode policy given by the system designer.

Definition 7. *AFM Schedulability. A task set Γ is AFM schedulable according to the scheduling algorithm Sched if and only if (1) all the tasks respect their deadlines when the criticality mode of the system is LO and (2) all the active jobs respect their deadlines when the criticality mode of the system is HI and the fault mode policy of the AFM strategy is applied.*

A task set Γ is AFM feasible if and only if there exists an AFM strategy and a scheduling algorithm such that Γ is AFM schedulable.

5   Feasibility and Schedulability Analysis

In this section, we present exact feasibility and schedulability tests for the AFM strategy. These tests are based on CTL model checking for timed game automata.

5.1   A Timed Game Model

We model a task set Γ as a network of timed game automata T_Γ as presented in this section. A complete model using the tool Uppaal-Tiga [8] is available in [1].

A Timed automaton (TA) [3] is a model extending the automaton model with a set of real variables, called clocks, evolving synchronously with time. Transitions of a time automaton can be labelled by a clock constraint, clocks can be reset in transitions and states can be constrained by a staying condition.

A network of timed automata is the parallel composition of a set of timed automata, the parallel composition uses an interleaving semantic. Synchronous communication between timed automata is done using input actions denoted $a?$ and output actions denoted $a!$, this notation is used in the tool Uppaal [15]. A configuration of a network of n timed automata T_i is a pair (q, v) where $q = (q_1, \ldots, q_n)$ is a vector of states, with q_i is a state of T_i, and v a vector of clock valuations.

A Timed game automaton (TGA) [16] is an extension of the time automaton model where the set of transitions is split into controllable (Δ_c) and uncontrollable (Δ_u) transitions. In Figs. 1 and 2, dashed lines represent uncontrollable transitions. This model defines the rules of a game between a controller (controllable transitions) and the environment (uncontrollable transitions).

Given a timed game automaton T and a logic formula ϕ, if T satisfies ϕ then there exists a strategy f, defining for every possible configuration, the controllable transition to execute, s.t. T supervised by f always satisfies ϕ whatever are the uncontrollable transitions chosen by the environment. A strategy is formally a partial mapping from the set of runs of the TGA to the set $\Delta_c \cup \{\lambda\}$ s.t. for a finite run ξ:

- if $f(\xi) = e \in \Delta_c$ then execute transition e from the last configuration of ξ,
- if $f(\xi) = \lambda$ then wait in the last configuration of ξ.

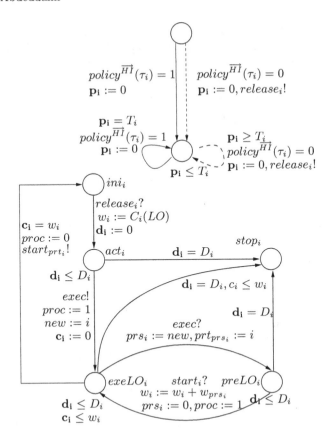

Fig. 1. $TaskLO_i$ and $PeriodLO_i$ of a LO criticality task τ_i

We model a LO criticality task $\tau_i \in \Gamma^{LO}$ using a timed game automaton $TaskLO_i$ with a set $\mathcal{Q} = \{ini_i, act_i, exeLO_i, preLO_i, stop_i\}$ of states and two clocks c_i and d_i. This automaton is synchronized with a timed game automaton $PeriodLO_i$ using the action $release_i$. If the fault mode policy defined by the designer states that the activation of task τ_i is not stopped for the current criticality configuration ($policy^{\overrightarrow{HI}}(\tau_i) = 0$), the action $release_i$ is launched by the automaton $PeriodLO_i$. In this case, the automaton $TaskLO_i$ is synchronized with the uncontrollable transition of automaton $PeriodLO_i$. This transition is an uncontrollable transition as the task set Γ is sporadic i.e. a job is triggered at the earliest every T_i time unit, but we don't control the time by which the job is triggered. Otherwise, if the fault mode policy states that jobs of task τ_i are no more activated ($policy^{\overrightarrow{HI}}(\tau_i) = 1$), the action $release_i$ is not launched.

The automaton $TaskLO_i$ starts its execution at state ini_i, when an action $release_i$ is captured, the automaton moves to state act_i and the clock d_i is reset to zero, when the clock d_i reaches the deadline D_i the automaton moves to state

$stop_i$. When the task starts its execution, the automaton moves to state $exeLO_i$, the global variable $proc$ is reset to one indicating that the processor is not idle and the clock c_i is reset to zero. The clock c_i is used to measure w_i the response time of task τ_i. The response time w_i of a task is set initially to $C_i(LO)$.

To be able to handle preemption using timed automata, we restrict ourselves to job level fixed priority scheduling algorithms. In a job level fixed priority scheduling algorithm, the priority does not change during time between two jobs $\tau_{i,k}$ and $\tau_{j,r}$ of two different tasks. This restriction does not limit much the generality of our work as most of the commonly known scheduling algorithms, as fixed priority (FP) and earliest deadline first (EDF), respect this restriction. As a consequence, in our model, a task can be preempted only if a new job is executed, this is done by using a global action $exec?$ synchronizing every preemption with the execution of a new job. When a task is preempted, the automaton moves to state $preLO_i$, the variable prs_i records the identifier of the preempting task and prt_i the identifier of the task preempted by τ_i. When the preempting task τ_{prs_i} completes, the response time w_i of the task τ_i is augmented by the response time of the preempting task τ_{prs_i}. This method has been used in [2] to model preemptions.

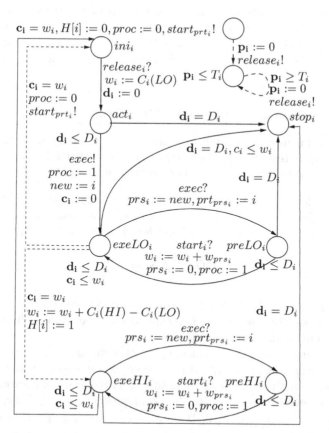

Fig. 2. $TaskHI_i$ and $PeriodHI_i$ of a HI criticality task τ_i

We model a HI criticality task $\tau_i \in \Gamma^{HI}$ using a timed game automaton $TaskHI_i$, this automaton is synchronized with a timed game automaton $PeriodHI_i$. In addition to states of automaton $TaskLO_i$, automaton $TaskHI_i$ has two specifics states $exeHI_i$ and $preHI_i$ used to model the behavior of a critical job. The transitions from $exeLO_i$ to ini_i and from $exeLO_i$ to $exeHI_i$ are uncontrollable as we don't know in advance if the job will terminate after the execution of $C_i(LO)$ time unit or exhibits an erroneous behavior.

If the job is critical, the variable w_i is augmented by $C_i(HI) - C_i(LO)$. In the transition from $exeLO_i$ to $exeH_i$, the criticality configuration $\overrightarrow{HI}$ is updated by adding the task τ_i to the set of tasks with a critical job ($HI[i] = 1$). When the critical job terminates its execution, transition from $exeHI_i$ to ini_i, the criticality configuration $\overrightarrow{HI}$ is updated by removing the task τ_i from the set of tasks with a critical job ($HI[i] = 0$).

Given a task set Γ, the network of TGA T_Γ is the parallel composition of $TaskLO_1, \ldots, TaskLO_{nLO}$ and $TaskHI_1, \ldots, TaskHI_{nHI}$ and $PeriodLO_i \ldots PeriodLO_{nLO}$ and $PeriodHI_i \ldots PeriodLO_{nHI}$. The TGA T_Γ is augmented with a clock t, this clock is never reset and is used to measure the total elapsed time.

We say that a configuration (q, v, t) is equivalent to an execution scenario $Exec(t)$ iff if a task is active, or preempted, or executed in q then it is also active or preempted, or executed in $Exec(t)$ and the current response times and duration since the activation of every active task are identical in $Exec(t)$ and (q, v, t).

5.2 Exact Feasibility and Schedulability Tests

We use CTL [14] model checking for time game automata to build feasibility and schedulability tests for the AFM strategy. The task set Γ is modeled using a network of timed game automata as presented in Sect. 5.1.

Theorem 1 (Exact AFM Feasibility test). *The task set Γ is AFM feasible (w.r.t. a fault mode policy) according to a job level fixed priority algorithm iff the network of timed game automata T_Γ modeling Γ satisfies the CTL Formula 1.*

$$AG\neg(\bigvee_{\tau_i \in \Gamma} stop_i) \tag{1}$$

Proof. An automaton T satisfies the formula $AG\phi$ iff there exists a wining strategy s.t. whatever the execution of uncontrollable transitions, there exists an execution where all the states satisfy ϕ.

Suppose that Formula 1 is satisfied, then, whatever is the execution of uncontrollable transitions in T_Γ, there exists an execution where $stop_i$ is not reached, meaning that no active job misses its deadline. This execution respects an AFM strategy as in every possible criticality configuration, a LO criticality job is active, iff $policy^{\overrightarrow{HI}}(\tau_i) = 0$, i.e. iff the fault mode policy states that the activation of jobs of task τ_i are not stopped for the current criticality configuration. A scheduling algorithm where no active job misses its deadline and respecting an AFM strategy can be computed using the wining strategy where $\forall t \in \mathbb{R}^+$

$Sched(Exec(t), t) = \tau_{i,k}$,with k the kth active job of τ_i, if there exists ξ in T_Γ with a last configuration $(q, \boldsymbol{v}, t)$ equivalent to $Exec(t)$ and $f(\xi) = e$ with e is a transition from act_i to $exeLO_i$.

Now, suppose that the task set Γ is AFM feasible according to a job level fixed priority algorithm and an AFM strategy. Thus, there exists a job level fixed priority scheduling algorithm $Sched$ respecting the AFM strategy where no active job misses its deadline. Using the scheduling algorithm, we can compute a wining strategy f for the timed game defined by T_Γ and Formula 1 with

$f(\xi) = $ take transition from act_i to $exeLO_i$ if the last configuration $(q, \boldsymbol{v}, t)$ of ξ is equivalent to $Exec(t)$, $act_i \in q$ and $Sched(Exec(t), t) = \tau_{i,k}$.

Theorem 2 (Exact FP AFM schedulability test). *The task set Γ is fixed priority (FP) AFM schedulable (w.r.t. a fault mode policy) iff the network of timed game automata modeling Γ satisfies the CTL Formula 2.*

$$AG\neg(\bigvee_{\tau_i \in \Gamma} stop_i) \bigwedge \neg(\bigvee_{\tau_i \in \Gamma} \bigvee_{\tau_j \in lp(i)} (act_i \wedge exeLO_j) \bigvee_{\tau_i \in \Gamma} \bigvee_{\tau_j \in lp(i)} (act_i \wedge exeHI_j)$$
$$\bigvee_{\tau_i \in \Gamma} \bigvee_{\tau_j \in lp(i)} (preLO_i \wedge exeLO_j) \bigvee_{\tau_i \in \Gamma} \bigvee_{\tau_j \in lp(i)} (preLO_i \wedge exeHI_j) \quad (2)$$
$$\bigvee_{\tau_i \in \Gamma^{HI}} \bigvee_{\tau_j \in lp(i)} (preHI_i \wedge exeLO_j) \bigvee_{\tau_i \in \Gamma^{HI}} \bigvee_{\tau_j \in lp(i)} (preHI_i \wedge exeHI_j))$$

Proof. Formula 2 is satisfied iff there exists an execution where no job misses its deadline (state $stop_i$ not reached) and if a job of a task τ_i cannot be active (state act_i) or preempted (state $preLO_i$ or $preHI_i$) if a job of a task τ_j of lower priority is executed (state $exeLO_j$ or $exeHI_j$). Thus the execution of jobs is done according to a fixed priority scheduling algorithm.

Theorem 3 (Exact EDF AFM schedulability test). *The task set Γ is earliest deadline first (EDF) AFM schedulable (w.r.t. a fault mode policy) iff the network of timed game automata modeling Γ satisfies the CTL Formula 3.*

$$AG\neg(\bigvee_{\tau_i \in \Gamma} stop_i) \bigwedge \neg(\bigvee_{\tau_i \in \Gamma} \bigvee_{\tau_j \in \Gamma} (act_i \wedge exeLO_j \wedge p_{ij}) \bigvee_{\tau_i \in \Gamma} \bigvee_{\tau_j \in \Gamma^{HI}} (act_i \wedge exeHI_j$$
$$\wedge p_{ij}) \bigvee_{\tau_i \in \Gamma} \bigvee_{\tau_j \in \Gamma} (preLO_i \wedge exeLO_j \wedge p_{ij}) \bigvee_{\tau_i \in \Gamma} \bigvee_{\tau_j \in \Gamma^{HI}} (preLO_i \wedge exeHI_j \wedge p_{ij}) \quad (3)$$
$$\bigvee_{\tau_i \in \Gamma^{HI}} \bigvee_{\tau_j \in \Gamma} (preHI_i \wedge exeLO_j \wedge p_{ij}) \bigvee_{\tau_i \in \Gamma^{HI}} \bigvee_{\tau_j \in \Gamma^{HI}} (preHI_i \wedge exeHI_j \wedge p_{ij}))$$

where p_{ij} is a state of an observer automaton reachable when $d_i - d_j > D_i - D_j$.

Proof. In earliest deadline first (EDF) schedulability, priorities are assigned to jobs dynamically according to their absolute deadlines. Formula 3 is satisfied iff there exists an execution where no job misses its deadline (state $stop_i$ not reached) and a job of a task τ_i cannot be active (state act_i) or preempted (state $preLO_i$ or $preHI_i$) if a job of a task τ_j with an absolute deadline less close to its deadline ($D_j - d_j > D_i - d_i$) is executed (state $exeLO_j$ or $exeHI_j$). Thus the execution of jobs is done according to EDF algorithm.

6   Illustrative Example

We illustrate our approach using a task set Γ_1 with $\tau_1 = (HI, 3, 10, 10, (1, 2))$, $\tau_2 = (HI, 2, 8, 8, (2, 4))$ and $\tau_3 = (LO, 1, 4, 4, (2, 2))$. The CPU utilization of Γ_1 in the case where no LO criticality job is stopped and all the jobs of HI criticality tasks are critical is equal to 1.2, i.e. the task set is not feasible if we don't scarifice some LO criticality jobs when the criticality of the system is HI.We present in Sects. 6.1 and 6.2 two benefits of using the AFM strategy.

6.1   Decrease the Number of Sacrificed LO Criticality Jobs

The first benefit of our approach is that the designer can define a fault mode policy with the aim of reducing the number of jobs that are sacrificed, i.e. that are no more activated when the criticality of the system is HI.

We compare our method with the classical adaptive mixed criticality strategy (AMC) [4]. In AMC, when the criticality of the system is HI, LO criticality tasks are no more activated, and all jobs of HI criticality tasks are supposed to have an execution time equal to their HI WCET. We consider, also for AMC, that the system returns to LO mode at the first instant where no active job is critical.

Using the sufficient schedulability test, AMC_{rtb}, of AMC presented in [4], we cannot conclude that Γ_1 is schedulable using AMC, however we use the AFM strategy with a policy, $policy_{AMC}$, defined by $policy_{AMC}^{\overrightarrow{HI}}(\tau_3) = 1$ for all the criticality configurations where τ_1 or τ_2 have a critical job. Using the exact schedulability test of Formula 2, we prove that the task set Γ_1 is fixed priority AFM schedulable according to $policy_{AMC}$.

We compare the AMC strategy to an AFM strategy using the fault mode policy, $policy_1$, where $policy_1^{\overrightarrow{HI}}(\tau_3) = 1$ for all the criticality configurations where τ_2 has a critical job and $policy_1^{\overrightarrow{HI}}(\tau_3) = 0$ otherwise. In other words, LO criticality jobs are sacrified only when a job of task τ_2 is critical. Using the exact schedulability test of Formula 2, we prove that the task set Γ_1 is fixed priority AFM schedulable according to $policy_1$. We conclude that there is no need to scarify jobs of task τ_3 when only jobs of tasks τ_1 are critical, this will reduce the number of LO criticality jobs to stop to ensure the schedulability of the system.

To illustrate this, we consider an execution scenario in a time window from 0 to 44. In this scenario, the two last jobs of τ_1 are critical and the three last jobs of τ_2 are critical. We remind that a job is critical if the job does notify its

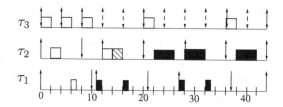

Fig. 3. FP Scheduling of Γ_1 using AMC

completion after the execution of its WCET at the LO criticality level. Critical jobs are presented in black in the figures.

Figure 3 represents the schedule using AMC. At time 12, the system moves to HI criticality mode until time 17. As a consequence, the forth job of task τ_3 is not activated. Note that even if the second job of task τ_2 is not critical, its WCET is equal to 4 as all jobs of HI criticality tasks are supposed to have an execution time equal to their HI WCET when the criticality of the system is HI in AMC. The criticality of the system is again HI in the time interval $[24, 26]$, $[28, 33]$ and $[40, 42]$. Using AMC, we can see that 6 jobs of τ_3 are not activated.

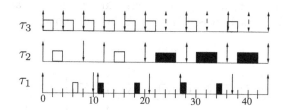

Fig. 4. FP Scheduling of Γ_1 using AFM and *policy*$_1$

In Fig. 4, we can see that for the same execution scenario, using AFM strategy with FP scheduling and *policy*$_1$, only 3 jobs of τ_3 are not activated. In fact, the others job of task τ_3 are activated as even if the criticality mode of the system is HI, only the task τ_1 has a critical job.

6.2 LO Criticality Jobs Are Necessary in Some HI Criticality Configurations

Even if some tasks are classified as LO criticality tasks by the designer, they may be necessary for the good functioning of the system. For example, in the case where task τ_3 is a non critical task performing some image analysis that are necessary when both jobs of τ_1 and τ_2 are critical to help the system to recover.

In this case, even if the task set is schedulable using *policy*$_1$, the system is unsafe because, as we can see in Fig. 4, at instant 32, jobs of both τ_1 and τ_2 are critical, and the job of τ_3 is not executed, while it is necessary to ensure the safety of the system.

To test if the task set is schedulable if LO criticality tasks are not stopped when jobs of both τ_1 and τ_2 are critical, we use the fault mode policy $policy_2$. This policy is defined with, $policy_2^{\overrightarrow{HI}}(\tau_3) = 1$ for all the criticality configurations where "only" τ_2 has a critical job and $policy_2^{\overrightarrow{HI}}(\tau_3) = 0$ otherwise.

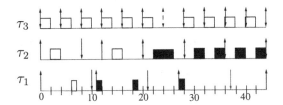

Fig. 5. FP Scheduling of Γ_1 using AFM and $policy_2$

Using the exact schedulability test of Formula 2, we prove that the task set is not FP AFM schedulable according to $policy_2$. As we can see in Figure 5, at time 32 the two tasks τ_1 and τ_2 have a critical job, in this case, according to the policy, the job of task τ_3 is activated. Using this policy, the third job of τ_1 misses its deadline.

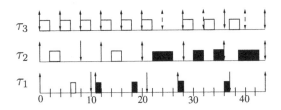

Fig. 6. EDF Scheduling of Γ_1 using AFM and $policy_2$

However, using the exact schedulability test of Formula 3, we prove that the task set is EDF AFM schedulable according to $policy_2$. As we can see in Fig. 6, at time 32 the two tasks τ_1 and τ_2 have a critical job, in this case, according to the policy, the job of task τ_3 is activated.

7 Conclusion

In this paper, we introduce a scheduling strategy for the mixed criticality real-time scheduling problem where the designer can define his own policy to deal with low criticality tasks when the criticality of the system is high. For this model, we propose exact feasibility and schedulability tests for job level fixed priority algorithms based on CTL model checking for timed game automata. Using an example, we illustrate the benefits of the proposed strategy. We are aware that our exact tests face the state explosion problem, since the upper bound complexity of model checking on time game automata is EXPTIME, however, as future work, we plan to propose a more specific game model taking into account the characteristics of our real-time scheduling problem, our intuition is that only a subset of configuration is needed to prove the feasibility of the problem. An other direction, is to be able to generate the fault mode policy.

References

1. Abdeddaïm, Y.: Link to upload UPPAAL-TIGA files. https://perso.esiee.fr/~abdedday/VECOS2020.tar.gz
2. Abdeddaïm, Y., Masson, D.: Real-time scheduling of energy harvesting embedded systems with timed automata. In: 2012 IEEE International Conference on Embedded and Real-Time Computing Systems and Applications, RTCSA, pp. 31–40 (2012)
3. Alur, R., Dill, D.L.: Automata for modeling real-time systems. In: 17th International Colloquium on Automata, Languages and Programming, ICALP, pp. 322–335 (1990)
4. Baruah, S.K., Burns, A., Davis, R.I.: Response-time analysis for mixed criticality systems. In: Proceedings of the 32nd IEEE Real-Time Systems Symposium, RTSS, pp. 34–43 (2011)
5. Baruah, S.K., Vestal, S.: Schedulability analysis of sporadic tasks with multiple criticality specifications. In: 20th Euromicro Conference on Real-Time Systems, ECRTS, pp. 147–155 (2008)
6. Bate, I., Burns, A., Davis, R.I.: A bailout protocol for mixed criticality systems. In: 27th Euromicro Conference on Real-Time Systems, ECRTS, pp. 259–268 (2015)
7. Bate, I., Burns, A., Davis, R.I.: An enhanced bailout protocol for mixed criticality embedded software. IEEE Trans. Software Eng. **43**(4), 298–320 (2017)
8. Behrmann, G., Cougnard, A., David, A., Fleury, E., Larsen, K.G., Lime, D.: UPPAAL-TIGA: Time for playing games! In: 19th International Conference on Computer Aided Verification, CAV, pp. 121–125 (2007)
9. Burn, A.: Mixed criticality - a personal view. vol. 5. Schloss Dagstuhl-Leibniz-Zentrum fuer Informatik (2015)
10. Burns, A., Davis, R.I.: Mixed criticality systems - a review. https://www-users.cs.york.ac.uk/~burns/review.pdf
11. Fleming, T., Burns, A.: Incorporating the notion of importance into mixed criticality systems. In: 3rd Workshop on Mixed Criticality Systems, WMC (2013)
12. Gettings, O., Quinton, S., Davis, R.I.: Mixed criticality systems with weakly-hard constraints. In: Proceedings of the 23rd International Conference on Real Time Networks and Systems, RTNS, pp. 237–246 (2015)
13. Huang, P., Kumar, P., Stoimenov, N., Thiele, L.: Interference constraint graph - A new specification for mixed-criticality systems. In: Proceedings of 2013 IEEE 18th Conference on Emerging Technologies & Factory Automation, ETFA. pp. 1–8 (2013)

14. Clarke, E.M., Emerson, E.A.: Design and synthesis of synchronization skeletons using branching time temporal logic. In: Kozen, D. (ed.) Logic of Programs 1981. LNCS, vol. 131, pp. 52–71. Springer, Heidelberg (1982). https://doi.org/10.1007/BFb0025774
15. Larsen, K.G., Pettersson, P., Yi, W.: UPPAAL in a nutshell. STTT 1(1–2), 134–152 (1997)
16. Maler, O., Pnueli, A., Sifakis, J.: On the synthesis of discrete controllers for timed systems (an extended abstract). In: STACS, 12th Annual Symposium on Theoretical Aspects of Computer Science, Munich, Germany, Proceedings, pp. 229–242 (1995)
17. Tamas-Selicean, D., Pop, P.: Design optimization of mixed-criticality real-time applications on cost-constrained partitioned architectures. In: Proceedings of the 32nd IEEE Real-Time Systems Symposium, RTSS, pp. 24–33 (2011)
18. Tindell, K., Alonso, A.: A very simple protocol for mode change. In: Technical report, Universidad Politecnica de Madrid (1996)
19. Vestal, S.: Preemptive scheduling of multi-criticality systems with varying degrees of execution time assurance. In: Proceedings of the 28th IEEE Real-Time Systems Symposium RTSS, pp. 239–243 (2007)

Formal Modeling and Verification, Testing

Model-Based Design of Resilient Systems Using Quantitative Risk Assessment

Braham Lotfi Mediouni[1(✉)], Iulia Dragomir[2], Ayoub Nouri[1], and Saddek Bensalem[1,3]

[1] Huawei Technologies France, Grenoble, France
{brahamlotfi.mediouni,ayoub.nouri}@huawei.com
[2] GMV Aerospace and Defence, Madrid, Spain
idragomir@gmv.com
[3] Univ. Grenoble Alpes, CNRS, Grenoble INP (Institute of Engineering Univ. Grenoble Alpes), VERIMAG, 38000 Grenoble, France
saddek.bensalem@univ-grenoble-alpes.fr

Abstract. Fault detection, isolation and recovery (FDIR) subsystems are an accepted technique to make safety-critical systems resilient against faults and failures. Yet, these subsystems should be devised only for those faults that violate the system's requirements, while providing a correct approach such that requirements are met again. As a consequence, the obtained system is minimal, although complete, and robust both with respect to safety and performance requirements. In this paper, we propose a systematic and automated approach based on formal methods that includes (1) the evaluation of the relevance of faults based on quantitative risk assessment, and (2) the validation of system robustness by statistical model checking. We apply this approach on an excerpt of a real-life autonomous robotics case study, and we report on the implementation and results obtained with the $\mathcal{S}$BIP framework.

Keywords: Model-based design · FDIR · Risk assessment · Statistical model-checking · Real-time · $\mathcal{S}$BIP framework · Robotics case study

1 Introduction

An accepted method for achieving real-time systems resilience is to rely on *fault detection, isolation and recovery (FDIR)* components. These aim to detect at runtime whether a fault has occurred and apply a recovery strategy that brings the system in a safe mode. Formal methods have been recently leveraged for correct-by-construction FDIR components [28] in the frame of untimed [10] and real-time [16] systems. In this context, synthesis algorithms are used for building the two parts of the FDIR component: the diagnoser for fault detection and the controller for recovery. The method proposed in [16] suffers from two

This work has been supported by the EU's H2020 research and innovation programme under grant agreement #730080 (ESROCOS) and #700665 (CITADEL).

B. Ben Hedia et al. (Eds.): VECoS 2020, LNCS 12519, pp. 149–164, 2020.
https://doi.org/10.1007/978-3-030-65955-4_11

main limitations: (i) a diagnoser is devised for each detectable fault, and (ii) the controller is manually modeled, being left for verification by model-checking techniques. First, synthesizing a diagnoser for each detectable fault has disadvantages since not all faults have an impact on the system requirements, and the system performance can be greatly degraded due to the large number of unnecessary components. Therefore, it is important to synthesize diagnosers only for those faults that are relevant with respect to the system requirements and objectives. Second, the controller validation problem is hard and often unfeasible since model-checking techniques suffer from scalability issues.

In this paper, we tackle the above limitations by proposing to (i) quantify the impact of potential faults on the system requirements –quantitative risk assessment–, and to (ii) use scalable techniques to validate the manually designed controllers. Quantifying the faults impact on the system allows to prioritize them such that only the most relevant ones are considered for diagnoser synthesis. Scalable techniques help to automate the controllers validation and to make it more systematic. To do so, we propose a formal model-based approach for designing resilient systems incrementally by iterative and sound model transformations.

In our approach, we consider *risk* to be any system-related change that may alter its nominal behavior or performance. A risk is induced either explicitly by introducing faults or implicitly by integrating new FDIR capabilities in the system. Quantitative risk assessment is used to study the impact of such changes on the system requirements and to improve the FDIR design accordingly. We use probabilities to measure risk and to evaluate whether it is acceptable. When deemed unacceptable, a risk should be mitigated. This can be through synthesizing a new FDIR component or by simply improving the existing ones, e.g., using a more appropriate recovery strategy.

In this work, we automate risk measurement by using *statistical model-checking (SMC)* [17,29] and we leverage its scalability to validate manually designed controller components. To the best of our knowledge, this is the first time SMC is used for safety risk assessment and FDIR subsystems design. The contributions of the paper are the following:

- We propose an iterative and incremental process for the design of resilient systems equipped with FDIR capabilities. We follow the well-known spiral development process in a model-based context and we extend it with a quantitative risk assessment and system validation procedures. These are partially automated using SMC as described in Sect. 4.
- We apply this design process on a real-life robotics case study. We devise three system designs at different levels of granularity on which we perform quantitative risk assessment. For each design, we propose FDIR behavior that we validate against the system's requirements. We use the SBIP formal framework [22,24] described in Sect. 3 for modeling and quantitative analysis. The obtained results are presented in Sects. 5 and 6.
- We discuss advantages and limitations of this process for the design of industrial applications in Sect. 7.

2   Related Work

Safety risk assessment is studied in the literature from two points of view: qualitative or quantitative. Qualitative safety assessment determines which scenarios lead the system from a nominal mode to a degraded mode where safety requirements do not hold. The practice consists of building safety artifacts such as fault trees or timed failures propagation graphs and analyzing them in order to certify the system safety. For example, automated safety analysis for fault trees is described in [8] for the AltaRica dataflow language. In [9], the xSAP tool is presented for the analysis of fault trees and timed failure propagation graphs in the context of symbolic transition systems from NUXMV [12].

Quantitative safety assessment allows to probabilistically measure risks. Probabilities computation is usually done manually on safety artifacts built a priori during the qualitative assessment, without considering the detailed system behavior. These artifacts are then decorated by probability distributions for individual faults. Some safety assessment tools automate this analysis, such as xSAP [9] for probabilistic fault trees, yet without taking into account timing aspects. Our work contributes to this class of risk assessment with a focus on timed systems and their detailed behaviors, and by investigating the use of SMC.

SMC was successfully applied in various domains such as biology [13], communication protocols [6], multimedia [27] and avionics [7]. It has several advantages compared to other formal verification techniques such as (probabilistic) model checking [5]. First, SMC is inherently *scalable* as only a subset of the system's executions are explored. Furthermore, its underlying statistical algorithms can be easily parallelized. Even though the obtained results are only estimations, their accuracy is controlled with confidence parameters that bound the estimation error, which distinguish it from pure simulation techniques. Also, rare events which might be missed by pure simulations can be caught with SMC. Specific techniques such as *importance sampling* and *importance splitting* [18,19] have been recently adapted to SMC in order to efficiently deal with this class of events. Another advantage of using SMC is the ability to handle both models and implementations, provided that the latter are obtained from formally defined models with a purely stochastic [22].

Safety risk assessment can be seen as an optimization in the design of FDIR components in general and diagnosers in particular. The correct design of FDIR components from complete system specifications has been studied from a methodological point of view in [10,16,28]. Implementations are provided in [16] for timed systems with partial observability and in [10] for untimed systems modeled with NUXMV. While [10] includes the safety assessment mechanism implemented in [9] for user-modeled discrete time failure propagation graphs, this question is left open in [16]. Our contribution completes the work from [16] by defining and automating a quantitative safety assessment method for (stochastic) timed systems allowing the efficient design of FDIR components. A more detailed comparison with existing risk assessment approaches and other formal methods used for robotics systems can be found in [21].

3   A Rigorous Framework for Modeling and Analyzing Stochastic Timed Systems

Stochastic models are of paramount importance in system design as they allow to capture uncertainties, a key concept for reasoning about risk. Besides, models of real-time behavior are mandatory when designing critical applications. To take full advantage of these models, formal and quantitative analysis techniques that handle real-life system models are primordial.

In this work, we consider SMC which takes as input an executable model of the system of interest and a formal specification of the requirement to verify, usually given in some logic. Using SMC, it is possible to (1) estimate the probability that the system satisfies the requirement, and (2) position the probability of satisfying the requirement with respect to a given threshold. Answering the first type of query relies on well known probability estimation techniques [17], while the second is handled using a hypothesis testing approach [29]. Concretely, SMC explores a sample of finite execution traces which are iteratively generated and monitored against the desired requirement. Monitored traces produce local verdicts {*true, false*} which are consumed sequentially by the statistical algorithms to compute a final probability estimation or a global Boolean verdict.

In this section, we briefly present the $\mathcal{S}$BIP framework [22,24] that represents the foundation on which relies our approach. It includes a stochastic real-time modeling formalism and a statistical model checking engine.

3.1   Stochastic Real-Time BIP

The stochastic real-time BIP[1] [24] is a component-based modeling formalism that allows one to design complex system models compositionally. This formalism is sufficiently expressive to model systems from various application domains, including different behaviors such as real-time, uncertainties and faults. It further enables the integration of external code in order to model complex computations.

In the stochastic real-time BIP formalism, components are designed as extended timed automata [4] and are composed through multi-party interactions, i.e., n-ary synchronizations among components actions. Formally, a component is a finite automaton enriched with data variables and real-valued clocks that allow us to measure time. System events can be associated with guards, that is, constraints describing when they are enabled. Guards can be expressed on variables and clocks and their corresponding events can further be annotated with an urgency level [11], namely *eager (ϵ)*, *delayable (d)* or *lazy (λ)*. Eager indicates that time elapse is disabled once the transition is enabled. Delayable states that the transition can be delayed at most to the upper bound of the time interval. Finally, lazy specifies that the event can be fired at any moment while enabled or never.

Uncertainty in the stochastic real-time BIP formalism concerns mainly events scheduling. It is expressed by associating events guards with probability density functions. Hence, the precise moment of executing an event is scheduled

---

[1] BIP stands for Behavior - Interaction - Priority.

according to that density. We consider two types of events, namely *timed* and *stochastic*. The former are associated with timing constraints on transitions. These events are implicitly scheduled according to a uniform or an exponential probability distribution as it is the case in several existing modeling formalisms, e.g., UPPAAL [14]. *Stochastic events* can be associated with arbitrary density functions, e.g., Normal or Poisson, and scheduled accordingly. The underlying semantics of a stochastic real-time BIP model is a Generalized Semi-Markov Process (GSMP) [20] where the interpretation of time is dense[2].

3.2 The SMC-BIP Engine

SMC-BIP [22] considers as input stochastic real-time BIP models and requirements expressed in Linear-time Temporal Logic (LTL) [26] and Metric Temporal Logic (MTL) [3]. It implements both types of SMC queries in addition to advanced features such as automatic parameter exploration, useful for system dimensioning, and rare events analysis, important for risk assessment. The tool offers an integrated development environment including a graphical user-interface with model edition, compilation and simulation features, and automates the different statistical analyses. In this work, we mainly use the probability estimation (PE) and parameter exploration (PX) capabilities offered by the tool. PX is an automated technique to identify the optimal parameter value of a parametric property such that the latter is satisfied with probability 1.

4 A Model-Based Approach Integrating Quantitative Risk Assessment

The proposed approach, illustrated in Fig. 1, is based on the idea of iterative and incremental transformation of models (Γ). Each model transformation can introduce new risks, for example due to relaxing environment assumptions. The idea depicted in this approach is to perform at each step of the development two assessments. First, *quantitative risk assessment* allows one to measure the impact that different risks have with regard to the system requirements, and perform a model upgrade if deemed necessary. Second, *validation* ensures that the upgrade is consistent with respect to the system requirements. The proposed approach is general and can be applied to different types of systems, e.g., untimed, real-time. Moreover, the notion of *risk* can have different interpretations, e.g., safety, security. Our setting consists of stochastic real-time systems designed and analyzed with the $\mathcal{S}$BIP framework, where risks are understood as the impact of faults on the system with respect to the desired properties.

Initially, system specifications and informal requirements are analyzed to build a nominal application model and a set of formal requirements. At this step, only validation is performed to ensure that the obtained model satisfies the formalized requirements. While this condition is not satisfied, the model is

---

[2] We refer the readers to [24] for the formal definition of the stochastic real-time BIP.

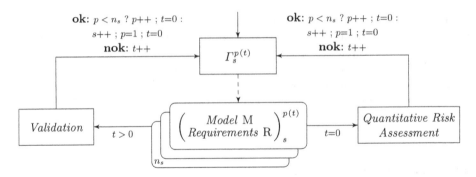

Fig. 1. Design approach based on formal methods integrating quantitative risk assessment where: Γ denotes, possibly manual, model transformation, s the number of performed steps, p the number of explored models within a step bounded by n_s, and t the number of iterative transformations for given s and p. Initially s, p, t are 0, 1, 1.

iteratively corrected (denoted by label **nok** and index t in Fig. 1). When the model is judged valid, one can proceed with the next transformation step.

The model is incrementally transformed towards the concrete implementation (represented by the s index). Transformation concerns different aspects of the system and may introduce new risks. Examples include integrating new behavior, correction of modeling or implementation errors, legacy code, instantiation of the model's parameters. For the latter, one obtains a family of models (indexed with $p \in \{1, \ldots, n_s\}$ in Fig. 1). Similarly, the system requirements are modified based on the purpose of the realized model transformations. By system requirements we denote those expected to be fulfilled by the model in the current stage of the design, that is, during the step s the exploration p and the iteration t.

The first analysis to perform on the transformed system model is the *quantitative risk assessment*. It implies computing the probability for the requirements to hold on the model. Based on this measurement, risks can be appreciated. If they are acceptable (represented with label **ok**) one can continue with inspecting a new model either from the same family if $p < n_s$ or by moving to the next step s. If the risks are high (represented with label **nok**) a *decision* on how to mitigate is taken, which usually involves the transformation of the model architecture and/or behavior. Once all the risks have been dealt with, the obtained model and its requirements are subject to the iterative validation described above.

Note that in this approach, a transformation can be applied either manually or automatically. Often, when iterating through steps, the transformations are automated, as they usually correspond to the model refinement towards implementation. However, when dealing with unacceptable risks, the transformation is manual as it depends on the corrections deemed necessary by the system designer.

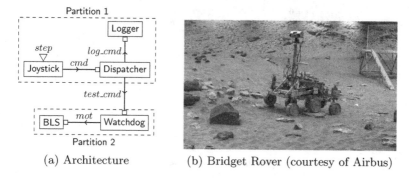

(a) Architecture   (b) Bridget Rover (courtesy of Airbus)

Fig. 2. Overview of the case study system architecture and rover running it.

In the following sections, we apply this approach on a robotics control system[3]. We distinguish several levels of granularity, where faults and FDIR behavior are integrated into the nominal behavior, and evaluated with respect to safety and performance. All models are described in the stochastic real-time BIP formalism where time evolves probabilistically following uniform density functions. These models are derived through manual transformations, where the added behavior is explicitly modeled by the designer. The quantitative risk assessment and validation activities are automated using the SMC-BIP engine.

5   Planetary Robotics Case Study

The case study considered in this paper is an excerpt of a Bridget Rover demonstrator control system [1] developed for the validation of the ESROCOS environment [2]. This control system aims to remotely drive the rover using a joystick in order to acquire images. We focus, in the following, on the drive with a joystick functionality, and more precisely on the software chain between the joystick and the locomotion software (BLS) [15], illustrated in Fig. 2.

Model. The Joystick component, depicted in Fig. 3a, regularly sends a motion command denoted cmd^4 to the rover locomotion software to be executed. The command sending is activated by the *step* trigger. The *cmd* request has several fields that characterize the motion to be executed: the id of type integer recording the package number, and the actual locomotion data consisting of translation, rotation and heading of type float. The trigger, depicted by ∇ in Fig. 2, activates the component with a period P. Additionally, it imposes that the activation does

---

[3] The contribution presented in this paper has been used for the development of the FDIR components in the robotics systems scenarios presented in [23,25].

[4] The suffixes *out*, *in* and *return* used in Fig. 3 are modeling the directionality of the requests. *Out* models that the component sends the request. *In* models that the component receives the request. *Return* models that the action associated with the request has finished executing.

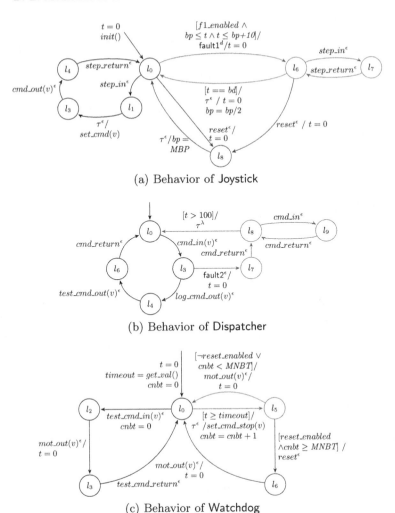

(a) Behavior of Joystick

(b) Behavior of Dispatcher

(c) Behavior of Watchdog

Fig. 3. Behavior of the main components from Fig. 2 represented as timed automata in $\mathcal{S}$BIP, where faults, fault detection and standard recovery action are represented in red, and more complex recovery strategy in blue (Color figure online)

not exceed a deadline D. In our model, the violation of this condition induces a timelock, i.e., the considered design cannot be scheduled and needs to be corrected. For the *step* trigger, P and D are set to 100 ms and 15 ms respectively.

The *cmd* action is first sent to a Dispatcher, provided in Fig. 3b. The Dispatcher transfers this request to two software components: first to the Logger via the *log_cmd* request and then to the Watchdog via the *test_cmd* request. Both *log_cmd* and *test_cmd* contain the data package received from the Joystick. The Logger records the received requests such that they can be reused later for the validation of the system through replaying.

The Watchdog, illustrated in Fig. 3c, interfaces the Dispatcher (and subsequently the Joystick) with a wrapper of the locomotion control software called BLS. Whenever a *test_cmd* request is received, the Watchdog transfers the data package in the *mot* request to the BLS. For simplicity, we do not consider the detailed behavior of the BLS: it is abstracted to consume all received requests.

All these components communicate asynchronously via queues depicted by □ in Fig. 2. The behavior of the queue is to store incoming requests up to a certain *size* and then to transfer them to the receiver component with a periodicity called *minimal inter-arrival time (MIAT)*. Additionally, the queue imposes that the handling of a request does not exceed a deadline D. Similarly to *step*, violating this condition, induces a timelock. All queues in the model are instantiated with $MIAT = 50$ ms, $D = 15$ ms and $size = 1^5$.

The parametric BIP model consists of 10 components, mostly timed. Each component has at least 3 variables in addition to clocks. External data types were implemented and used to model the complex data structure of the robot. The BIP description is about 536 lines of code plus 298 of external C++[6].

Requirements. The main prerequisite for the system is that the rover is moving according to the requested motions. This feature is formalized by many requirements based on the granularity of the system design and assumptions made over the environment, as listed in Table 1. For the nominal model, we define requirements ϕ_0 and ϕ_1. ϕ_0 states that all *cmd* requests sent by the Joystick, modeled with the Boolean variable *is_sent*, are received within 100 ms by the locomotion system (BLS component), modeled with the Boolean variable *is_received_c*. ϕ_1 describes that the BLS regularly receives a request *mot*, within 100 ms.

Validation. We use SMC-BIP tool with the confidence parameters $\alpha = 0.005$ and $\delta = 0.05$ for all our experiments (at this step and after). These confidence parameters require the evaluation of almost 1200 system executions to come up with a global verdict, using PE. The computed satisfaction probability is 1 for both ϕ_0 and ϕ_1. The computations took for each requirement approximately 4 min. This analysis time is due to: (1) the large number of simulations required by SMC to ensure high precision δ and confidence level α, (2) the collection of long system runs (representing 100 cycles), and (3), in the case of PX, the size of the instantiation domain of property parameters. As the requirements on the nominal model are satisfied, we can proceed with the next step of the approach.

---

[5] Notice that the values for P, D, $MIAT$, and *size* are part of the system specification.
[6] Model sources are available at https://drive.google.com/file/d/1oN90ZraClQx AH5hHE2tl7t2IMsZVzo7L/view?usp=drivesdk.

Table 1. Requirements of the planetary robotics case study. ϕ_{1-6} are subject to PE, while ϕ_{7-13} are analyzed with PX.

ID	Formal specification
Requirements on the nominal system	
ϕ_0	$\Box_{[0,10000]}\ (is\_sent\ \Rightarrow \Diamond_{[0,100]}\ is\_received\_c)$
ϕ_1	$\Box_{[0,10000]}\ (is\_received\_c\ \Rightarrow \Diamond_{[1,100]}\ is\_received\_c)$
Requirements on the FDIR behavior	
ϕ_2	$\Box_{[0,10000]}\ (is\_sent\ \Rightarrow \Diamond_{[0,110]}\ is\_received)$
ϕ_3	$\Box_{[0,10000]}\ (is\_received\ \Rightarrow (\Diamond_{[1,110]}\ is\_received) \lor (\Diamond_{[110,200]}\ is\_timeout))$
ϕ_4	$\Box_{[0,10000]}\ (\Diamond_{[0,200]}\ nb\_received = nb\_sent\ +\ nb\_timeout)$
ϕ_5	$\Box_{[0,10000]}\ (cnbt \geq MNBT\ \Rightarrow \Diamond_{[0,200]}\ is\_reset)$
ϕ_6	$\Box_{[0,10000]}\ (is\_reset\ \Rightarrow \Diamond_{[0,100]}\ is\_received)$
Requirements on the system performance	
$\phi_7(n)$	$\Box_{[0,10000]}\ (\Diamond_{[0,200]}\ nb\_timeout - (nb\_received - nb\_sent) \leq n)$
$\phi_8(n)$	$\Box_{[0,10000]}\ (nb\_timeout \leq n)$
$\phi_9(n)$	$\Box_{[0,10000]}\ (cnbt \leq n)$

6   Three Steps of the Approach on the Case Study

In the following, we describe an enhanced model of the planetary rover at the application level with faults and recovery. Due to the lack of space, we present here the results obtained only on a subset of the faults we studied. We refer the reader to [21] for further details.

6.1   On Robustness to Faults

Model with Faults. Starting from the validated nominal behavior, the first transformation we consider introduces faults in the model. Hence, it relaxes the assumptions made previously: the Joystick issues periodically a *cmd* request and no requests are lost. Two faults are considered independently. The fault1 affects the Joystick (represented in red in Fig. 3a). It stops the request sending for a given break duration *bd*. The fault activation is controlled by a break period *bp*, which decreases each time the fault occurs. Note that this fault is persistent since *bd* converges to 0 ms, and the Joystick will eventually be continuously failing. The fault2 provokes requests losses in the Dispatcher (represented in red in Fig. 3b). Once the Dispatcher fails, incoming requests are not dispatched for at least 100 ms. Even if the component recovers, the fault can happen again.

For the experiments, we set the fault1 *bd* to 20 ms and *bp* to 190 ms. Moreover, in order to have a systematic way of evaluating the impact of faults, we define Boolean variables (e.g., *f1_enabled*) to control their activation.

Table 2. Results obtained with the $\mathcal{S}$BIP framework on the different system designs with respect to requirements from Table 1. n_ϕ^* refers to the parameter value for which $\phi(n)$ is satisfied with probability 1.

	fault1 (without recovery)		fault1 (with recovery)		fault2	
	Probability/ Parameter	Time (sec)	Probability/ Parameter	Time (sec)	Probability/ Parameter	Time (sec)
Nominal system						
ϕ_0	1	240	1	196	0	14
ϕ_1	0	14	0	13	0	14
System with FDIR behavior						
ϕ_2	1	240	1	196	0	14
ϕ_3	1	202	1	242	1	180
ϕ_4	1	192	1	246	0	14
ϕ_5	-	-	1	194	-	-
ϕ_6	-	-	1	217	-	-
$\phi_7(n)$	$n_{\phi_7}^* = 0$	140	$n_{\phi_7}^* = 0$	140	$n_{\phi_7}^* = 88$	900
$\phi_8(n)$	$n_{\phi_8}^* = 88$	720	$n_{\phi_8}^* = 61$	197	$n_{\phi_8}^* = 88$	900
$\phi_9(n)$	$n_{\phi_9}^* = 88$	720	$n_{\phi_9}^* = 5$	180	$n_{\phi_9}^* = 30$	1080

Risk Assessment. For risk assessment, we use requirements ϕ_0 and ϕ_1 defined in Sect. 5. These are sufficient to assess the impact of the faults on the nominal behavior. The complete results obtained with the $\mathcal{S}$BIP framework are given in Table 2. ϕ_0 is satisfied on the model with fault1, but not on the one with fault2. In case of fault1, if a command is not sent it is not received, hence the implication in ϕ_0 evaluates to *true*. However, for fault2, commands are actually sent then lost (not received by the BLS). Property ϕ_1 is not satisfied regardless of the occurring fault. Given these results, we conclude that the risk associated with these faults is high and that an FDIR behavior needs to be added so that the rover operates safely. Otherwise, the locomotion system could continue executing the last command and lead the rover in a harmful situation.

Model with FDIR Behavior. In order to operate safely when risks are present, we equip the Watchdog with a data package validity checking: *motion commands must be received before a* timeout *event*. This corresponds to the diagnoser part of FDIR. When this condition is not satisfied (due to faults in the system), the Watchdog will ensure that *the rover stops* by issuing a special command. This corresponds to the controller part and is achieved by setting the data to stop – translation, rotation and heading are set to 0 – before being sent to the BLS.

For simplicity[7], we model this FDIR behavior directly in the Watchdog as illustrated in Fig. 3c. For our experiments, we configure the value of the timeout to 110 ms. Consequently, requirements formalization listed in Table 1 for the FDIR component use this new time interval.

Validation. We define a new set of requirements ϕ_{2-4} specific to the FDIR behavior of the Watchdog. ϕ_2 is based on ϕ_0 and states that whenever a motion command is sent by the Joystick (modeled with variable *is_sent*), it is received by the Watchdog (modeled with variable *is_received*). ϕ_3 is a refinement of ϕ_1 to account for the FDIR behavior. It checks that the BLS receives *mot* requests periodically, where the data package either comes from the Joystick (modeled with the *is_received* variable) or is the stop data package sent by the Watchdog (modeled with the *is_timeout* variable). Requirement ϕ_4 models the consistency of the data package reception: all the packages received by the BLS (modeled with variable *nb_received*) are either generated by the Joystick (modeled with variable *nb_sent*) or by the Watchdog (modeled with variable *nb_timeout*).

The Watchdog is robust with respect to both faults: ϕ_3 is satisfied with probability 1, as shown in Table 2. In addition, ϕ_2 and ϕ_4 are also satisfied when considering fault1. However, in the presence of fault2, the probability of satisfaction for ϕ_2 and ϕ_4 is 0. This result is expected since fault2 models request losses. We consider the obtained results satisfying and move to step 2.

6.2   On System Performance

Model for Performance Measurement. We are now interested in the impact of the FDIR behavior on the system performance. Therefore, we do not perform any model transformation. Instead, we enrich the set of requirements with new ones concerning the system performance ϕ_{7-9} described below.

Risk Assessment. ϕ_7 explores the different bounds for inconsistency in the number of packages. ϕ_8 explores the maximal number of stop commands the Watchdog issues in a given time period, while ϕ_9 considers the number of consecutive stop commands (modeled with variable *cnbt*). We are interested to have a low number of stop commands such that the rover operates smoothly.

We remark from the results in Table 2 that both faults lead the Watchdog to issue a large number of stop commands (ϕ_8), i.e., 88. In addition, while the number of consecutive stop commands (ϕ_9) is bounded to 30 in case of fault2, the persistent nature of fault1 is critical, obtaining the higher bound 88. Therefore, we introduce a reset mechanism for the Joystick to optimize the FDIR behavior.

Model with reset Mechanism for the Joystick. We introduce a reset mechanism in the Joystick that will allow this component to go back to its nominal behavior, illustrated in blue in Fig. 3a. This action additionally resets the break period

---

[7] The system architecture and specification, Watchdog included, have been provided in the frame of this case study such that the used resources (e.g., number of components and threads) are minimal.

bp to the maximal allowed duration *MBP* for future occurrences of fault1. The reset mechanism is activated by the Watchdog whenever enabled (modeled with *reset_enabled* variable) and the consecutive number of stop commands *cnbt* is above the *MNBT* threshold. We configure the Watchdog to tolerate a maximum number of 5 consecutive timeouts before triggering a Joystick reset ($MNBT = 5$).

Validation. The reset mechanism is additionally validated by requirements ϕ_5 and ϕ_6. ϕ_5 states that whenever fault1 is detected, the Watchdog triggers the *reset* action (modeled with variable *is_reset*). ϕ_6 validates the efficiency of the reset mechanism modeled by the reception of a command by the Watchdog after a reset is triggered. We also check and compare the performance of the reset mechanism with requirements ϕ_{7-9}.

The results are given in the second column of Table 2. We remark that the reset mechanism is robust: both ϕ_5 and ϕ_6 being satisfied. Additionally, this mechanism reduces the overhead due to FDIR on the system performance. More specifically, the total number of issued stop commands is reduced by 31%, whereas the number of consecutive stop commands is bounded to 5 (the *MNBT* bound of the reset mechanism).

7   Discussion

In this paper, we proposed a model-based design approach that relies on formal methods to develop real-time resilient systems. The method is incremental: it starts from the nominal model, then transformations are applied to take into account different sources of risks. The impact of the considered risks is evaluated using a quantitative risk assessment method and FDIR components are introduced accordingly. These are then validated against safety and performance properties. The approach was successfully used for the design and validation of the control software of a planetary rover.

Approach. Following a model-based approach for the design and validation of FDIR components provides a lot of flexibility and allows rapid exploration of various situations. Combined with formal methods, it provides more confidence in the obtained results given that the built models are faithful, which is not trivial and requires some expertise. Finally, the use of statistical model checking automates quantitative risk analysis, and helps to deal with real-life system models. However, both the identification and the evaluation of risks remain manual and subject to the designer's interpretation.

Case Study. The presented results are part of the work realized for the validation of the ESROCOS environment [23] with a real-life robotics case study. Although the approach was successfully applied, and the FDIR component described here was deployed on the rover and tested during field trials, we wish to share some of the challenges we faced. Building faithful models is by far the most challenging. The choice of the appropriate abstractions to perform and the probability distributions to use require a deep knowledge of the system under analysis. Besides, using risk assessment helped to take well founded decisions in order to build robust FDIR components. However, the notion of risk is wide and several times we found ourselves analyzing risks at different levels, such as risks due to faults, risks due to adding new FDIR behavior, etc. Moreover, managing the transformed models and the associated requirements can quickly become cumbersome if not methodically performed.

Tools. Risk analysis automation is primordial for the design of complex systems as the design space is substantial and proceeding manually is not feasible. In our case, once we built a model it becomes almost straightforward to analyze it using the SMC-BIP engine. Nevertheless, some difficulties remain to use the tool properly, like the correct formalization of requirements in MTL or the instrumentation of the model in order to perform SMC.

Future Work. In this paper, we only considered quantitative risk assessment. Using qualitative assessment before may help a lot in filtering irrelevant risks with respect to the requirements of interest. Moreover, risk identification could be done in a knowledge-based manner by using machine-learning techniques for instance. Finally, we are also interested to evaluate the applicability of the approach to security risk assessment. Indeed, we believe that our approach is general enough to also handle other types of requirements, such as security.

References

1. ESROCOS Planetary Exploration Demonstrator. https://github.com/ESROCOS/plex-demonstrator-record
2. ESROCOS Project Github Repository. https://github.com/ESROCOS
3. Alur, R., Henzinger, T.: Real-time logics: complexity and expressiveness. Inf. Comput. **104**(1), 35–77 (1993)
4. Alur, R., Dill, D.L.: A theory of timed automata. Theor. Comput. Sci. **126**(2), 183–235 (1994)
5. Baier, C., Katoen, J.P.: Principles of Model Checking (Representation and Mind Series). The MIT Press, Cambridge (2008)
6. Basu, A., Bensalem, S., Bozga, M., Caillaud, B., Delahaye, B., Legay, A.: Statistical abstraction and model-checking of large heterogeneous systems. In: Hatcliff, J., Zucca, E. (eds.) FMOODS/FORTE -2010. LNCS, vol. 6117, pp. 32–46. Springer, Heidelberg (2010). https://doi.org/10.1007/978-3-642-13464-7_4
7. Barringer, H., et al. (eds.): RV 2010. LNCS, vol. 6418. Springer, Heidelberg (2010). https://doi.org/10.1007/978-3-642-16612-9
8. Batteux, M., Prosvirnova, T., Rauzy, A., Kloul, L.: The AltaRica 3.0 project for model-based safety assessment. In: 11th IEEE International Conference on Industrial Informatics, INDIN 2013, Bochum, Germany, 29–31 July 2013, pp. 741–746. IEEE (2013). https://doi.org/10.1109/INDIN.2013.6622976
9. Bittner, B., et al.: The xSAP safety analysis platform. In: Chechik, M., Raskin, J.-F. (eds.) TACAS 2016. LNCS, vol. 9636, pp. 533–539. Springer, Heidelberg (2016). https://doi.org/10.1007/978-3-662-49674-9_31
10. Bittner, B., et al.: An integrated process for FDIR design in aerospace. IMBSA **2014**, 82–95 (2014)

11. Bornot, S., Sifakis, J., Tripakis, S.: Modeling urgency in timed systems. In: de Roever, W.-P., Langmaack, H., Pnueli, A. (eds.) COMPOS 1997. LNCS, vol. 1536, pp. 103–129. Springer, Heidelberg (1998). https://doi.org/10.1007/3-540-49213-5_5

12. Cavada, R., et al.: The NUXMV symbolic model checker. In: Biere, A., Bloem, R. (eds.) CAV 2014. LNCS, vol. 8559, pp. 334–342. Springer, Cham (2014). https://doi.org/10.1007/978-3-319-08867-9_22

13. David, A., Larsen, K., Legay, A., Mikucionis, M., Poulsen, D.B., Sedwards, S.: Statistical model checking for biological systems. Int. J. Softw. Tools Technol. Transf. (STTT) 17(3), 351–367 (2015)

14. David, A., Larsen, K.G., Legay, A., Mikucionis, M., Poulsen, D.B.: Uppaal SMC tutorial. STTT 17(4), 397–415 (2015)

15. Dragomir, I.: ESROCOS planetary exploration demonstrator: the watchdog component in TASTE and BIP. https://github.com/ESROCOS/control-mc_watchdog

16. Dragomir, I., Iosti, S., Bozga, M., Bensalem, S.: Designing systems with detection and reconfiguration capabilities: a formal approach. In: Margaria, T., Steffen, B. (eds.) ISoLA 2018. LNCS, vol. 11246, pp. 155–171. Springer, Cham (2018). https://doi.org/10.1007/978-3-030-03424-5_11

17. Hérault, T., Lassaigne, R., Magniette, F., Peyronnet, S.: Approximate probabilistic model checking. In: Steffen, B., Levi, G. (eds.) VMCAI 2004. LNCS, vol. 2937, pp. 73–84. Springer, Heidelberg (2004). https://doi.org/10.1007/978-3-540-24622-0_8

18. Jegourel, C., Legay, A., Sedwards, S.: Importance splitting for statistical model checking rare properties. In: Sharygina, N., Veith, H. (eds.) CAV 2013. LNCS, vol. 8044, pp. 576–591. Springer, Heidelberg (2013). https://doi.org/10.1007/978-3-642-39799-8_38

19. Kahn, H., Marshall, A.W.: Methods of reducing sample size in Monte Carlo computations. J. Oper. Res. Soc. Am. 1(5), 263–278 (1953). http://www.jstor.org/stable/166789

20. Kulkarni, V.G.: Brownian motion. Introduction to Modeling and Analysis of Stochastic Systems. STS, pp. 247–280. Springer, New York (2011). https://doi.org/10.1007/978-1-4419-1772-0_7

21. Mediouni, B.L., Dragomir, I., Nouri, A., Bensalem, S.: Quantitative risk assessment in the design of resilient systems. Technical report TR-2018-10, VERIMAG (2018). http://www-verimag.imag.fr/TR/TR-2018-10.pdf

22. Mediouni, B.L., Nouri, A., Bozga, M., Dellabani, M., Legay, A., Bensalem, S.: SBIP 2.0: statistical model checking stochastic real-time systems. In: Lahiri, S.K., Wang, C. (eds.) ATVA 2018. LNCS, vol. 11138, pp. 536–542. Springer, Cham (2018). https://doi.org/10.1007/978-3-030-01090-4_33

23. Munoz, M., et al.: ESROCOS: a robotic operating system for space and terrestrial applications. In: Symposium on Advanced Space Technologies in Robotics and Automation (ASTRA) 2017, Leiden, Netherlands, 20–22 June 2017 (2017)

24. Nouri, A., Mediouni, B.L., Bozga, M., Combaz, J., Bensalem, S., Legay, A.: Performance evaluation of stochastic real-time systems with the SBIP framework. Int. J. Crit. Comput.-Based Syst. 8(3–4), 340–370 (2018)

25. Ocon, J.,et al.: The ERGO framework and its use in planetary/orbital scenarios. In: International Astronautical Congress (IAC) 2018, Bremen, Germany, 1–5 October 2018 (2018)

26. Pnueli, A.: The temporal logic of programs. In: 18th Annual Symposium on Foundations of Computer Science, Providence, Rhode Island, USA, 31 October - 1 November 1977, pp. 46–57 (1977). https://doi.org/10.1109/SFCS.1977.32

27. Raman, B., et al.: Stochastic modeling and performance analysis of multimedia SoCs. In: International Conference on Systems, Architectures, Modeling and Simulation, SAMOS 2013, pp. 145–154 (2013)
28. Wander, A., Forstner, R.: Innovative Fault Detection, Isolation and Recovery Strategies On-Board Spacecraft: State of the Art and Research Challenges. Deutscher Luft- und Raumfahrtkongress (2012)
29. Younes, H.L.S.: Verification and Planning for Stochastic Processes with Asynchronous Events. Ph.D. thesis, Carnegie Mellon (2005)

An Evaluation of Estimation Techniques for Probabilistic Verification

Mariia Vasileva and Paolo Zuliani$^{(\boxtimes)}$

School of Computing, Newcastle University, Newcastle upon Tyne, UK
{m.vasileva2,paolo.zuliani}@newcastle.ac.uk

Abstract. Formal techniques for verifying stochastic systems (e.g..,
probabilistic model checking) do not generally scale well with respect
to the system size. Therefore, simulation-based techniques such as statistical model checking are often used in practice. In this paper, we focus
on stochastic hybrid systems and evaluate Monte Carlo and Quasi-Monte
Carlo (QMC) methods for computing probabilistic reachability. We compare a number of interval estimation techniques based on the Central
Limit Theorem (CLT), and we also introduce a new approach based on
the CLT for computing confidence intervals for probabilities near the
borders of the [0,1] interval. We empirically show that QMC techniques
and our CLT approach are accurate and efficient in practice. Our results
readily apply to any stochastic system and property that can be checked
by simulation, and are hence relevant for statistical model checking.

1 Introduction

Verification techniques for stochastic systems such as probabilistic model checking can be very precise and can deal with a variety of stochastic systems (*e.g.*,
discrete-time Markov chains [21], continuous-time Markov chains [6] and Markov
decision processes [7]). However, as for standard, non-probabilistic model checking, these techniques suffer from the state explosion problem, which limits their
applicability in many practical cases. Statistical model checking [22] is often used
in practice on stochastic systems that exceed the limits of probabilistic model
checking, or for which no formal technique is available (*e.g.*, *nonlinear* stochastic
hybrid systems). In this paper, we focus on the probabilistic reachability problem for hybrid systems that depend on random parameters, which amounts to
computing the probability that the system reaches a *goal* state.

Checking reachability in hybrid discrete/continuous systems is an undecidable problem for all but the simplest systems (timed automata) [2]. (See [12] for
an up to date survey.) Formal verification of hybrid systems can include checking
the satisfiability of formulas involving real variables, which is known to be an
undecidable problem when, *e.g..*, trigonometric functions are involved [20]. The
notion of δ-complete decision procedure was introduced to combat the undecid-
ability of general sentences over the reals [13]. This approach has been extended
to a bounded probabilistic reachability method with statistically valid enclosures [18]. Essentially, this technique amounts to computing (multi-dimensional)

© Springer Nature Switzerland AG 2020
B. Ben Hedia et al. (Eds.): VECoS 2020, LNCS 12519, pp. 165–179, 2020.
https://doi.org/10.1007/978-3-030-65955-4_12

integrals of indicator functions, which can be done in three possible ways: rigorous, Monte-Carlo (MC) and Quasi-Monte Carlo (QMC). The computational complexity of rigorous (*i.e.*, numerically precise) computation of integrals grows exponentially with respect to the number of dimensions [20]. This motivates the use of QMC methods, which are asymptotically more efficient than MC methods. While MC methods are based on the Law of Large Numbers and random sampling, QMC methods are based on *deterministic* sampling from so-called quasi-random sequences. A drawback of QMC methods is that their integration error is difficult to estimate in practice, so one instead estimates the error of *Randomised* Quasi-Monte Carlo methods via confidence interval techniques based on the Central Limit Theorem (CLT). However, a problem of many such techniques is that the actual coverage probability of the interval near the boundaries (0 and 1) can be poor [9,16].

To summarise, in this paper we make the following contributions:

- we compare several confidence interval techniques for estimating probabilities in MC and QMC methods, and we show that QMC methods are more efficient in general;
- we propose a simple but effective modification of the CLT interval for estimating probabilities close to 0 or 1, and we empirically show that it performs well in practice.

While we focus on hybrid systems and reachability, our results readily apply to any stochastic system and property whose truth can be checked by simulation.

Probabilistic Reachability. Hybrid systems provide a framework for modelling real-world systems that combine continuous and discrete dynamics [2]. In particular, parametric hybrid systems (PHS) [18] represent continuous and discrete dynamic behaviour dependent on parameters that remain unchanged during the system evolution. Such systems can flow, described by differential equations, and jump, described by difference equations or control graphs.

In this paper, we consider *stochastic* PHS, which introduce random parameters to an otherwise deterministic PHS. Bounded k-step reachability in stochastic PHS aims at finding the **probability** that for the given initial conditions the system reaches a goal state in k discrete transitions within a given finite time. It can be shown that this probability can be computed as an integral of the form $\int_G d\mathbb{P}$, where G denotes the set of all random parameter values for which the system reaches a goal state in k steps, and $\mathbb{P}$ is the probability measure of the random parameters [18].

2 Integral Estimation Methods

Monte Carlo Method. Consider an integrable function f, the integral $I = \int_a^b f(y)dy < \infty$, and a random variable U on $[a, b]$ with density φ. The expectation of $f(U)$ is $\mathbb{E}[f(U)] = \int_a^b f(y)\varphi(y)dy$. If U is uniformly distributed on $[a, b]$, then the integral becomes: $I = \int_a^b f(y)dy = (b-a)\mathbb{E}[f(U)]$. Now, if we take N samples $\{u_1, \ldots, u_N\}$ from U and compute the sample mean $\frac{1}{N}\sum_{i=1}^N f(u_i)$, we obtain the MC estimate:

$$\int_a^b f(y)dy \approx (b-a)\frac{1}{N}\sum_{i=1}^N f(u_i). \tag{1}$$

The Strong Law of Large Numbers states that this approximation is convergent to I with probability 1 (for $N \to \infty$). The variance of the MC estimator (1) is:

$$Var(MC) = \int_a^b \cdots \int_a^b \left(\frac{1}{N}\sum_{i=1}^N f(u_i) - I\right)^2 du_1...du_N = \frac{\sigma_f^2}{N} \tag{2}$$

where σ_f^2 is the integrand variance, which is assumed to exist. In practice, the integrand variance is often unknown, and that is why the next estimation for the CI is instead used: $\widehat{\sigma}_f^2 = \frac{1}{N-1}\sum_{i=1}^N (f(u_i) - \frac{1}{N}\sum_{j=1}^N f(u_j))^2$, which enjoys the unbiasedness property $\mathbb{E}[\widehat{\sigma}_f^2] = \sigma_f^2$.

Quasi-Monte Carlo Method. QMC methods can be regarded as a deterministic counterpart to classical MC methods. Unlike MC integration, which uses estimates (1) with randomly selected points, QMC methods use (1) but select the points u_i *deterministically*. In particular, QMC techniques produce deterministic sequences of points that provide the best-possible spread over the integration domain. These deterministic sequences are often referred to as low-discrepancy sequences, of which the Sobol sequence [19] is a well-known example. An effective way to use the QMC method is by performing a change of variables to reduce the integration to the $[0, 1]$ domain. When we need to integrate over a large domain $[a, b]$, that avoids multiplying the QMC estimate by a large factor $(b-a)$ as required by (1).

A QMC advantage with respect to MC is that its error is $O(1/N)$, while the MC error (see Eq. (2)) is $O(1/\sqrt{N})$, where N is the sample size. The Koksma-Hlawka inequality bounds the error of QMC estimates, but in practice it is very hard to estimate [14], thereby hampering the use of QMC methods. As such, other methods for estimating the QMC error need to be developed. For example, Ermakov and Antonov [5] have recently introduced the *Qint* method for QMC variance estimation, based on a set of random quadrature formulas (see below).

Randomised Quasi-Monte Carlo. As discussed earlier, the practical application of QMC is limited by the difficulty of computing an estimate of the integration error. However, allowing randomisation into the deterministic QMC procedure enables constructing confidence intervals. A *Randomised* QMC (RQMC) procedure can be described as follows. Suppose that $\mathfrak{X} = \{x_1, ..., x_n\}$ is a deterministic low-discrepancy set. By means of a transformation $\tilde{\mathfrak{X}} = \Gamma(\mathfrak{X}, \epsilon)$ a finite set $\tilde{\mathfrak{X}}$ is generated by the random variable ϵ with the same quasi-random properties of the set $\mathfrak{X}$ (see Fig. 1). For a randomised set $\tilde{\mathfrak{X}}_i$ of size n we construct a RQMC estimate similar to (1):

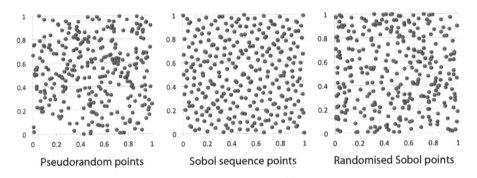

Fig. 1. Uniform pseudorandom, Sobol sequence and randomised Sobol sequence points (obtained by transformation $\Gamma = (\mathfrak{X} + \epsilon) \mod 1$, where ϵ is a random sample from MC sequence and $\mathfrak{X}$ is low-discrepancy sample from Sobol sequence) distribution in the 2-dimensional unit space. The comparison is based on the first 300 points of sequences.

$$RQMC_j = \frac{1}{n} \sum_{i=1}^{n} f(\tilde{\mathfrak{X}}_{i,j}) \tag{3}$$

for $0 < j \leqslant r$, where r is the total number of different pseudo-random sequences. Then, we take their average for overall RQMC estimation (3):

$$RQMC = \frac{1}{r} \sum_{j=1}^{r} RQMC_j \tag{4}$$

which is then built out of rn samples in total. If we choose the Γ transformation in such a way that each of the estimates $RQMC_j$ has the unbiasedness property, i.e., $\mathbb{E}[RQMC_j] = I$ for all j, (e.g., $\Gamma = (\mathfrak{X} + \epsilon) \mod 1$), then the estimator (4) will also be unbiased, i.e., $\mathbb{E}[RQMC] = I$. By independence of the samples used in (3) and (4), we have that:

$$Var(RQMC) = \frac{Var(RQMC_j)}{r}.$$

Thus, we have the following variance estimation: $\widehat{Var}(RQMC) = \frac{1}{r(r-1)} \sum_{j=1}^{r} \left(RQMC_j - RQMC\right)^2$.

3 Confidence Interval Estimation and Error Analysis

In the following we shall use the notation below:

- $\tilde{X} = \frac{1}{n} \sum_{i=1}^{n} x_i$ - sample mean;
- $C_a = Quant(1 - \frac{a}{2})$ - inverse cumulative distribution function (quantile function) of a normal random variable with mean 0 and standard deviation 1; parameter a defines the confidence level at $1 - a$;

- $\hat{p} = n_s/n$ - the binomially-distributed proportion, where: n_s - number of successes and n_f - number of failures in a Bernoulli trial process; n - total number of Bernoulli trials;
- $\hat{q} = 1 - \hat{p}$, and CI = confidence interval.

3.1 Intervals Based on the Standard CLT Interval

Modified Central Limit Theorem (CLT) interval. First, we consider the case when the sample x_i is extracted from the normal distribution $N(\mu, \sigma^2)$ with unknown mean μ and known variance σ^2. Here, μ can be approximated by the sample mean: $\mu \approx \tilde{X}$. To clarify this approximation, the standard CI for μ with confidence level $1 - a$ is:

$$CI_{CLT} = \left(\tilde{X} - C_a \frac{\sigma}{\sqrt{n}}; \tilde{X} + C_a \frac{\sigma}{\sqrt{n}} \right). \tag{5}$$

In practice, the variance σ^2 is often unknown, but one can use the same CI by replacing σ with the sample standard deviation $s = \sqrt{\frac{1}{n-1} \sum_{i=1}^{n} (x_i - \tilde{X})^2}$. This method is widely used for estimating the distribution of binomially-distributed proportions. A number of works (*e.g.*, [8–10]) note that the CI_{CLT} approximation can be poor when applied to Bernoulli trials with $\hat{p}$ close to 0 or 1. Indeed, in the Bernoulli case, when $\hat{p}$ is 0 (or 1) the CLT interval (5) cannot be constructed, since $s = 0$ (recall that the sample standard deviation s substitutes σ, which is most often unknown). In order to address this problem, we simply overapproximate the sample standard deviation with $\frac{1}{n^2}$ if $\hat{p}$ is equal to 0 (or 1).

Wilson Interval. It was introduced by Wilson in 1927 in his fundamental work [11] and uses the inversion of the CLT interval. The interval is:

$$CI_W = \left(\frac{n_s + \frac{C_a^2}{2}}{n + C_a} - \frac{C_a \sqrt{n}}{n + C_a^2} \sqrt{\hat{p}\hat{q} + \frac{C_a^2}{4n}}; \frac{n_s + \frac{C_a^2}{2}}{n + C_a} + \frac{C_a \sqrt{n}}{n + C_a^2} \sqrt{\hat{p}\hat{q} + \frac{C_a^2}{4n}} \right) \tag{6}$$

This interval has some obvious advantages - it can not exceed probability boundaries, and it can be easily calculated even if $\hat{p}$ is 0 or 1. At the same time, CI_W has downward spikes when $\hat{p}$ is close to 0 and 1, because it is formed by an inverted CLT approximation.

Agresti-Coull Interval. This method was introduced by Agresti and Coull in 1998 [1]. One of the most interesting features of this CI is that it makes a crucial assumption about n_s and n_f. This interval formally adds two successes and two failures to the obtained values in case of 95% confidence level and then uses the CLT method. The interval can be constructed as follows:

$$CI_{AC} = \left(\tilde{X} - \frac{1}{n + C_a^2} (n_s + \frac{1}{2} C_a^2); \tilde{X} + \frac{1}{n + C_a^2} (n_s + \frac{1}{2} C_a^2) \right) \tag{7}$$

Additionally, this interval can be modified by using the center of the Wilson interval (6) in place of $\hat{p}$:

$$CI_{ACw} = \left(\frac{n_s + \frac{C_a^2}{2}}{n + C_a} - C_a\sqrt{\hat{p}\hat{q}(n + C_a^2)}; \left(\frac{n_s + \frac{C_a^2}{2}}{n + C_a} - C_a\sqrt{\hat{p}\hat{q}(n + C_a^2)} \right) \right. . \quad (8)$$

Logit Interval. The Logit interval is based on a transformation of the standard interval [10]. It uses the empirical logit transformation: $\lambda = ln(\frac{\hat{p}}{1-\hat{p}}) = ln(\frac{n_s}{n-n_s})$. The variance of λ is: $\widehat{Var}(\lambda) = \frac{n}{n_s(n-n_s)}$ and the Logit interval is estimated as:

$$CI_L = \left(\frac{e^{\lambda_L}}{1 + e^{\lambda_L}}, \frac{e^{\lambda_U}}{1 + e^{\lambda_U}} \right) \quad (9)$$

where the lower bound transformation is $\lambda_L = \lambda - C_a\sqrt{\widehat{Var}(\lambda)}$ and the upper bound transformation is $\lambda_U = \lambda + C_a\sqrt{\widehat{Var}(\lambda)}$.

Anscombe Interval. This interval was proposed by Anscombe in 1956 [4] and is based on the Logit interval (9). The key difference is in λ and $\widehat{Var}(\lambda)$ estimation, where λ is defined as $\lambda = ln(\frac{n_s + \frac{1}{2}}{n - n_s + \frac{1}{2}})$ and the variance is $\widehat{Var}(\lambda) = \frac{(n+1)(n+2)}{n(n_s+1)(n-n_s+1)}$. On this basis, the Anscombe interval CI_{Anc} is estimated in the same way as Logit interval (9).

Arcsine Interval. It uses a variance-stabilising transformation of $\hat{p}$. In 1948, Anscombe introduced an improvement [3] for achieving better variance stabilisation by replacing $\hat{p}$ to $p^{\dagger} = \frac{n_s+3/8}{n+3/4}$, obtaining

$$CI_{Arc} = \left(\sin(\arcsin(\sqrt{p^{\dagger}}) - \frac{C_a}{2\sqrt{n}})^2, \sin(\arcsin(\sqrt{p^{\dagger}}) + \frac{C_a}{2\sqrt{n}})^2 \right) . \quad (10)$$

3.2   Alternative Intervals Based on the Beta-Function

Bayesian Interval. This method is based on the assumption that the (unknown) probability p to estimate is itself a random quantity [23]. The Bayesian interval is also called *credible*, because it is based on the posterior distribution of the unknown quantity computed by using its prior distribution and the Bayes theorem. The prior distribution can be constructed by means of the *Beta* distribution. If p has a prior distribution $Beta(\alpha, \beta)$ then after n Bernoulli trials with n_s successes, p has posterior distribution $Beta(n_s + \alpha, n - n_s + \beta)$. We can construct a Bayesian equal-tailed interval by the formula:

$$CI_B = \left(Beta^{-1}(\frac{a}{2}, n_s + \alpha, n - n_s + \beta), Beta^{-1}(1 - \frac{a}{2}, n_s + \alpha, n - n_s + \beta) \right) \quad (11)$$

where, $Beta^{-1}(a, \alpha, \beta)$ is the inverse of the cumulative distribution function of $Beta(\alpha, \beta)$. The probability density function of the $Beta$ distribution is $f(x; \alpha, \beta) = \frac{1}{B(\alpha,\beta)} x^{\alpha-1}(1-x)^{\beta-1}$, where $0 \le x \le 1$, $\alpha, \beta > 0$ and B is the beta function. In our experiments we used $\alpha = \beta = 1$, which gives the uniform distribution.

Jeffreys Interval. The Jeffreys interval is a Bayesian interval and uses the Jeffreys prior [15], which is a non-informative prior given by the $Beta$ distribution with parameters $(\frac{1}{2}, \frac{1}{2})$. We can form Jeffreys' equal-tailed interval by (11) with parameters $(\alpha = \frac{1}{2}, \beta = \frac{1}{2})$.

Clopper-Pearson Interval. This method was introduced by Clopper and Pearson in 1934 [8] and is based on the inversion of the binomial test, rather than on approximations. The Clopper-Pearson interval is:

$$CI_{CP} = \left(Beta^{-1}(\frac{a}{2}, n_s, n - n_s + 1), Beta^{-1}(1 - \frac{a}{2}, n_s + 1, n - n_s)\right) . \quad (12)$$

The interval states that the computed coverage probability is always above or equal to the $1 - a$ confidence level. In practice, it can be achieved in cases when n is large enough, while in general the actual coverage can exceed $1 - a$. We can conclude from Eq. (12) that due to the absence of the α and β parameters, a tighter CI can be achieved only by increasing the number of trials. We report this interval only for completeness, although we will not use it in our experiments as it is similar to the Bayesian and Jeffreys intervals.

4   Results

We evaluate confidence interval (CI) estimation methods based on the CLT interval with the RQMC and MC techniques and the Bayesian CI estimation method with the MC technique. In the RQMC case $r = 10$ quasi-random sequences were obtained by changing the pseudo-random points ϵ of the transformation $\Gamma = (\mathfrak{X} + \epsilon) \mod 1$, while the Sobol sequence points $\mathfrak{X}$ remained the same. In the MC case we used the same 10 pseudo-random points sequences that were used for RQMC calculations. (The high confidence levels used (up to 0.99999) motivates our choice of ten repetitions.) The samples used in Sects. 4.1 and 4.2 were obtained by sampling Bernoulli's, *i.e.*, no model simulation was performed.

4.1   Border Probability Cases

Intervals Based on CLT and Bayesian Interval. The comparison of the different CI estimation techniques for low probability cases is presented in Fig. 2. It shows that all intervals except the Arcsin interval (10) (see plot $c = 0.99$ of Fig. 2 for probability=0.001) contain the true probability value. The Bayesian method tends to overestimate the true probability values as they increase while CI_{CLT} tends to underestimate them. Also, it is interesting to note that the most accurate center value is returned by the Agresti-Coull interval. The reason why CI_{CLT} tends to include the true probability value near the upper bound of the interval is directly related to the number of samples. As shown in Fig. 2, for true probability values between 0.007–0.01, the CI_{CLT} center is moving up evenly to the true probability value with the increase of the confidence value. For the other true probability values (0.001–0.006) this drift remains, but it can not be seen from the figure because of the small difference in the number of samples for all confidence levels, which causes the CI center to move wave-like.

In Fig. 3 we plot the number of samples that the CI estimation techniques used to return intervals for four confidence levels. It can be clearly seen that when increasing the confidence level the CIs based on the CLT interval outperform the Bayesian CI. The plot with $c = 0.99999$ in Fig. 3 illustrates that the best techniques in the number of samples are (best to worst): CI_{CLT}, $Qint$, CI_{Arc}, CI_W, CI_L, CI_{Ans}, CI_{ACw} and CI_B. The CI_L and CI_{Anc} techniques always show almost the same results near the bounds, because of the modification of the CI_L. Initially, CI_L is not able to deal with probability values near the bounds according to its λ formula (see Sect. 3.1). It has been modified to use the Anscombe estimation formula in cases when $\hat{p} = 0$ or $\hat{p} = 1$. It is also important to note that the difference in the number of samples between CI_{CLT}, CI_{Arc} and CI_B for extreme probability cases is relevant. For example in the plot with $c = 0.9999$ of Fig. 3 the number of samples used to obtain interval for $p = 0.005$ equals to 1,078 for CI_{CLT}, 2,662 for the CI_{Arc} and 4,440 for CI_B.

Summarising, for probability values near the bounds (0 or 1) our modified CLT method achieves better results with fewer samples in comparison with the other techniques. For probability values away from the bounds, the CLT, Wilson, Agresti–Coull, Logit and Anscombe methods are all very similar, and so for such probabilities we come to the conclusion that the CLT interval should be recommended, due to its simplest form. Meanwhile for smaller sample sizes, the CI_{CLT} is strongly preferable to the others and so might be the choice where sampling cost is paramount.

Qint Method Results. In Figs. 2 and 3 we also plotted the results of the $Qint$ algorithm (see Sect. 2). In our research we used $Qint$ with $n = k \times 2^s$, where $k = 2$. These parameters were used to form n points of the Sobol sequence x_i with numbers $i \in I_{k,s} = \{1, 2, ..., k \times 2^s\}$. These parameters were chosen on the basis of the original study of the $Qint$ method [5]. As mentioned in Sect. 2, $Qint$ uses a cubature randomization method and provides an integral estimation variance that we used to obtain a CI by our modified CLT interval (5).

In Fig. 2 we display the $Qint$ intervals for border probability values. We can see from the plots that the $Qint$ CI always contains the true probability value. At the same time for all confidence levels from 0.99 to 0.99999 and for true probability values 0.006-0.01, $Qint$ shows better centration than CI_B and CI_{CLT}. For example, the greatest difference between the $Qint$ CLT center result and the true probability values is 0.00245 for $c = 0.99$ (p = 0.004), while this difference for CI_B reaches 0.00518 for $c = 0.99$ (= 0.007). We can see in Fig. 3 that, as expected, $Qint$ uses fewer samples than other CIs but CI_{CLT}. Our modification

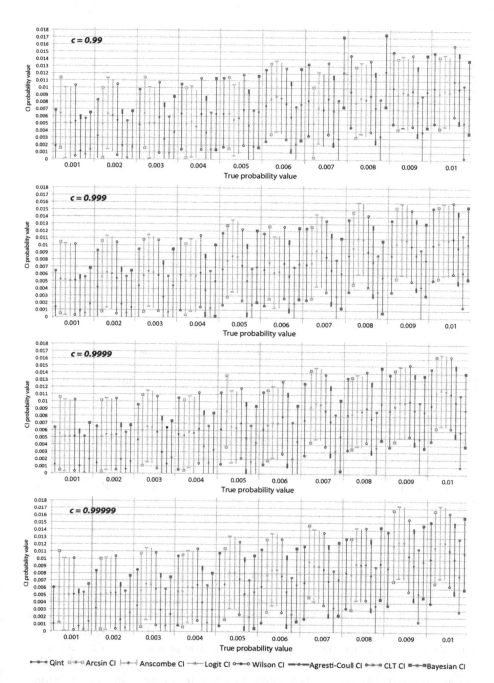

Fig. 2. Comparison of confidence intervals for probability values near 0, interval size equal to 10^{-2} and c - confidence level.

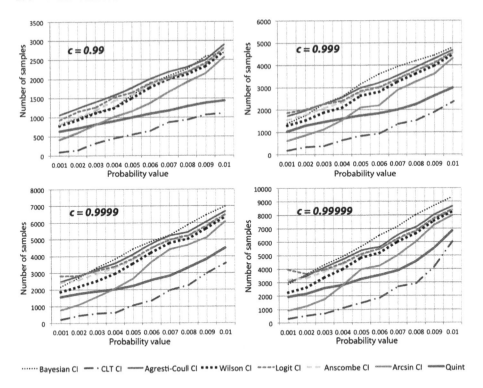

Fig. 3. Comparison of sample size for probability values near 0, interval size equal to 10^{-2} and c - confidence level.

allows the $Qint$ algorithm to return intervals even if $n_s = 0$, which significantly decreases the final sample size.

The fact that with the chosen parameters $Qint$ can not outperform our modified CI_{CLT} leads us to the conclusion that our use of the standard deviation formula with $\frac{1}{n^2}$ lower bound is a rather effective and simple solution. However, the deep range of the possible parameters variation of the $Qint$ algorithm lead us to believe that further research towards their comparison is needed.

4.2 MC and QMC Error Comparison

As mentioned in Sect. 2, the aymptotic QMC advantage over MC on the integration error holds in general. When the true probability value is extremely close to 0 (*i.e.*, $n_s = 0$ is obtained), we have that both the MC and QMC produce a 0 estimate and hence their error equals the true probability value. Also, the chaotic coverage properties of the MC method are far more persistent than they are appreciated. The chaotic behaviour does not disappear even when n is quite large and the true probability p is not near the boundaries. For instance, in Fig. 4 (a) it is visible that even when n is quite large (*i.e.*, tends to 10,000 samples) the actual error value of the MC method reaches 0.005. Hence we conclude that

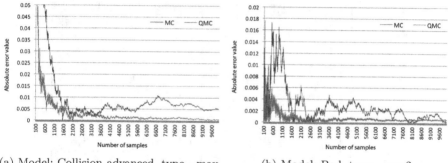

(a) Model: Collision advanced, type - max.   (b) Model: Bad, type - max2.

Fig. 4. MC (blue line) and QMC (red line) absolute error with respect to the number of samples. (Color figure online)

MC-based CI estimation techniques can be misleading and defective in several respects and their *point estimates* should be used with care [9].

A notable phenomenon, which was noticed for both the MC and QMC methods is that the actual error contains non-negligible oscillations as both p and n vary. There exist some "unlucky" pairs (p, n) such that the corresponding absolute error is much greater than the results for smaller n. The phenomenon of oscillation is both in n, for fixed p, and in p, for fixed n. Furthermore, drastic changes in coverage probability can occur in nearby p for fixed n and in nearby n for fixed p [9]. We can see it on the simple example in Fig. 4 (b). However, the same figure shows that error of QMC is more "stable" than the MC error.

4.3   Hybrid Systems Results

The results in this Section have been obtained via the ProbReach tool [17] for computing bounded reachability in stochastic parametric hybrid systems. Five models[1], including nonlinear systems, were chosen for our experiments to give use cases representative of real applications. Based on our model set, we provide in Table 1 a comparison of the CIs described in Sect. 3. The true probability value P is either an analytically computed single probability value or a rigorously computed absolute (non-statistical) interval [17].

As it can be seen in Table 1, all the intervals computed overlap with each other. The key difference in the interval size can be found in the results of the Bad model Type min and the Collision (Basic) model Type min. From the results we can conclude that the true probability value is very close to 0. This allows the Bayesian, CLT and Agresti-Coull methods to form intervals that are actually half of the required width 10^{-2}, while the other techniques return fully sized intervals. That happens because CI_B is using the posterior distribution to form the interval, which is always defined on the whole [0,1] interval. At the same time, the CI_{CLT} and CI_{AC_W} calculations of the mean value are quite close

---

[1] Available at https://github.com/dreal/probreach/tree/master/model.

Table 1. Confidence interval computation obtained via ProbReach, with solver precision $\delta = 10^{-3}$ and interval size equal to 10^{-2}, Type - extremum type and P - true probability value, where single point values were analytically computed and interval values are numerically guaranteed enclosures (computed by ProbReach).

Confidence level $c=0.99$

Model	Type	P	CI_B	CI_{CLT}	CI_{ACW}	CI_w	CI_L	CI_{Ans}	CI_{Arc}	Q_{mt}
Good	max	0.1	[0.09671, 0.10671]	[0.09564, 0.10564]	[0.09632, 0.10632]	[0.09574, 0.10574]	[0.09575, 0.10575]	[0.09577, 0.10577]	[0.09559, 0.10559]	[0.09147, 0.10147]
	min	0.1	[0.09529, 0.10529]	[0.09566, 0.1056]	[0.09666, 0.10666]	[0.09679, 0.10679]	[0.09678, 0.10678]	[0.0968, 0.1068]	[0.09639, 0.10639]	[0.09164, 0.10164]
Bad	max	0.95001	[0.94416, 0.95416]	[0.94495, 0.95493]	[0.94422, 0.95422]	[0.94397, 0.95397]	[0.94396, 0.95396]	[0.94392, 0.95392]	[0.94735, 0.95735]	[0.94459, 0.95459]
	max2	0.88747	[0.88215, 0.89215]	[0.88055, 0.89055]	[0.88031, 0.88031]	[0.88136, 0.89019]	[0.8803, 0.8902]	[0.88019, 0.89019]	[0.88325, 0.89325]	[0.88136, 0.89136]
	min	4×10^{-7}	[0, 0.00517]	[0, 0.00319]	[0, 0.00483]	[0, 0.00955]	[0, 0.00959]	[0, 0.00959]	[0, 0.00959]	[0, 0.005]
Deceleration	max	[0.08404, 0.08881]	[0.08471, 0.09471]	[0.08802, 0.09802]	[0.08817, 0.09817]	[0.08685, 0.09685]	[0.08614, 0.09614]	[0.0863, 0.0963]	[0.08963, 0.09932]	[0.08852, 0.09852]
	min	[0.04085, 0.04275]	[0.03835, 0.04835]	[0.03919, 0.04919]	[0.03854, 0.04854]	[0.03884, 0.04884]	[0.03886, 0.04886]	[0.03889, 0.04889]	[0.03873, 0.04873]	[0.03337, 0.04337]
Collision (Basic)	max	[0.96567, 0.97254]	[0.96371, 0.97381]	[0.96873, 0.97873]	[0.9684, 0.9784]	[0.96851, 0.97851]	[0.96875, 0.97875]	[0.96853, 0.97853]	[0.96851, 0.97851]	[0.96301, 0.97301]
	min	[0, 0.00201]	[0, 0.00525]	[0, 0.005]	[0, 0.00483]	[0, 0.00955]	[0, 0.00959]	[0, 0.00959]	[0, 0.00959]	[0, 0.005]
Collision (Extended)	max	[0.35751, 0.49961]	[0.42267, 0.43675]	[0.42418, 0.4342]	[0.42187, 0.43187]	[0.42345, 0.43345]	[0.42463, 0.43463]	[0.42457, 0.43457]	[0.42385, 0.43385]	[0.42342, 0.43342]
	min	[0.04296, 0.06311]	[0.0482, 0.0582]	[0.04772, 0.05772]	[0.04785, 0.05785]	[0.04823, 0.05823]	[0.04812, 0.05812]	[0.0481, 0.0581]	[0.04757, 0.05772]	[0.04618, 0.05618]
Collision (Advanced)	max	[0.14807, 0.31121]	[0.2072, 0.2172]	[0.20873, 0.21872]	[0.21872, 0.2185]	[0.20854, 0.21854]	[0.20854, 0.21854]	[0.20855, 0.21855]	[0.20111, 0.21111]	[0.20167, 0.21166]
	min	[0.02471, 0.05191]	[0.02631, 0.03631]	[0.03045, 0.04045]	[0.03016, 0.04016]	[0.03001, 0.04]	[0.03001, 0.04]	[0.03016, 0.04016]	[0.03164, 0.04164]	[0.0304, 0.0404]
Anesthesia	n/a	[0.00916, 0.04222]	[0.01361, 0.02361]	[0.01339, 0.02332]	[0.01374, 0.02374]	[0.01373, 0.02373]	[0.01318, 0.02318]	[0.01311, 0.02311]	[0.01592, 0.02592]	[0.01815, 0.02815]

Confidence level $c=0.99999$

Model	Type	P	CI_B	CI_{CLT}	CI_{ACW}	CI_w	CI_L	CI_{Ans}	CI_{Arc}	Q_{mt}
Good	max	0.1	[0.09499, 0.10499]	[0.09378, 0.10378]	[0.09386, 0.10386]	[0.09389, 0.10389]	[0.09391, 0.10391]	[0.09392, 0.10392]	[0.09405, 0.10405]	[0.09512, 0.10512]
	min	0.1	[0.09419, 0.10419]	[0.09667, 0.10667]	[0.09668, 0.10668]	[0.09677, 0.10677]	[0.09671, 0.10671]	[0.09679, 0.10679]	[0.09675, 0.10675]	[0.09525, 0.10525]
Bad	max	0.95001	[0.94525, 0.95525]	[0.94579, 0.95579]	[0.94564, 0.95548]	[0.94548, 0.95548]	[0.94545, 0.95545]	[0.94543, 0.95543]	[0.94735, 0.95735]	[0.94543, 0.95543]
	max2	0.88747	[0.88215, 0.89215]	[0.88055, 0.89055]	[0.88057, 0.89057]	[0.88046, 0.89046]	[0.88046, 0.89046]	[0.88046, 0.89046]	[0.88325, 0.89325]	[0.88052, 0.89052]
	min	4×10^{-7}	[0, 0.00517]	[0, 0.00319]	[0, 0.00494]	[0, 0.00984]	[0, 0.00992]	[0, 0.00992]	[0.00445, 0.0139]	[0, 0.005]
Deceleration	max	[0.08404, 0.08881]	[0.08613, 0.09613]	[0.08624, 0.09624]	[0.08312, 0.09312]	[0.08725, 0.09725]	[0.08725, 0.09725]	[0.08726, 0.09726]	[0.08746, 0.09746]	[0.08737, 0.09735]
	min	[0.04085, 0.04275]	[0.03514, 0.04514]	[0.03919, 0.04919]	[0.03918, 0.04918]	[0.03942, 0.04942]	[0.03943, 0.04943]	[0.03944, 0.04944]	[0.039, 0.049]	[0.03377, 0.04377]
Collision (Basic)	max	[0.96567, 0.97254]	[0.96359, 0.97359]	[0.96241, 0.97241]	[0.96767, 0.9767]	[0.96892, 0.96892]	[0.96689, 0.97589]	[0.96683, 0.97583]	[0.96863, 0.97863]	[0.96462, 0.97462]
	min	[0, 0.00201]	[0, 0.00517]	[0, 0.00319]	[0, 0.00494]	[0, 0.00984]	[0, 0.00992]	[0, 0.00992]	[0.00445, 0.0139]	[0, 0.005]
Collision (Extended)	max	[0.35751, 0.49961]	[0.42651, 0.43652]	[0.42719, 0.43724]	[0.42757, 0.43757]	[0.42656, 0.43656]	[0.41774, 0.42774]	[0.41779, 0.42779]	[0.42745, 0.43745]	[0.42875, 0.43875]
	min	[0.04296, 0.06311]	[0.04979, 0.05979]	[0.04766, 0.05766]	[0.04764, 0.05764]	[0.04748, 0.05748]	[0.04745, 0.05745]	[0.04776, 0.05776]	[0.05776, 0.05673]	[0.04576, 0.05576]
Collision (Advanced)	max	[0.14807, 0.31121]	[0.20515, 0.21519]	[0.20558, 0.21563]	[0.20533, 0.21533]	[0.20531, 0.21531]	[0.20547, 0.21547]	[0.20547, 0.21547]	[0.20385, 0.21385]	[0.20453, 0.21453]
	min	[0.02471, 0.05191]	[0.03011, 0.04015]	[0.02902, 0.03902]	[0.02954, 0.03945]	[0.03956, 0.04956]	[0.03861, 0.04861]	[0.03887, 0.04887]	[0.0363, 0.04363]	[0.03031, 0.04031]
Anesthesia	n/a	[0.00916, 0.04222]	[0.01284, 0.02284]	[0.01513, 0.02511]	[0.01623, 0.02623]	[0.01545, 0.02545]	[0.01557, 0.02557]	[0.01562, 0.02562]	[0.01385, 0.02385]	[0.01852, 0.02852]

to zero, thus cutting out the negative part of the interval. This trend holds for all probability values within $[0, 0.001]$. Table 1 also shows that with the increase of the confidence level the interval's precision grows, which in turn is directly related to the usage of the inverse cumulative distribution function for normal random variable with given confidence level in formulas for CI_{CLT} (5), CI_W (6), CI_{AC_W} (8) and CI_{Arc} (10). It also results in the increase of the sample size for CI_L and CI_{Anc}.

The comparison of the obtained intervals (see Table 1) with the true probability value or interval P shows that all CIs contain the single probability values but CI_{Acr} (see Bad type min model of Table 1), and all CIs overlap with the true probability intervals. We also note that the true probability intervals of the Collision Extended, Collision Advanced, and Anesthesia models contain all confidence intervals for all confidence levels. The Collision Basic and Deceleration models' true probability intervals do not contain CIs due to their size (< 0.01). The original $Qint$ algorithm was not able to provide results for the Bad type min and Collision Basic type min models, because of the very small probability involved (4×10^{-7} and $[0, 0.00201]$) it could not detect $n_s > 0$ for the chosen confidence level and interval size. Therefore, $Qint$ was used in conjunction with our CLT method described in Sect. 3.1. In conclusion, all the CI techniques examined returned reliable intervals.

Table 2 reports the number of samples required to compute the CIs obtained via ProbReach. As it was noted earlier for Fig. 4, the number of samples needed for CI computation grows from the bounds to the center of the $[0,1]$ interval (this is because the variance of a Bernoulli is largest at $p = 0.5$). The most important outcome is that all CIs (except CI_{Arc}) show better result in number

Table 2. Sample size comparison for confidence interval computation obtained via ProbReach, with solver δ precision equal to 10^{-3} and interval size equal to 10^{-2}, **Type** - extremum type; confidence level = 0.99999.

Model	Type	CI_B	CI_{CLT}	CI_{AC_W}	CI_W	CI_L	CI_{Ans}	CI_{Arc}	$Qint$
Good	max	70422	69484	69582	69496	69530	69529	77262	68456
	min	71898	71286	71339	71293	71321	71321	79369	68994
Bad	max	37388	36518	36771	36629	36687	36868	60006	36164
	max2	79306	79097	79125	79101	79118	79118	96442	77892
	min	5797	124	2766	1963	4136	4136	572	94
Deceleration	max	65248	65233	65330	65299	65320	65319	72114	59882
	min	33147	32969	33133	33018	33060	33060	34231	29096
Collision (basic)	max	25279	24711	24834	24789	24934	24933	26045	23016
	min	5797	124	2766	1963	4136	4136	572	94
Collision (extended)	max	191466	190776	191253	190894	191485	191472	376294	185456
	min	41153	38942	39745	39473	39537	39541	47923	37608
Collision (advanced)	max	131517	129746	131185	129845	129934	129933	183405	127486
	min	27305	25657	25835	25736	25792	25791	29362	24569
Anesthesia	n/a	16197	15453	15834	15634	15734	15733	17845	15314

of samples with respect to CI_B. Overall, $Qint$ showed the best result for every model, closely followed by CI_{CLT}: from Table 2 we see that $Qint$ used on average between 1,850 and 24,802 fewer samples than other CI techniques.

5 Conclusions

In this paper, we have provided a comprehensive evaluation of confidence interval calculation techniques for Monte Carlo (MC) and Quasi-Monte Carlo (QMC) methods. We have shown that:

- the Central Limit Theorem (CLT) interval generally performs best, in particular for small sample sizes;
- when estimating probabilities near the borders (*i.e.*, close to 0 or 1), our simple CLT modification has proved to be very effective, while other techniques cannot form intervals;
- QMC methods are more efficient than MC methods by providing precise estimates with fewer samples.

Based on our analysis, we suggest that our results can be used as guidelines for statistical model checking of time-bounded properties beyond reachability.

References

1. Agresti, A., Coull, B.A.: Approximate is better than "exact" for interval estimation of binomial proportions. Am. Stat. **52**(2), 119–126 (1998)
2. Alur, R., Courcoubetis, C., Henzinger, T.A., Ho, P.-H.: Hybrid automata: an algorithmic approach to the specification and verification of hybrid systems. In: Grossman, R.L., Nerode, A., Ravn, A.P., Rischel, H. (eds.) HS 1991-1992. LNCS, vol. 736, pp. 209–229. Springer, Heidelberg (1993). https://doi.org/10.1007/3-540-57318-6_30
3. Anscombe, F.J.: The transformation of Poisson, binomial and negative-binomial data. Biometrika **35**(3/4), 246–254 (1948)
4. Anscombe, F.J.: On estimating binomial response relations. Biometrika **43**(3/4), 461–464 (1956)
5. Antonov, A.A., Ermakov, S.M.: Empirically estimating error of integration by Quasi-Monte Carlo method. Vestnik St. Petersburg Univ. Math. **47**(1), 1–8 (2015)
6. Baier, C., Haverkort, B.R., Hermanns, H., Katoen, J.: Model-checking algorithms for continuous-time Markov chains. IEEE Trans. Softw. Eng. **29**(6), 524–541 (2003)
7. Bianco, A., de Alfaro, L.: Model checking of probabilistic and nondeterministic systems. In: Thiagarajan, P.S. (ed.) FSTTCS 1995. LNCS, vol. 1026, pp. 499–513. Springer, Heidelberg (1995). https://doi.org/10.1007/3-540-60692-0_70
8. Clopper, C.J., Pearson, E.S.: The use of confidence or fiducial limits illustrated in the case of the binomial. Biometrika **26**(4), 404–413 (1934)
9. Brown, L.D., Cai, T.T., DasGupta, A.: Interval estimation for a binomial proportion. Stat. Sci. **16**(2), 128–133 (2001)
10. Dean, N., Pagano, M.: Evaluating confidence interval methods for binomial proportions in clustered surveys. J. Surv. Stat. Methodol. **3**(4), 484–503 (2015)

11. Edwin, W.B.: Probable inference, the law of succession, and statistical inference. J. Am. Stat. Assoc. **22**(158), 209–212 (1927)
12. Fränzle, M., Chen, M., Kröger, P.: In memory of Oded Maler: automatic reachability analysis of hybrid-state automata. ACM SIGLOG News **6**(1), 19–39 (2019)
13. Gao, S., Avigad, J., Clarke, E.M.: Delta-decidability over the reals. In: LICS, pp. 305–314 (2012)
14. Gnewuch, M., Srivastav, A., Winzen, C.: Finding optimal volume subintervals with k points and calculating the star discrepancy are NP-hard problems. J. Complex. **25**(2), 115–127 (2009)
15. Mahajan, K.K., Arora, S., Kaur, K.: Bayesian estimation for Gini index and a poverty measure in case of Pareto distribution using Jeffreys' prior. MASA **10**(1), 63–72 (2015)
16. Pradhan, V., Banerjee, T.: Confidence interval of the difference of two independent binomial proportions using weighted profile likelihood. Commun. Stati. - Simul. Comput. **37**(4), 645–659 (2008)
17. Shmarov, F., Zuliani, P.: ProbReach: verified probabilistic δ-reachability for stochastic hybrid systems. In: HSCC, pp. 134–139. ACM (2015)
18. Shmarov, F., Zuliani, P.: Probabilistic hybrid systems verification via SMT and Monte Carlo techniques. In: Bloem, R., Arbel, E. (eds.) HVC 2016. LNCS, vol. 10028, pp. 152–168. Springer, Cham (2016). https://doi.org/10.1007/978-3-319-49052-6_10
19. Sobol', I.M.: On the distribution of points in a cube and the approximate evaluation of integrals. USSR Comput. Math. Math. Phys. **7**(4), 86–112 (1967)
20. Traub, J.F., Wasilkowski, G.W., Woźniakowski, H.: Information-Based Complexity. Academic Press, Cambridge (1988)
21. Vardi, M.Y.: Automatic verification of probabilistic concurrent finite-state programs. In: FOCS, pp. 327–338 (1985)
22. Younes, H.L.S., Simmons, R.G.: Statistical probabilistic model checking with a focus on time-bounded properties. Inf. Comput. **204**(9), 1368–1409 (2006)
23. Zuliani, P., Platzer, A., Clarke, E.M.: Bayesian statistical model checking with application to stateflow/simulink verification. Formal Methods Syst. Des. **43**(2), 338–367 (2013)

Formal Verification of a Certified Policy Language

Amir Eaman$^{(\boxtimes)}$ and Amy Felty

School of Electrical Engineering and Computer Science, University of Ottawa,
Ottawa, Canada
{aeama028,afelty}@uottawa.ca

Abstract. Access control is an information security process which guards protected resources against unauthorized access, as specified by restrictions in security policies. A variety of policy languages have been designed to specify security policies of systems. In this paper, we introduce a certified policy language, called TEpla, with formal semantics and simple language constructs, which we have leveraged to express and formally verify properties about complex security goals. In developing TEpla, we focus on security in operating systems and exploit *security contexts* used in the *Type Enforcement* mechanism of the SELinux security module. TEpla is certified in the sense that we have encoded the formal semantics and machine-checked the proofs of its properties using the Coq Proof Assistant. In order to express the desired properties, we first analyze the behavior of the language by defining different ordering relations on policies, queries, and decisions. These ordering relations enable us to evaluate how algorithms for deciding whether or not requests are granted by policies will react to changes in policies and queries. The machine-checked mathematical proofs guarantee that TEpla behaves as prescribed by the semantics. TEpla is a crucial step toward developing certifiably correct policy-related tools for Type Enforcement policies.

Keywords: Access control · Policy languages · Formal methods

1 Introduction

Access control as a security mechanism is concerned with the management of access requests to resources. To determine if a request is allowed, it is checked against a set of authorization rules which are written in a particular policy language dependent on the type of access control available in the underlying computer system. Access control policy languages have an essential role in expressing the intended access authorization to regulate requests to resources. Security policy languages used to develop security policies significantly affect this process, mainly because the policy developers' understanding of the semantics of the languages has a direct effect on the way they write policies. Formal semantics can tremendously improve the use of a language by constructing a precise

© Springer Nature Switzerland AG 2020
B. Ben Hedia et al. (Eds.): VECoS 2020, LNCS 12519, pp. 180–194, 2020.
https://doi.org/10.1007/978-3-030-65955-4_13

reference for the underlying language. Semantic-related tools which analyze or reason about specifications written in the language require formal semantics to process the language correctly. Moreover, the implementation of such tools can be verified, which is another important consequence of formal semantics.

We propose a small and certifiably correct policy language, TEpla. TEpla can provide ease of use, analysis, and verification of its properties. By *certified* policy language, we mean a policy language with formal semantics and formally verified mathematical proofs of important properties, which reflects the concept of certification in formal methods communities and programming languages [6]. One of our goals is to avoid language-introduced errors (i.e., errors that are introduced to IT systems due to multiple contradictory interpretations of policies). Ease of reasoning and analysis of policies is facilitated by a clear specification of TEpla's behavior and semantics as it satisfies important formal properties designed for this purpose [22]. In addition to these properties, TEpla is flexible enough for defining complex security constraints through introducing user-defined predicates. This enables security administrators to define various security goals in security policies. We analyze the language's behavior by defining different ordering relations on policies, queries, and decisions. These ordering relations enable us to evaluate how language decisions react to changes in policies and queries. See, for example, the *non-decreasing* property of TEpla policies discussed in Sect. 4.

In order to keep the core of the language simple, in this study, we focus on developing a new certified policy language for the *Type Enforcement* mechanism, which is a subset of the SELinux security module [16] implemented in Linux distributions. Type Enforcement exploits the *security context* of resources to regulate accesses. The security context is a set of allowable values for particular attributes assigned to system resources.

SELinux is a Linux Security Module (LSM) that enables security developers to define security policies. It implements the Mandatory Access Control (MAC) [20] strategy, which allows policy writers to express whether a *subject* can perform an operation on an *object*, e.g.., whether an SELinux process can perform a read or write on a file or socket.

We carried out a study [8] on policy languages, which proposes solutions for dealing with the many gaps for using policy languages with informal semantics, mainly focusing on the SELinux policy language in particular, and gaps in developing verified security policies in general. TEpla is an important step in closing these gaps. We believe that the same development paradigm used for TEpla can be adopted to develop other verified policy languages, such as one for AppArmor [13] or one for full SELinux, thus providing higher-trust policy languages for Linux.

As mentioned earlier, TEpla also provides additional language constructs that allow security administrators to encode different security goals in policies as user-defined predicates. Using this mechanism, administrators can express a variety of conditions, thus significantly increasing the flexibility over the language's built-in conditions. However, there are some conditions that policy writers need to verify about their predicate definitions in order to ensure that their defined predicates are compatible with TEpla properties. Note that our proof development uses no axioms; we require all conditions to be proved.

We use the Coq proof assistant [3, 21] (version 8.12) to write machine-checked mathematical proofs for TEpla's properties. The Coq development of TEpla contains approximately 4700 lines of script and is available at http://www.site.uottawa.ca/~afelty/vecos20/. This online appendix also contains a mapping from names used in this paper to names used in the Coq code.

In Sect. 2, we present most of the infrastructure of TEpla, including rules, decisions, queries, and policies, and present the part of the semantics that involves evaluating queries against the rules. This section also defines ordering relations on TEpla decisions, policies, and queries. In Sect. 3, we present the syntax and semantics of constraints. A constraint can be considered as an additional form of a policy rule, which takes a user-defined predicate as an argument. We present the syntax and semantics of constraints, and discuss the conditions that must hold for predicates. In Sect. 4, we discuss the main properties that we have proved about TEpla, and Sect. 5 concludes the paper.

The work presented here appears in the Ph.D. thesis of the first author [7], and the reader is referred there for details, including a BNF grammar of TEpla's syntax. Here, in Sects. 2 and 3, we informally describe the TEpla language structures and their meaning, and present parts of the Coq encoding to illustrate.

2   Rules, Decisions, Queries, and Policies

The main element in a system is a *resource*, which can be either a subject or an object, as described in the previous section. In fact, a resource can act as a subject in some contexts and an object in others. In many policy languages, including TEpla, resources have attributes. As mentioned, the values of these attributes form the security context of the resources. In TEpla, the security context is the values of an attribute called *basic type*. Each resource is assigned one basic type, providing it with an identity in the same way as done in SELinux. For example, consider two resources of a system called *file_web* and *port_protocol*. We can assign, for instance, the values of the basic type attribute to be `mail_t` and `http_t`, respectively.

TEpla allows policy developers to group basic types of resources together to form a *group type*, providing a single identifier for a group of resources. We group together basic types when there exists a conceptual relationship among them. Basic and group types together form the notion of a *type*, which is the main building block of TEpla.

SELinux uses the terminology *source* and *destination* to mean subjects and objects, and *domain* and *type* to classify their types, respectively. Here, we continue to use *subject* and *object* and we use *type* to classify both.

Two other central data types in TEpla include *object class* and *permitted action*. Object classes specify possible instances of all resources of a certain kind, such as files, sockets, and directories. Permitted actions specify the actions that subjects are authorized to perform on objects. Permitted actions can range from being as simple as reading data, sharing data, or executing a file [15].

2.1   Syntax in Coq

We start by defining the basic data types. Here, $\mathbb{C}$, $\mathbb{P}$, basicT, which represent object classes, permitted actions, and basic types, respectively, are all defined as nat ($\mathbb{N}$), which is the datatype of natural numbers in Coq [21]. These definitions plus some examples are below.

```
Definition C := N.     Definition P := N.     Definition basicT := N.
Definition File : C := 600.
Definition mail_t : basicT := 300.    Definition http_t : basicT := 301.
Definition networkManager_ssh_t : basicT := 302.
Definition Read : P := 702.           Definition Write : P := 703.
```

We encode a group type as a list of basic types, i.e., we represent them using Coq's built-in datatype for lists. For example, the code below introduces $\mathbb{G}$ to define group types and program_G, which represents the example set {mail_t, http_t}. A group type should contain at least 2 elements.

```
Definition G : Set := list basicT.
Definition program_G : G := [mail_t; http_t].
```

We can now encode our principle entity, the *type* structure; we define the inductive datatype $\mathbb{T}$ with two constructors singleT and groupT. These constructors take arguments of type basicT and $\mathbb{G}$ respectively to produce a term belonging to $\mathbb{T}$.

```
Inductive T : Type:=
  | singleT : basicT → T
  | groupT : G → T.
```

As an example, consider two subjects whose security contexts are represented by the values http_t and mail_t, and a third subject that is allowed to access objects of both types. These are represented by (singleT http_t), (singleT mail_t), and (groupT program_G) respectively.

The access control rules that are used to form policies are defined inductively as type $\mathbb{R}$. These rules consist of *Allow* and *Type Transition* rules. The definition of $\mathbb{R}$ below is followed by an example *Allow* rule.

```
Inductive R : Set :=
  | Allow : T * T * C * P * B → R
  | Type_Transition : T * T * C → R.
Definition R_A : R :=
  Allow (groupT program_G, singleT mail_t, File, Read, true).
```

Rules are implemented using tuples. *Allow* rules enable policy writers to express eligible access from subjects (whose type is expressed by the first component) to objects (whose type is expressed by the second component). The third component specifies the object class of the object. The fourth component expresses possible actions that the object can perform on the subject. The last component is a Boolean condition, which we do not use here; it is discussed in future work.

The second kind of rule provides support for transition of types in security contexts from one value to another. *Type Transition* rules in policies express which types can switch to other types, which is an important feature that we adapt from SELinux but do not discuss further in this paper.

TEpla has a three-valued decision set for access requests including *NotPermitted*, *Permitted* and *UnKnown*. In TEpla, queries are denied by default, i.e., every access request that should be granted must be expressed explicitly by rules in a policy. Decisions are defined by the inductive type $\mathbb{DCS}$:

```
Inductive DCS: Set := Permitted | NotPermitted | UnKnown.
```

The *UnKnown* decision arises from conflicts in policies. Conflicts are caused by rendering a decision for access requests in a part of security policies that is different from an already taken decision according to other policy statements. For example, a specific *Allow* rule may permit a particular query, but there is a constraint in the policy that is not satisfied by the query. Such conflicts signify errors that must be corrected by a policy administrator.

Access requests or *queries* are inquiries into the policy to check the possibility that the subject is allowed to perform the specified action on the object. In TEpla, they consist of four components: the types of the subject and object, the object class of the object, and an action. The definition of queries ($\mathbb{Q}$) is shown below, along with an example.

```
Definition Q : Set := T * T * C * P.
Definition sampleQ : Q := (singleT mail_t, singleT http_t, File, Write).
```

Processing of a query with respect to a policy involves an attempt to check the authorization of a subject with the given type to carry out a specific action on an object having the given type and class. In the example, a subject of type `mail_t` is requesting to write to an object whose class is `File` and whose type is `http_t`.

Policies, defined below as the type $\mathbb{TEPLCY}$, consist of a sequence of rules and a sequence of constraints. Constraints, denoted as $\mathbb{CSTE}$, will be defined in the next section.

```
Inductive TEPLCY: Set := TEPolicy : list R * list CSTE → TEPLCY.
```

2.2   Evaluating Queries Against Policy Rules

We define the semantics of TEpla as a mapping from policies and access requests to decisions, in the form of five translation functions implemented in Coq, which together act as the decision-making chain that evaluates a query against a policy, taking into account all the various parts of the policy.

The first function, shown in Listing 1, evaluates a query against a single rule leading to a decision of either `Permitted` or `NotPermitted`.

```
Definition R_EvalTE (R_policy:R) (q:Q) : DCS:=
 match R_policy with
 |Allow (alw_srcT,alw_dstT,alw_C,alw_P,alw_B) ⇒
     match q with
          |(qsrcT, qdsT, qC, qP) ⇒
```

```
            if ((TSubset qsrcT alw_srcT) && (TSubset qdsT alw_dstT) &&
                (Nat.eqb qC alw_C) && (Nat.eqb qP alw_P) && (alw_B)
            then Permitted else NotPermitted
        end
  |Type_Transition (trn_srcT, trn_dstT, trn_C) ⇒ ...
  end.
```

Listing 1. Evaluation of a rule and a query

For *Allow* rules, the first four conditions of the if statement check to see if the rule applies to the query. The first two check that the types of the subject and object in the query are a subset of the corresponding types in the rule. (TSubset performs this check.) The next two conditions check that the object class and permitted action are the same, using the built-in function Nat.eqb. The last condition checks that the last component of rule, which is a Boolean condition, evaluates to *true*. If all conditions are satisfied, the result is Permitted. A *Type Transition* rule is similar (details omitted), but only the types of the subject and object need to be checked in order to determine that the rule applies.

The second function for evaluating queries against constraints is presented in Sect. 3; we also discuss the other three functions there.

2.3 Ordering Relation on Decisions, Queries and Policies

We define a *Partially Ordered Set (poset)* [9] called $(\mathbb{DCS}, <::)$ on TEpla's three-valued set of decisions as *NotPermitted* $<::$ *Permitted* $<::$ *UnKnown*. The lowest decision in this ordering is *NotPermitted*, which means that all accesses are first denied by default. To permit an access query, a relevant rule in the first component of policies must authorize the access. If the query is not granted at this stage, TEpla denies the access, which means that the ultimate access decision is *NotPermitted*. In the case that the query is granted (with decision *Permitted*), TEpla proceeds to check whether or not the query satisfies the constraint component of policies. The decision for the query continues to be *Permitted* as long as it satisfies the constraints; if not, that is the query fails to satisfy some constraints, the decision changes to *UnKnown*. We allow composition of policies in which decisions never go from *UnKnown* or *Permitted* to *NotPermitted* when TEpla checks the sub-policies of the composed policy (see Sect. 4 for more details about this property).

Additionally, we define a relation on queries $(\mathbb{Q}, <<=)$. Two queries $\mathbb{Q}_1 = (\text{SourceT\_}\mathbb{Q}_1, \text{DestT\_}\mathbb{Q}_1, \mathbb{C}_1, \mathbb{P}_1)$ and $\mathbb{Q}_2 = (\text{SourceT\_}\mathbb{Q}_2, \text{DestT\_}\mathbb{Q}_2, \mathbb{C}_2, \mathbb{P}_2)$ are in relation $\mathbb{Q}_1 <<= \mathbb{Q}_2$ if and only if $(\text{TSubset SourceT\_}\mathbb{Q}_2 \text{ SourceT\_}\mathbb{Q}_1)$ and $(\text{TSubset DestT\_}\mathbb{Q}_2 \text{ DestT\_}\mathbb{Q}_1)$ hold.

Finally, we define the binary relation $(\mathbb{TEPLCY}, \lesssim)$ on policies, where $p_1 \lesssim p_2$ whenever p_2 has more information that p_1. More formally:

$$\forall (p_1, p_2 \in \mathbb{TEPLCY}), p_1 \lesssim p_2 \text{ iff } length(p_1) \leqslant length(p_2) \wedge p_1 \subseteq p_2.$$

In this definition, *length* means the sum of the lengths of the rule component and the constraint component of a policy. We call the combined list *authorization rules.* Here $p_1 \subseteq p_2$ means that p_2 has more authorization rules and it contains all the authorization rules in p_1.[1]

3   Constraints and Predicates

As discussed earlier, rules alone cannot always accommodate the security requirements of systems precisely enough. TEpla's *constraints* and *predicates*, described in this section, represent one of the powerful features of TEpla, which distinguish it from other languages that lack this feature. TEpla constraints allow policy writers not only to rely on conditions or constraints defined in the language but also to define their complementary security logic.

3.1   Syntax in Coq

Constraints are defined below as the type $\mathbb{CSTE}$.

```
Inductive CSTE: Set :=
  | Constraint : C * P * T * T * list T *
       (list R → list T → C → P → T → T → T → T → B) → CSTE.
```

Constraints have six arguments. When a constraint is checked against a query or access request, the values of the first two arguments are compared to the values of the $\mathbb{C}$ and $\mathbb{P}$ components of a query, and the constraint is only applicable when the values of these components match. A constraint takes a function as its last argument, which returns a Boolean; these functions act as predicates that express when the constraint is satisfied. To express specific security goals, administrators can define different predicates by using various arguments provided for the function. These arguments supply a comprehensive set of values by which policy developers can define the required security criteria. To illustrate constraints and predicates, we use a "separation of duty" running example, which includes the constraint $\mathbb{CSTE}\_SoD$, defined below, and the predicate Prd_SoD, defined later.

```
Definition CSTE_SoD : CSTE:= Constraint(File, Read, groupT program_G,
                              singleT networkManager_ssh_t , [], Prd_SoD ).
```

This constraint only allows subjects whose types are elements of program_G to perform the action Read on objects whose basic type is networkManager_ssh_t and whose object class is File as long as the additional requirement is met that objects of types program_G and networkManager_ssh_t are never permitted to be acted upon by subjects of the same type. Prd_SoD will formally express what is meant by this additional requirement, and it will be defined after presenting the implementation of the function for evaluating constraints against queries. Informally, whenever two *Allow* rules permit subjects of the same type to perform actions, if the object in one of the rules has a type in program_G, then the object in the other rule cannot have type networkManager_ssh_t. Similarly, if the object in

---

[1] In the Coq implementation, we do not have a separate definition for $\lesssim$. Instead, we express it directly when needed using list operators.

one rule has type `networkManager_ssh_t`, then the object in the other rule cannot have a type that is a subset of `program_G`. The constraint $CSTE\_SoD$ is applicable to all queries whose $\mathbb{C}$ and $\mathbb{P}$ are `File` and `Read`, respectively.

Using the above example constraint, and the example rule in Sect. 2.1, we can define the example policy below, where both the rule component and the constraint component are lists of length 1.

```
Definition TEPLCY_example : TEPLCY:= TEPolicy ([R_A], [CSTE_SoD]).
```

3.2   Evaluating Queries Against Constraints

The function $CSTE\_EvalTE$ implemented in Listing 2 evaluates a query against a constraint. It takes a single constraint, a query, and a list of rules (all the rules in the rule component of a policy) as arguments. The rules argument can be used to extract access information required for expressing security goals encoded in predicates.

```
Definition CSTE_EvalTE
  (constraint_rule:CSTE) (Q_to_constr:Q) (listR:list R) : DCS:=
match constraint_rule with
 |Constraint (cstrn_C, cstrn_P,cstrn_T_arg1,cstrn_T_arg2,
                cstrn_listT,cstrn_PRDT) ⇒
   match Q_to_constr with
     |(Q_srcT, Q_dstT, Q_C, Q_P) ⇒
        if (Nat.eqb Q_C cstrn_C && Nat.eqb Q_P cstrn_P) then
        match (cstrn_PRDT listR cstrn_listT cstrn_C cstrn_P
               Q_srcT Q_dstT cstrn_T_arg1 cstrn_T_arg2) with
         |true ⇒ Permitted
         |false ⇒ UnKnown
        end
        else NotPermitted
   end
end.
```

Listing 2. Evaluation of a constraint

In order to check whether or not the constraint is applicable to the query, the object class and permitted action components are compared and must be the same. If applicable, the constraint predicate is checked. Note that the arguments passed to `cstrn_PRDT` include the list of rules as well as all the other components of the constraint and query, except the two that are used to check the applicability of the constraint. If the evaluation of the predicate returns `true`, then the decision is `Permitted`. Otherwise, the decision is `UnKnown`. Note that if the constraint is not applicable, the default value `NotPermitted` is returned.

A query must be evaluated against all the rules and constraints in a policy. We omit the other three functions that are defined to complete this task, and just remark that the main function that calls the others is called $TEPLCY\_EvalTE$. We note that they are implemented so that $CSTE\_EvalTE$ is always passed the complete list of rules in a policy as its third argument, and thus the complete list of rules is always passed as input (as the first argument) when the constraint predicate is called inside $\mathbb{CSTE}\_$EvalTE.

It is often useful to view various kinds of information in the list of rules as *sets* of values, and so we provide several general operators that support this view, such as *intersection, union,* as well as *set comparison* operators such as *subset* and *set equality.* Here, we follow the general approach in [11], where it is shown that such operators form a suitable formalism for expressing security conditions and goals formulated as constraints. Those that are useful for our separation of duty example include *selector functions,* which retrieve various kinds of information from a list of rules, and *operator functions,* which apply certain operations on the results of selector functions along with other arguments of the predicate. We include a selector function called $\texttt{list}\mathbb{R}\texttt{Search\_subject}\mathbb{T}\texttt{s}$, which receives a list of rules and an object type as inputs and returns a list of types, which we view here as a set. This function searches all the Allow rules of input rules to find all types of subjects that are allowed to access (i.e., perform any kind of action on) objects of the type specified by the object type argument. The result is a list (set) containing these subject types. The definition of Prd_SoD uses this function along with operator functions called IntersectionList, which returns the set of common elements of two lists, and $\texttt{is\_emptylist}\mathbb{T}$, which checks whether or not a list of types is empty. The Prd_SoD predicate is defined in Listing 3.

```
Fixpoint Prd_SoD (listℝ:list ℝ) (ListT:list T) (sClass:ℂ) (perm:ℙ)
                 (QSrcT:T) (QDesT:T) (PRDTsrcT:T) (PRDTDesT:T) : 𝔹:=
  if (TSubset QSrcT PRDTsrcT && TSubset QDesT PRDTDesT)
  then is_emptylistT ( IntersectionList
                          (listℝSearch_subjectTs listℝ PRDTsrcT)
                          (listℝSearch_subjectTs listℝ PRDTDesT))
  else true.
```

Listing 3. The predicate Prd_SoD

Returning to our example constraint $\mathbb{CSTE}\_$SoD in Sect. 3.1, we have now completed the definition of its last component, and thus we can now see how a query is evaluated against this constraint by $\mathbb{CSTE}\_$EvalTE in Listing 2. When Prd_SoD is called inside $\mathbb{CSTE}\_$EvalTE, it first checks whether or not the predicate is applicable to the query, by checking that the subject and object types of the query (arguments $\mathbb{Q}\texttt{Src}\mathbb{T}$ and $\mathbb{Q}\texttt{Des}\mathbb{T}$) are subsets of the input arguments $\mathbb{PRDT}\texttt{src}\mathbb{T}$ and $\mathbb{PRDT}\texttt{Des}\mathbb{T}$, respectively. The predicate returns true if this condition is false. When the condition is true, it gathers all the types of subjects in rules that act on objects of types mail_t and/or http_t, and gathers all the types of subjects in rules that act on objects of type networkManager_ssh_t, and ensures that there is no overlap. It checks all rules in a policy, which can be seen by the fact that the first argument to Prd_SoD is passed on directly to both calls to $\texttt{list}\mathbb{R}\texttt{Search\_subject}\mathbb{T}\texttt{s}$.

Recall that in the definition of $\mathbb{CSTE}$, a predicate takes eight arguments, but there is of course no requirement that the predicate uses them all. In Prd_SoD, note that the second, third, and fourth arguments are not relevant for expressing separation of duty. The fact that the second argument is not used is why an empty list [] appears as the fifth component of $\mathbb{CSTE}\_$SoD.

We have used Prd_SoD and some other predicates to develop a security policy called TEpla_policy as a case study, which can be found in the Coq code. This example policy has twenty rules and five constraints. All the predicates used there satisfy the conditions on predicates that we now present in the next section.

3.3   Conditions on Predicates

Policy writers have to verify three conditions on predicates using a library of lemmas we provide for this purpose. The conditions express that given two queries related by $<<=$ or two policies related by $\lesssim$, the evaluation of a query against a policy preserves the defined order on decisions $<::$. We describe them briefly here.[2] The first one is about queries (involving the $<<=$ relation) and the other two are about policies (involving the $\lesssim$ relation).

Two of the conditions involve a relation on Boolean values called transition_Verify_Decision that relates the Boolean results of applying a predicate twice with some argument or collection of arguments differing between the two calls.

The first condition is one of the two that uses this relation on Booleans. It is called predicate_query_condition and the specific arguments that differ in the two calls are the query subject and object types. This condition is used in a lemma called predicate_query_condition_implication, which simply states that whenever a predicate $\mathbb{P}$ satisfies predicate_query_condition, then given any two queries $\mathbb{Q}_1$ and $\mathbb{Q}_2$ such that $\mathbb{Q}_1 <<= \mathbb{Q}_2$, a constraint $\mathbb{C}$ whose last argument is $\mathbb{P}$, and any list of rules listR, if d_1 and d_2 are the decisions resulting from evaluating $\mathbb{Q}_1$ and $\mathbb{Q}_2$, respectively, against $\mathbb{C}$ and listR (i.e., applying function $\mathbb{CSTE}$_EvalTE in Listing 2), then $d_1 <:: d_2$.

The second and third conditions involve evaluating a predicate in a constraint on a single query but with two sets of rules (the first argument of the predicate). The second condition simply states that the same result is obtained from applying the predicate on the two lists of rules, whenever the two lists differ only in the order of the rules. This condition is called Predicate_plc_cdn.

The third condition, called Predicate_plc_cdn_Transition, is the other condition that uses the relation transition_Verify_Decision on Booleans. The condition states that given two lists of rules, listR and listR', the transition_Verify_Decision relation holds between the results of applying the predicate to listR and listR ++ listR'. This condition and the second condition are used in a lemma called constraintEvalPropSnd. This lemma states that whenever all the constraints in a given listC of constraints satisfy both conditions, then given a query $\mathbb{Q}$ and two lists of rules listR and listR', if d_1 and d_2 are the decisions resulting from evaluating $\mathbb{Q}$ against listC and the two rule lists listR, and listR ++ listR', respectively, (i.e., applying function listCSTE_EvalTE), then $d_1 <:: d_2$.

---

[2] The lemmas stating that these three conditions hold for our running example Prd_SoD are called qry_condition_SoDpredicate, plc_conditionS_SoDpredicate, and plc_-conditionF_SoDpredicate in the Coq code.

The expressive power of predicates is limited by the conditions that they have to satisfy, as presented in this section. Alternatively, however, we propose two methods to extend the expressive power of predicates. The first is simply to relax the restriction and not require these conditions to be verified, which would allow constraints to violate the ordering on decisions by changing an UnKnown to a Permitted. Allowing this freedom provides policy developers with the same expressive power as the studies that use sets to express security goals, such as [11], which empirically illustrates that practical binary constraints can be expressed by comparisons of two sets. The second method is to replace the above constraints with a structural restriction on policies that requires that the rule component never changes. Such a situation can occur, for example, when different departments in an organization have different security goals, but they all have the same set of rules defined by a central security administrator. With this change, some formal properties we present in Sect. 4 will still hold. This solution eliminates the need for an expert in Coq to verify conditions.

4   Properties of TEpla and Their Formalization

Determinism is one of the important properties of policy languages discussed in [22]. A deterministic language always produces the same decision for the same policies and queries. Recall that the function $\mathbb{TEPLCY\_EvalTE}$ evaluates a query against a policy. The behavior of this function specifies the overall semantics of TEpla. Thus TEpla satisfies determinism simply because evaluation is defined as a function.

4.1   Order Preservation of TEpla Queries

TEpla has in fact been designed so $\mathbb{TEPLCY\_EvalTE}$ is *order-preserving* for the relation $\lesssim$ on policies, $<<=$ on queries, and $<::$ on decisions. This means that $\mathbb{TEPLCY\_EvalTE}$ acts as a *homomorphism* [9] on the posets we defined on $\mathbb{TEPLCY}$, $\mathbb{Q}$, and $\mathbb{DCS}$.

Of particular importance is the preservation of order on decisions with respect to queries: if $q_1 <<= q_2$, then the decisions d_1 and d_2 that result applying function $\mathbb{TEPLCY\_EvalTE}$ on q_1 and q_2, respectively, are in the relation $d_1 <:: d_2$. The $<<=$ relation is defined (see Sect. 2.3) to be as general as possible; it involves only subject and object types, which are elements that queries in any language must have. When policies are large, verifying policies often involves testing a number of queries against the policy. Having an unambiguous ordering facilitates sorting, filtering, and optimizing query evaluation by administrators. This property can be compared to the *safety* property defined in [22].

Theorem Order_Preservation_TEpla (in Listing 4) expresses this order preservation on queries.

Theorem Order_Preservation_TEpla :
   $\forall$ (listR:list $\mathbb{R}$) (listCSTE:list $\mathbb{CSTE}$) (q q' : $\mathbb{Q}$),
       (q $<<=$ q') $\wedge$ const_imp_prd_List listCSTE $\rightarrow$

$((TEPLCY\_EvalTE (TEPLCY (listR, listCSTE)) q) <::$
$(TEPLCY\_EvalTE (TEPLCY (listR, listCSTE)) q')) = true.$

Listing 4. Order preservation of decisions with respect to queries

The const_imp_prd_List predicate in this theorem expresses that all the predicates of the input list of constraints listCSTE satisfy the first condition from Sect. 3.3 (predicate_query_condition).

4.2   Non-decreasing Property of TEpla Policies

It is common to add new policy statements as new regulations arise. The next property states that when adding new rules, policies do not change their decisions in the reverse direction of the order on decisions (i.e., $<::$). When adding new rules, changing decisions, for example, from *Permitted* to *NotPermitted*, is impossible. Thus granted requests will never be revoked. Revoking access from already granted requests is problematic because once the information has been revealed, there is no way to reverse the effect of revealing this information. This property is aligned with *monotonicity* defined in [22]. We state and prove the property in Listing 5, which expresses that TEpla is *non-decreasing*.

Theorem Non_Decreasing_TEpla :
$\forall$ (Pol_list: list TEPLCY) (Single_pol:TEPLCY) (q:Q) (d d': DCS),
  validCnstrtListPolicy Pol_list $\wedge$ validConstrt Single_pol $\rightarrow$
  (TEPLCY_EvalTE ($\oplus$ (Pol_list)) q) = d $\rightarrow$
  (TEPLCY_EvalTE ($\oplus$ (Single_pol::Pol_list)) q) = d' $\rightarrow$
  (d $<::$ d') = true.

Listing 5. Theorem Non_Decreasing_TEpla

This theorem states that adding a policy Single_pol, to any list of policies Pol_list can change the decisions only according to the order relation $<::$ on decisions. The predicate validCnstrt expresses that the constraints in Single_pol satisfy the second and third conditions from Sect. 3.3 (Predicate_pl_cdn and Predicate_plc_cdn_Transition). The predicate validCnstrtListPolicy applies this check to every policy in Pol_list. The $\oplus$ operator extracts the rule lists of all the policies in its argument list of policies and combines them into one list, and similarly for constraints, forming a single policy from these rules and constraints. Note that in this theorem, ($\oplus$ Pol_list) $\lesssim$ ($\oplus$ (Single_pol :: Pol_list)).

4.3   Independent Composition of TEpla Policies

It is important to be able to analyze the behavior of access control policies based on their components or *sub-policies*, as the decisions for the combined policies can be determined from the decisions of included policies. Similar to *independent composition* in [22], we codify the following property of TEpla.

Theorem Independent_Composition :
∀ (PLCY_DCS_pair : list (TEPLCY * DCS)) (q : ℚ) (dstar : DCS),
 Foreach q (map fst PLCY_DCS_pair) (map snd PLCY_DCS_pair) ∧
 (TEPLCY_EvalTE (⊕ (map fst PLCY_DCS_pair)) q) = dstar →
 (maximum (map snd PLCY_DCS_pair) <:: dstar) = true.

Listing 6. Theorem Independent_Composition

In this statement, PLCY_DCS_pair is a list of policies and a list of decisions of the same length such that for each index i into these lists, if p_i and d_i are the policy and decision at this index, respectively, then (p_i, d_i) is an *evaluation pair on q*, which means that $(\text{TEPLCY\_EvalTE } p_i \ q) = d_i$, i.e., that d_i is the decision returned from evaluating policy p_i on query q. Although we do not show its definition, the Foreach predicate is defined to express this property. It also expresses that all the constraints in each policy satisfy the second and third conditions from Sect. 3.3 (Predicate_pl_cdn and Predicate_plc_cdn_Transition). The independent composition theorem states that whenever a pair of lists satisfies this property, then the decision obtained by evaluating the combined policy on q is the maximum of the decisions resulting from evaluating each policy independently. The function maximum takes a list of decisions and returns the maximum according to the binary relation $<::$.

5 Conclusion

We have presented the infrastructure of the TEpla Type Enforcement policy language, and formally verified some of its important properties in Coq. TEpla, with formal semantics and verified properties, is an essential step toward developing certifiably correct policy-related tools for Type Enforcement policies.

The properties that we have considered here, namely *determinism, order preservation, independent composition*, and *non-decreasing*, analyze the behavior of the language by defining different ordering relations on policies, queries, and decisions. These ordering relations enabled us to evaluate how language decisions react to changes in policies and queries.

Moreover, we provide the language constructs (in particular, the integration of user-defined predicates) for allowing security administrators to encode different security goals in policies. This makes the language flexible because policy developers are not limited to built-in conditions to express their intended predicates.

In related work, ACCPL (A Certified Core Policy Language) [19] represents some preliminary work using our approach, i.e., building in formal semantics from the start, but in the domain of web services and digital resources, with some very basic properties proved, which include determinism, but not the other properties considered here. In other work, a variety of other studies have included the formalization of various aspects of access control policies using different and sometimes quite complex logics and algorithms, e.g.., [1, 2, 4, 23]. In our approach, we start with a simple language, and some simple notions of orderings and relations on sets, and show that it is possible to express fairly complex access control requirements. We were inspired, for example, by the work in [11], which shows that complex access control constraints such as separation of duty [12, 20] can be expressed using set operators. Additionally, although we began with the particular domain of policies for operating systems, one of our goals is to develop general ideas that can be adapted to other domains such as the web and distributed platforms. Future work will include exploring such extensions. Eventually, we plan to use the program extraction feature of Coq to generate a certified program from the algorithms used to express TEpla semantics, similar to what was done in [5] for firewall policy evaluation.

With regard to work on SELinux in particular, different studies have been carried out that put forward some possible tools for helping policy writers write policies that are more easily understood and reasoned about. Languages such as Lobster [10], Seng [14], Please [17], and CDSFramework [18] are intended to enhance the SELinux policy language by providing easier syntax and more language features, such as defining object-oriented policy syntax, for example. Despite their attempt to help users to specify SELinux security policies, as analyzed in [8], these languages give rise to limited results that cannot be verified due to a lack of formalized definition of semantics and language behavior, which results in potentially contradictory interpretations and precludes correct reasoning. These issues contribute to the ongoing development of numerous policy-related tools that try to model SELinux policies without proving the correctness of the results and analyses, as each tool attempts to cover more features rather than verifying their properties and results.

Our future work will also include addressing some of the current limitations of the language, including extending the kinds of constraints provided, as well as designing and developing certified tools for policy-related tasks such as automating various kind of policy analyses. We expect to be able to reuse many definitions and lemmas of the current Coq development.

References

1. Abadi, M., Burrows, M., Lampson, B., Plotkin, G.: A calculus for access control in distributed systems. In: Feigenbaum, J. (ed.) CRYPTO 1991. LNCS, vol. 576, pp. 1–23. Springer, Heidelberg (1992). https://doi.org/10.1007/3-540-46766-1_1
2. Archer, M., Leonard, E.I., Pradella, M.: Analyzing security-enhanced Linux policy specifications. In: Proceedings POLICY 2003, IEEE 4th International Workshop on Policies for Distributed Systems and Networks, pp. 158–169 (2003)
3. Bertot, Y., Castéran, P.: Interactive Theorem Proving and Program Development: Coq'Art: The Calculus of Inductive Constructions. Springer, Heidelberg (2004). https://doi.org/10.1007/978-3-662-07964-5
4. Brucker, A.D., Brügger, L., Wolff, B.: The unified policy framework (UPF). Archive of Formal Proofs (2014). https://www.isa-afp.org/entries/UPF.html
5. Capretta, V., Stepien, B., Felty, A., Matwin, S.: Formal correctness of conflict detection for firewalls. In: ACM Workshop on Formal Methods in Security Engineering (FMSE), pp. 22–30 (2007)
6. Chlipala, A.: Certified Programming with Dependent Types: A Pragmatic Introduction to the Coq Proof Assistant. The MIT Press, Cambridge (2019). https://mitpress.mit.edu/books/certified-programming-dependent-types

7. Eaman, A.: TEpla: a certified type enforcement access control policy language. Ph.D. thesis, University of Ottawa (2019). https://ruor.uottawa.ca/handle/10393/39876

8. Eaman, A., Sistany, B., Felty, A.: Review of existing analysis tools for SELinux security policies: challenges and a proposed solution. In: 7th International Multidisciplinary Conference on e-Technologies (MCETECH), pp. 116–135 (2017)

9. Harzheim, E.: Ordered Sets. Springer, Boston (2005). https://doi.org/10.1007/b104891

10. Hurd, J., Carlsson, M., Finne, S., Letner, B., Stanley, J., White, P.: Policy DSL: high-level specifications of information flows for security policies. In: High Confidence Software and Systems (HCSS) (2009)

11. Jaeger, T., Tidswell, J.: Practical safety in flexible access control models. ACM Trans. Inf. Syst. Secur. (TISSEC) **4**, 158–190 (2001)

12. Jaeger, T., Zhang, X., Edwards, A.: Policy management using access control spaces. ACM Trans. Inf. Syst. Secur. (TISSEC) **6**(3), 327–364 (2003)

13. Jang, M., Messier, R.: Security Strategies in Linux Platforms and Applications, 2nd edn. Jones and Bartlett Publishers Inc., Burlington (2015)

14. Kuliniewicz, P.: SENG: an enhanced policy language for SELinux (2006). Presented at Security Enhanced Linux Symposium. http://selinuxsymposium.org/2006/papers/09-SENG.pdf

15. Mayer, F., Caplan, D., MacMillan, K.: SELinux by Example: Using Security Enhanced Linux. Prentice Hall, Upper Saddle River (2006)

16. National Security Agency: Security-Enhanced Linux (2019). https://www.nsa.gov/what-we-do/research/selinux/

17. Quigley, D.P.: PLEASE: policy language for easy administration of SELinux. Master's thesis, Stony Brook University (2007). Technical report FSL-07-02. www.fsl.cs.sunysb.edu/docs/dquigley-msthesis/dquigley-msthesis.pdf

18. Sellers, C., Athey, J., Shimko, S., Mayer, F., MacMillan, K., Wilson, A.: Experiences implementing a higher-level policy language for SELinux (2006). Presented at Security Enhanced Linux Symposium. http://selinuxsymposium.org/2006/papers/08-higher-level-experience.pdf

19. Sistany, B., Felty, A.: A certified core policy language. In: 15th Annual International Conference on Privacy, Security and Trust (PST), pp. 391–393 (2017)

20. Stallings, W., Brown, L.: Computer Security, Principles and Practices. Pearson Education, London (2018)

21. The Coq Development Team: The Coq Reference Manual: Release. INRIA (2020). https://coq.inria.fr/refman/

22. Tschantz, M.C., Krishnamurthi, S.: Towards reasonability properties for access-control policy languages. In: 11th ACM Symposium on Access Control Models and Technologies (SACMAT), pp. 160–169 (2006)

23. Wu, C., Zhang, X., Urban, C.: A formal model and correctness proof for an access control policy framework. In: Gonthier, G., Norrish, M. (eds.) CPP 2013. LNCS, vol. 8307, pp. 292–307. Springer, Cham (2013). https://doi.org/10.1007/978-3-319-03545-1_19

Multi-path Coverage of All Final States for Model-Based Testing Theory Using Spark In-memory Design

Wilfried Yves Hamilton Adoni[1], Moez Krichen[2,3(✉)], Tarik Nahhal[1], and Abdeltif Elbyed[1]

[1] Hassan II University of Casablanca, Casablanca, Morocco
adoniwilfried@gmail.com, t.nahhal@fsac.ac.ma, abdeltif.elbyed@univh2c.ma
[2] FCSIT, Albaha University, Albaha, Saudi Arabia
[3] ReDCAD Laboratory, University of Sfax, Sfax, Tunisia
moez.krichen@redcad.org

Abstract. This paper deals with an efficient and robust distributed framework for finite state machine coverage in the field model based testing theory. All final states coverage in large-scale automaton is inherently computing-intensive and memory exhausting with impractical time complexity because of an explosion of the number of states. Thus, it is important to propose a faster solution that reduces the time complexity by exploiting big data concept based on Spark RDD computation. To cope with this situation, we propose a parallel and distributed approach based on Spark in-memory design which exploits A* algorithm for optimal coverage. The experiments performed on multi-node cluster prove that the proposed framework achieves significant gain of the computation time.

Keywords: Model-based testing · Coverage · Big data · Big graphs · Apache spark · Apache hadoop · Parallel and distributed computing

1 Introduction

Since we moved into the era of new technologies such as IoT, Cloud service, artificial intelligence, big data and 5G, the systems test attracts more interest for industrial and scientific communities. The reliability test of systems is based on a large number of compliance actions. The best known testing technique where the behavior of the system under test is checked against a prediction model is the Model-based testing [12,14].

The basic idea of this technique is to create a set of functional tests based on some requirements, input sequences and actions. Once created, this model can be applied for testing both software and hardware. In this context, several research projects have been conducted in the field of system testing [1,7,15,17].

Moez Krichen et al. [16] proposed a conformance testing for real-time systems. The proposed framework is designed for non-deterministic timed

© Springer Nature Switzerland AG 2020
B. Ben Hedia et al. (Eds.): VECoS 2020, LNCS 12519, pp. 195–204, 2020.
https://doi.org/10.1007/978-3-030-65955-4_14

automata. This allows the user to define essential features and assumptions based on variants of the tested model. Moreover, they proposed a dynamic testing for real-time applications [7], it is dynamic in the sense that it provides an automatic generation of observers of time automata.

The checking of advanced models emanated from concurrent, dynamic or real-time systems is challenging and has been the focus of several decades of research. Indeed, the main problem that affects checking of sophisticated software or hardware is the state explosion issue [18] because of large amount of stored and shared information between many components of the system under test. To cope with this situation, the most efficient techniques to cope with this problem are based on symbolic bounded model checking, partial order reduction and symbolic model [11]. These breakthrough approach have enabled the testing of complex systems with large amount of states. Nevertheless, these techniques present some drawbacks. In fact, symbolic techniques disable the computation for specific states like state probabilities of Petri nets [6].

Another idea consists of using a computer cluster to achieve verification of a system represented by very complex model. These computers are connected under a master-slave architecture. Each computer checks a part of the model in parallel. Thus the analysis of very complex systems is done in a fast way with a quasi-optimal verification of the system.

Although big data computation techniques have so far been poorly explored for verifying software systems. We believe that the main issues to be tackled in conformance testing can benefit a lot from big data concepts. In this paper, we are interested to the problem of multipath coverage of all final states in large-scale automaton with state explosion. This task is very expensive with impracticable computation time.

The major contributions of the paper are summarized below.

- Introduction of big data (big-graph) concepts in the area of model based testing.
- An optimized coverage technique for generating test cases based on the extended version of MapReduce-A* [3,4].
- A framework designed for large-scale finite automaton coverage.

The remainder of this paper is organized as follows. Section 2 gives background about Model-based testing and big data framework. In Sect. 3, we introduce the problem formulation. Section 4 presents the design of the proposed framework. Next, we expose our experiment results in Sect. 5 and conclude this work in Sect. 6.

2   Background

2.1   Model-Based Testing

Model-based testing (MBT) is a variant of test techniques that is based on explicit behavior models, describing expected behaviors of the system under test (SUT), or the behavior of its environment, constructed from functional requirements. The MBT is an evolving approach that aims to automatically generate from one of these models, test cases to play on the SUT. The interest of adopting model-based testing in projects is to improve the detection of SUT bugs and improve software quality. Moreover, Model-based testing can reduce the time and effort spent on testing since the test scripts are derived automatically from the model of the SUT.

2.2   Apache Spark

Hadoop is an Apache open source framework allowing parallel and distributed processing of large amoung of data sets. It is based on Google File System and works with multi-nodes connected under a master-slaves architecture. Hadoop consists of two main components: HDFS [9] and MapReduce [8,19]. HDFS component is the Hadoop Distributed File System, it is designed to store very big dataset. The dataset is split into many chunk files and distributed across the cluster nodes in order to prevent fault-tolerance. The MapReduce component is designed for intensive computation on large data file in parallel and distributed way. The MapReduce programming model consists of two main functions: $map()$ and $reduce()$ functions.

Spark [20] is an Apache open source framework allowing parallel and distributed processing of large amounts of data sets. Spark is designed to face some limitations of Hadoop [19] such as iterative processing, in-memory computing and near real-time processing. Spark works with multi-nodes cluster connected under a master-slaves architecture, it provides a Resilient Distributed Datasets (RDDs) that computing operations in memory. RDD is a immutable object which allows transformation and action operations to do parallel processing on multi-node cluster.

3   Problem of Coverage

Let $A = (Q, \sum, s_0, F, \delta)$ be a determinist finite automaton (DFA) such as:

- Q is a finite set of states.
- δ is a finite set of inputs.
- $s_0 \in Q$ is the initial state.
- $F \subseteq Q$ is a set of final states.
- $\delta : Q \times \sum \rightarrow Q$ is the transition function.

An automaton $A_i = (Q_i, \sum_i, s_0^i, F_i, \delta)$ with the initial state changed to s_0^i is a sub-automaton of A if $Q_i \subseteq Q$, $\sum_i \subseteq \sum$ and $F_i \subseteq F$. In MBT, the reliability test generally passes through the coverage of all states reachable by the system. This task consists of starting from the initial state of the system to find all the optimal or quasi-optimal paths that reach the set of final states of the system. However, in an automaton with state explosion, the search of all paths covering the final states of the system is very expensive.

Current coverage techniques [1,7,16,17] are not suitable for large-scale automaton. One technique that has attracted attention is the divide-to-conquer which consists of partitioning the automata into a sub-automata and then finding the intermediate coverage. However, the quality of the result of this technique depends on the partitioning technique adopted. There are parsing algorithms that allow for quick coverage while others are well suited for optimal coverage [2,5].

4   Proposed Spark Coverage Technique

The framework we propose is an adapted version of MRA* framework [3,4]. The cluster is setup so that the RDD is used for partitioned and distributed the automaton across the cluster. Then, we process the big automaton data using Spark in-memory computation. As shown in Fig. 1, the conceptual model of our framework works under a master-slaves architecture and it consists of three mains stages:

– Input stage which consists in the automaton partition.
– Map stage which consists in computing of intermediate states coverage.
– Reduce stage which consists in merging of all obtained coverages.

First of all, the coverage job submitted from the master is split into two sets of mapper and reducer tasks and distributed across the Spark cluster. The coverage job consists of finding all paths reaching the set of final states $s \in F$. The master manages the coverage job, assigns each map and reduces task to all workers. The coverage job is synchronized so that the output of the previous stage is taken as the input of the next stage. It is important to note that the input stage is the pre-processing stage and intermediate coverages are performed into map and reduce stages. Thus the total time t_σ required to find all paths that cover the set of final states $s_i \in F$ is calculated as follows:

$$t_\sigma = t_{map} + t_{red} \tag{1}$$

where t_{map} and t_{red} are respectively the time passed in the map and reduce stage.

4.1   Input Stage: Automaton Partition

First, we need to load the automaton graph A as a parallelized RDD object. Then, A is partitioned into k sub-automata. A good partitioning strategy allows to schedule all tasks so that the load balancing among all nodes in the cluster is balanced. Spark provides different types of partitioning scheme. But in our work, we have used the "Edge-partition" technique [10] which equally distributes the automaton among all the workers in the cluster.

As shown in Fig. 2 the automaton partition is defined as subsets of transitions such that each transition edge belongs to exactly one partition. The states

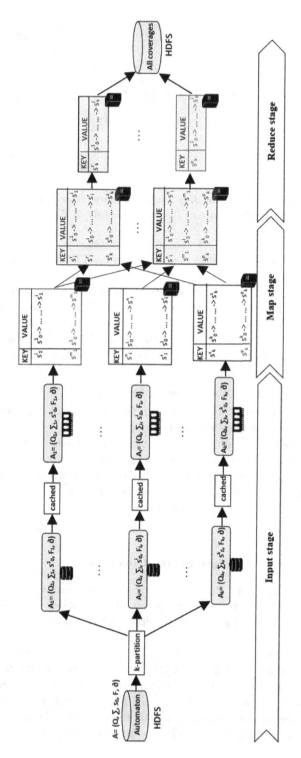

Fig. 1. Overview of our framework

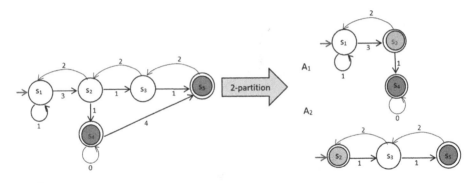

Fig. 2. Automaton partitioning under $k = 2$ sub-automata. State s_2 is a frontier (intermediate) state, it serves as start state of A_2 and final state of A_1

which appear in two sub-automata A_1 and A_2 are used for the communications channels [10].

From the initial state $s_0 \in Q$, we create the first sub-automaton, the intermediate states s_i located on each sub-automaton $A_i \subseteq A$ are used as initial and final states. The initial state s_i of the ith sub-automaton A_i is the final state of the $(i-1)^{th}$ sub-automaton A_{i-1} so on until the last k^{th} sub-automaton A_k. To avoid overload of the cluster and maintain load balancing, we assume that the number k of partitions is:

$$k = N_{node} \times N_{core} \tag{2}$$

where N_{node} is the number of workers within the cluster and N_{core} the number of core processors used by each worker node.

Finally, when the k-partitioning of the automaton is done, we create an object $RDD(A_i)$ of each sub-automaton A_i. This means that the input dataset remains in memory, but it can also be stored on the local disk if there is not enough RAM. The idea of creating RDD object of each sub-automaton is beneficial since at any time, we can correctly re-balance the number of states or transitions in each sub-automaton.

4.2   Map Stage: Intermediate States Coverage

The map stage consists of computing for a given sub-automaton object $RDD(A_i)$, all cover paths that reach the set of final states $s_i \in F_i$. For the cover computation, we used an adapted version of A* mapper algorithm proposed by [3,4].

This procedure takes as input an RDD object of a sub-automaton A_i, and maps it into key-value pairs of states RDDs and transitions RDDs. Then, it initializes the initial state s_0^i and the two main queues: the *open-list* O that contains all candidates states and the *close-list* S that contains promising states. The next step, consists of exploring in depth and selecting the most promising states until the final state found. Finally, the last step consists of extracting the coverage paths from the close-list S. Afterward, the result is submitted to the reduce stage.

The time complexity for all intermediate states coverage depends on the total number of worker nodes N_{node} within the cluster and the number of core processors N_{core} per node. To cover an intermediate final state s_i into the sub-automaton A_i, the time $m_{i,j,k}$ taken by the kth mapper assigned to the jth core processor of the ith worker is about $O(|Q_i| + |T_i|) \log(|Q_i|)$ where T_i is the set of transitions within A_i. The time $\Delta m_{i,j,k}$ taken to reach all final states during the map stage is calculated as follows:

$$t_{map} = \max(\sum_{i=1}^{N_{node}} \sum_{j=1}^{N_{map}} \sum_{k=1}^{|F_i|} \Delta m_{i,j,k}) \tag{3}$$

4.3 Reduce Stage: Merging All States Coverage

In the final stage, the intermediate coverages from the map stage are aggregated and merged based on all final states that share the same key. This is performed by an adapted version of A* reducer procedure presented in [3,4].

It takes as input the coverage path of each state, then group and merge all paths to build the final coverage path. The coverage $\sigma_{i,j}$ is extended to the coverage $\sigma_{j,k}$ if the intermediate state j is the final state reached by $\sigma_{i,j}$ and the start state of $\sigma_{j,k}$, so on until the last coverage.

When one worker node completes its tasks, it waits until the other reducers achieve their tasks before sending the result to the master. Then the time $\Delta r_{i,j,k}$ taken to merge all intermediate coverages from state s_0 to all final state $s_i \in F$ of the original automaton A is:

$$t_{red} = \max(\sum_{i=1}^{N_{node}} \sum_{j=1}^{N_{map}} \sum_{k=1}^{|F|} \Delta r_{i,j,k}) \tag{4}$$

5   Experiment Results

The experiments have been achieved on a Spark cluster composed of one master node and three worker nodes. The cluster configuration is given below in Table 1.

Table 1. Cluster setup

Software	Hardware
Linux Ubuntu 18.01.1.0 amd64	Processor Intel Core i5-2410M
Apache Spark 2.4.0	4 CPU @ 2.3 GHz
JDK 1.8	16 Gb RAM

We have used a dataset generated from the mutants-equiv-eval dataset[1]. This dataset corresponds to the finite behavioural models of equivalent mutant problem. Table 2 shows the number of states and transitions in different finite behavioural models, each automaton contains at least 5% of final states to cover.

Table 2. Mutant equivalence dataset

Automaton	States	Degree	Transitions	Type
ageRR	30,000	6	180,000	Directed
ageRRN	60,000	5	300,000	Directed
claroline	500,000	4	2,000,000	Directed
cpterminal	1,000,000	6	6,000,000	Directed
elsaRR	1,500,000	6	9,000,000	Directed

Table 3 shows the execution time between sequential and parallel approaches using single Spark node, the computation is parallelized on 4 cores processor. First, we remark that the sequential approach takes impracticable time to achieve the coverage of all final states.

On the other hand, by using our approach, the computational time is much faster and reduces significantly the execution time. For example, the task of finding all coverage paths within the claroline automaton takes on average 6 h with sequential approach, it is computationally expensive. In contrast, by using our approach we find all coverage paths in 25 min on disk computing versus 3 min in-memory computing.

Table 3. Runtime of sequential coverage versus Spark coverage using one node

	Sequential	Spark on disk	Spark in-memory
ageRR	26 min	2 min	9.58 s
ageRRN	1 h	4 min	26.85 s
claroline	6.17 h	25 min	2.63 min
cpterminal	11.24 h	45.3 min	4.7 min
elsaRR	21 h	1h24 min	8.17 min

Figure 3 shows the performance comparison of increasing workers on the computational time. In this case, the computation is distributed across the cluster and also parallelized on the 4 cores of each worker. We remark that the runtime varies strongly and is improved while extending the cluster from 1 to 4 workers. Adding new Spark worker reduces the computation time, however when we

---

[1] https://projects.info.unamur.be/vibes/mutants-equiv.html.

move from a very large dataset to a small automaton, the computational time is not affected by the addition of new Spark node into the cluster. This leads us to consider that there exists an optimal number of nodes that satisfies the full coverage of a given finite automaton.

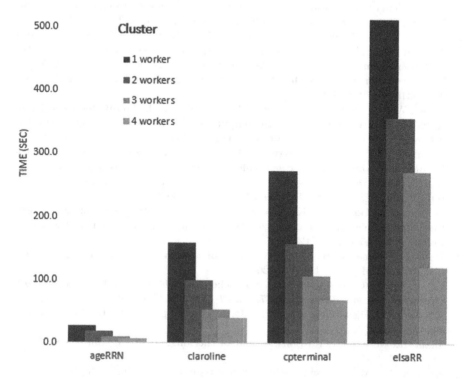

Fig. 3. Impact of number of workers on the computation time

6   Conclusion and Perspectives

In this paper, we have proposed a parallel and distributed framework for large-scale automaton coverage. The time complexity decreases from exponential to linear time. The experimental results prove that our approach is faster and works well with very large automaton. But our approach presents some limitations: 1) the path optimality depends on the partitioning strategy and the number of sub-automatons, see [4] for more details; 2) the computation is often memory expensive. For future work, we are interested in studying the impact of automaton partitioning on the time complexity and propose an extended version of the framework for the coverage of timed-automaton and distributed systems [13].

References

1. Adamatzky, A.: Computation of shortest path in cellular automata. Math. Comput. Modell. **23**(4), 105–113 (1996)
2. Adoni, H.W.Y., Nahhal, T., Krichen, M., Aghezzaf, B., Elbyed, A.: A survey of current challenges in partitioning and processing of graph-structured data in parallel and distributed systems. Distributed and Parallel Databases 1–36 (2019)
3. Adoni, W.Y.H., Nahhal, T., Aghezzaf, B., Elbyed, A.: MRA*: parallel and distributed path in large-scale graph using MapReduce-A* based approach. In: Sabir, E., García Armada, A., Ghogho, M., Debbah, M. (eds.) Ubiquitous Networking. LNCS, vol. 10542, pp. 390–401. Springer, Cham (2017). https://doi.org/10.1007/978-3-319-68179-5_34
4. Adoni, W.Y.H., Nahhal, T., Aghezzaf, B., Elbyed, A.: The MapReduce-based approach to improve the shortest path computation in large-scale road networks: the case of A* algorithm. J. Big Data **5**(1), 1–24 (2018)
5. Adoni, W.Y.H., Nahhal, T., Krichen, M., El byed, A., Assayad, I.: DHPV: a distributed algorithm for large-scale graph partitioning. J. Big Data **7**(1), 1–25 (2020)
6. Bell, A., Haverkort, B.R.: Sequential and distributed model checking of Petri nets. Int. J. Softw. Tools Technol. Transf. **7**(1), 43–60 (2004)
7. Bensalem, S., Bozga, M., Krichen, M., Tripakis, S.: Testing conformance of real-time applications by automatic generation of observers. Electron. Notes Theor. Comput. Sci. **113**, 23–43 (2005)
8. Dean, J., Ghemawat, S.: MapReduce: simplified data processing on large clusters. Commun. ACM **51**(1), 107–113 (2008)
9. Ghemawat, S., Gobioff, H., Leung, S.T.: The Google file system. In: ACM SIGOPS Operating Systems Review, vol. 37, pp. 29–43. ACM, New York (2003)
10. Guerrieri, A.: Distributed computing for large-scale graphs. Ph.D. thesis, University of Trento (2015)
11. Jhala, R., Majumdar, R.: Software model checking. ACM Comput. Surv. **41**(4), 21:1–21:54 (2009)
12. Krichen, M.: Model-based testing for real-time systems. Ph.D. thesis, Universit Joseph Fourier, December 2007
13. Krichen, M.: A formal framework for black-box conformance testing of distributed real-time systems. IJCCBS **3**(1/2), 26–43 (2012)
14. Krichen, M.: Contributions to model-based testing of dynamic and distributed real-time systems. Ph.D. thesis, ENIS, Sfax, Tunisia (2018)
15. Krichen, M., Tripakis, S.: Black-box conformance testing for real-time systems. In: Graf, S., Mounier, L. (eds.) SPIN 2004. LNCS, vol. 2989, pp. 109–126. Springer, Heidelberg (2004). https://doi.org/10.1007/978-3-540-24732-6_8
16. Krichen, M., Tripakis, S.: Conformance testing for real-time systems. Form. Methods Syst. Des. **34**(3), 238–304 (2009)
17. Maâlej, A.J., Lahami, M., Krichen, M., Jmaïel, M.: Distributed and resource-aware load testing of WS-BPEL compositions, pp. 29–38, February 2019
18. Rafe, V., Rahmani, M., Rashidi, K.: A survey on coping with the state space explosion problem in model checking, 6 (2013)
19. Vavilapalli, V.K., et al.: Apache hadoop YARN: yet another resource negotiator, pp. 1–16. ACM Press (2013)
20. Zaharia, M., et al.: Apache spark: a unified engine for big data processing. Commun. ACM **59**(11), 56–65 (2016)

Artificial Intelligence and Machine Learning

A C-IFGSM Based Adversarial Approach for Deep Learning Based Intrusion Detection

Yingdi Wang, Yixiang Wang, Endong Tong[✉], Wenjia Niu[✉],
and Jiqiang Liu

Beijing Key Laboratory of Security and Privacy in Intelligent Transportation,
Beijing Jiaotong University, Beijing 100044, China
{edtong,niuwj}@bjtu.edu.cn

Abstract. With the rapid development of machine learning algorithms, the security problem has gradually emerged. Most existing algorithms may be attacked by adversarial examples. An adversarial example is a slightly modified input sample that can lead to a false result of machine learning algorithms. This poses a potential security threat for many machine learning-based applications. Especially in the domain of intrusion detection, the intrusion adversarial examples may result in malicious attacks on intrusion detection classifiers. To our knowledge, all previous work only apply the adversarial examples generation methods in the field of image classification, which is not suitable for network traffic datasets. Aiming at generating more similar intrusion adversarial examples, this paper explores a Constraint-Iteration Fast Gradient Sign Method (C-IFGSM) that can adapt to complex network traffic datasets with multiple types of features and multiple relationship among features. Experiments show that the C-IFGSM based adversarial approach can achieve good performance on intrusion adversarial examples.

Keywords: Intrusion detection · Adversarial examples · Deep neural network · Feature constraints

1 Introduction

In recent years, with the increasing development of machine learning algorithms, they have helped humans solve problems in various fields, such as face recognition, robot path planning and so on [14,23]. At the same time, machine learning algorithms have also been increasingly applied to security fields, such as spam detection, malware detection and intrusion detection [1,12]. Machine learning models show superior performance in solving security problems [21]. Especially in the domain of intrusion detection, machine learning models can handle large amounts of network traffic data and adapt to new types of network attacks [2].

However, machine learning algorithms also have vulnerabilities. Researches have shown that an adversarial example will make the machine learning model

© Springer Nature Switzerland AG 2020
B. Ben Hedia et al. (Eds.): VECoS 2020, LNCS 12519, pp. 207–221, 2020.
https://doi.org/10.1007/978-3-030-65955-4_15

misclassified by adding a slight modification to real examples, which cannot be distinguished by humans [5]. With in-depth study of adversarial examples, researchers have found that the threat of adversarial examples exists in many applications, such as face recognition applications [16], speech recognition applications [3] and path planning applications [4]. Moreover, adversarial examples have transferability among neural networks [17], which means that adversarial examples generated for a particular neural network are likely to attack other neural network-based models.

Adversarial examples pose huge potential threats to security applications. Especially for neural network-based intrusion detection, the misclassified of anomaly network traffic by intrusion detection systems may lead to malicious attacks on systems [19]. So it is urgent to do the research on the adversarial example attack on intrusion detections. In recent years, many researchers have tried to generate adversarial examples for intrusion detection [13,20,22]. From these works, we can see that deep learning-based intrusion detection systems do suffer from the attacks by adversarial examples.

However, these existing researches only apply the basic generation methods for image adversarial examples, such as Fast Gradient Sign Method (FGSM) [5] and Jacobian-based Saliency Map Attak (JSMA) [11], without considering the data characteristics of network traffic dataset. For example, Zheng Wang used FGSM to generate adversarial examples for NSL-KDD dataset in his paper [19]. Although FGSM can add a slight noise to the original examples to cause deep neural networks misclassification, it cannot guarantee that the types and relationship of features in adversarial examples are consistent with original examples. Through generating the adversarial examples, the discrete features in original examples may be modified to continuous, and two features with a positive correlation may be added with noise in opposite directions. That is, the adversarial examples generated by basic methods cannot well simulate the network traffic data in the real world.

Aiming at the existing problems of generating adversarial examples for deep learning-based intrusion detection, this paper proposed a C-IFGSM based method which can adapt to complex datasets with multiple types of features and multiple relationship among features. We use NSL-KDD dataset as an example to analyze the type of features and the relationship among features. Based on analysis, we found that images and network traffic datasets have different characteristics. Taking NSL-KDD dataset as an example, it contains discrete features and continuous features, and there exists complex relationships among features, while the pixel of images are continuous and independent values. Therefore, we define the intrusion adversarial examples more specifically in this paper. That is, by making slight modifications to the original dataset, intrusion adversarial examples can not only make the intrusion detection classifiers go wrong, but also keep the feature types and feature relationships unchanged. The experiments on NSL-KDD dataset proved that the C-IFGSM based method can efficiently generate intrusion adversarial examples that can well simulate the network traffic data in the real world.

The rest of the paper is structured as follows: In Sect. 2, we discuss the related works in this field. This is followed in Sect. 3 by details about the analysis of network traffic dataset and FGSM. Section 4 describes the C-FGSM based method for crafting intrusion adversarial examples. Section 5 describes the experiments with NSL-KDD dataset. Finally, Sect. 6 concludes this paper.

2   Related Work

In order to craft intrusion adversarial examples that can satisfy the characteristices of network traffic data, we investigate the existing research work in the field of adversarial examples.

Crafting Images Adversarial Example. Most of prior work on adversarial examples is in the field of image classification. The most common algorithm is FGSM [5], which crafts image adversarial examples by modifying the pixel values along the gradient of loss function. After that, more researchers proposed advanced methods [8,11,17], all of which add continuous perturbation.

Crafting Text Adversarial Example. With the development of text sentiment analysis, crafting text adversarial examples attracted some interest [9,10, 15]. In these work, the methods for crafting text adversarial examples need to consider the discrete features of words and the grammar of sentences.

Crafting Malware Adversarial Example. There also exists adversarial attacks against malware detection [6,7]. In the paper [6], it only modified some specific discrete features to guarantee the malware functionality of adversarial examples. This reminds us that crafting malware adversarial examples needs to consider both the type of features and its functionality.

In summary, it is obvious that the methods of crafting adversarial examples in different applications are restricted by the specific characteristics of dataset. Therefore, in addition to misleading classification results, we also need to ensure the similarity between adversarial examples and original examples when crafting intrusion adversarial examples.

3   Background

In this section, we analyzed the most commonly used standard dataset NSL-KDD, and introduced the basic Iteration Fast Gradient Sign Method (IFGSM) algorithm in detail.

3.1   IFGSM

IFGSM is upgraded from FGSM, which is a common used algorithm for crafting image adversarial examples. The formula can be expressed as follows.

$$X^{adv} = X + \epsilon sign\big(\nabla_x J(X, y_{true})\big) \qquad (1)$$

where ϵ is the magnitude of the perturbation, $\nabla_x J(X, y_{true})$ calculates the gradient of loss function J of original example X and its real label y_{true}, and the *sign* function will get the sign of the gradient. By this formula, the adversarial example X^{adv} will be obtained by moving the original example X one step in the direction of the gradient of loss function J.

IFGSM is applied multiple iterations with small step size [8]. Because the FGSM algorithm only updates the gradient once, which sometimes is not enough to attack successfully. Therefore, we consider using the IFGSM to generate intrusion adversarial examples. The formula of IFGSM is as follows.

$$X_0^{adv} = X, X_{N+1}^{adv} = Clip_{X,\epsilon} X_N^{adv} + \alpha sign\big(\nabla_x J(X_N^{adv}, y_{true})\big) \qquad (2)$$

where X_0^{adv} is the first iteration initialized with the original example X. The $(N+1)_{th}$ iteration adversarial example $X_{(N+1)}^{adv}$ is obtained by adding perturbation on the N_{th} adversarial example X_N^{adv}. In order to ensure that the perturbation within the range of ϵ, there is a relationship between α and the number of iterations T, $\alpha = \epsilon/T$. Through iterations, IFGSM greatly increases the probability of crafting adversarial examples successfully.

3.2   Analysis of NSL-KDD Dataset

Table 1. The features of NSL-KDD dataset

No.	Feature	Type	No.	Feature	Type
1	duration	continuous	23	count	continuous
2	protocol_type	nominal	24	srv_count	continuous
3	service	nominal	25	serror_rate	continuous
4	flag	nominal	26	srv_serror_rate	continuous
5	src_bytes	continuous	27	rerror_rate	continuous
6	dst_bytes	continuous	28	srv_rerror_rate	continuous
7	land	discrete	29	same_srv_rate	continuous
8	wrong_fragment	continuous	30	diff_srv_rate	continuous
9	urgent	continuous	31	srv_diff_host_rate	continuous
10	hot	continuous	32	dst_host_count	continuous
11	num_failed_logins	continuous	33	dst_host_srv_count	continuous
12	logged_in	discrete	34	dst_host_same_srv_rate	continuous
13	num_compromised	continuous	35	dst_host_diff_srv_rate	continuous
14	root_shell	discrete	36	dst_host_same_src_port_rate	continuous
15	su_attempted	discrete	37	dst_host_srv_diff_host_rate	continuous
16	num_root	continuous	38	dst_host_serror_rate	continuous
17	num_file_creations	continuous	39	dst_host_srv_serror_rate	continuous
18	num_shells	continuous	40	dst_host_rerror_rate	continuous
19	num_access_files	continuous	41	dst_host_srv_rerror_rate	continuous
20	num_outbound_cmds	continuous			
21	is_hot_login	discrete			
22	is_guest_login	discrete			

We take NSL-KDD dataset as an example for feature analysis [18]. There are 41 features representing the statistical information of a connection, which can be divided into the following three parts.

Basic features (1–9): contain the basic connection information of data packets.

Content features (10–22): contain some useful payload in data packets.

Traffic features (23–41): show the traffic information presented by the current connection and its nearby connections. Traffic features are calculated based on two statistics method, a 2-second time window (23–31) and a 100-connection window (32–41).

Multi-type Features. As can be seen from Table 1, the NSL-KDD dataset contains 3 nominal features, 6 discrete features, and 32 continuous features. Unlike texts or images, it contains multiple types of features due to the complexity of network traffic. Therefore, when adding noise, it is necessary to consider the various feature types. Aiming at the problem that IFGSM adds continuous noise to all features, this paper proposes a discretizing processing step added to the basic algorithm to adapt to the multi-type features of NSL-KDD dataset.

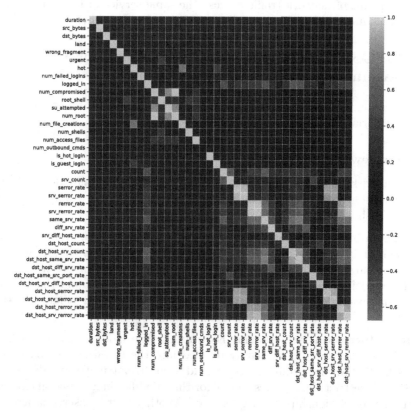

Fig. 1. The Pearson Correlation Coefficient matrix heat map of numeric and discrete features in NSL-KDD dataset.

Multi-relationship Features. In order to find the multi-relationship between features, we made a correlation matrix on NSL-KDD dataset, which was obtained by calculating the Pearson Correlation Coefficient (PCC) between any two numeric features. The formula of PCC is as follows.

$$\rho_{X,Y} = \frac{cov X, Y}{\sigma_X \sigma_Y} = \frac{E((X - \mu_X)(Y - \mu_Y))}{\sigma_X \sigma_Y} \qquad (3)$$

where $\rho_{X,Y}$ is the PCC between the two features, calculated by the ratio of the covariance to the standard deviation between the two features. The larger the absolute value of $\rho_{X,Y}$, the stronger the correlation between the two features.

In order to visualize the relationship between features, we draw a PCC matrix heat map of numeric features as Fig. 1, where the correlation between features mainly exists in the traffic features. Taking "rerror_rate" and "srv_rerror_rate" as an example, the meanings of them are as follows: in the past two seconds, the percentage of "REJ" errors in connections with the same target host as the current connection, and in the past two seconds, the percentage of "REJ" errors in connections with the same service as the current connection. We can make a superficial analysis of the correlation reasons from the feature meaning. That is, a large amount of network traffic attacks the same service of the same host for a period of time, so there is a positive correlation between them.

4   C-IFGSM Based Adversarial Approach for Deep Learning-Based Intrusion Detection

In this section, we will introduce the C-IFGSM based method for crafting intrusion adversarial examples in detail.

4.1   Framework

The framework of the method is mainly divided into two modules, deep learning-based intrusion detection module and intrusion adversarial examples crafting module, as shown in Fig. 2.

Deep Learning-Based Intrusion Detection. It can be seen from the formula that IFGSM is a white-box attack algorithm, which means that it needs to obtain the parameters of the model in advance. So we first need to train a deep learning model for intrusion detection. Moreover, we need to input the intrusion adversarial examples to the intrusion detection classifier to evaluate its attack effect.

Crafting Intrusion Adversarial Examples. The module of crafting intrusion adversarial examples is mainly divided into the following two parts. One part is C-IFGSM algorithm, where we add a constraint matrix to IFGSM. C-IFGSM will restrict the features that algorithm modifies in adversarial examples to satisfy the relationships among features. Another part is discretization. It is the discretization of continues and discrete features in adversarial examples after the C-IFGSM module to satisfy the feature types of original examples.

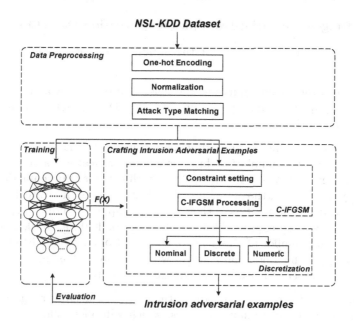

Fig. 2. The framework of C-IFGSM based adversarial approach for deep learning-based intrusion detection

4.2 Deep Learning-Based Intrusion Detection

After comparing several widely used neural network models (which will be mentioned in Sect. 5), we chose the Deep Neural Network (DNN) model that can achieve the highest accuracy. Its network structure is shown in Table 2.

We built a DNN based intrusion detection classifier. The model contains four hidden layers, with 512, 512, 256, 64 nodes in turn. The activation functions are ReLu and LeakyReLu. Because the adversarial examples have transferability in neural network structures, so it is representative to select basic DNN model as the attack target.

Table 2. The accuracy comparison of intrusion detection classifiers

Layers	Active function	Input $\times$ Output
INPUT		Batch size $\times$ 122
Hidden layer 1	ReLu	122×512
Hidden layer 2	LeakyReLu	512×512
Hidden layer 3	LeakyReLu	512×256
Hidden layer 4	LeakyReLu	256×64
OUTPUT	Softmax	64×5

4.3   Crafting Adversarial Examples for Intrusion Detection

The following is an explanation of the sub-modules for crafting adversarial examples.

Data Processing. Before building the classification model and crafting the adversarial examples, the data processing of NSL-KDD dataset is mainly divided into the following three steps.

One-Hot Encoding. One-hot encoding can convert the nominal features to numeric features. The number of bits for a numeric feature is the same with the number of its categories. In this way, for each category of a numeric feature, there is only one bit is 1, the others are all 0. For example, there are 4 categories (value) of feature "protocol_type", they are "tcp, udp, icmp, udp". After one-hot encoding, they will be expressed as "[1, 0, 0, 0], [0, 1, 0, 0], [0, 0, 1, 0], [0, 0, 0, 1]", where the number of this feature changes from 1 to 4. Therefore, after one-hot encoding, the 3 nominal features will be expressed as numeric features, and the total number of features will change from 41 to 122.

Normalization. After one-hot encoding, there will be a large range difference between discrete and continuous features, which will affect the weights of features. Therefore, we normalize the features to make them all in range of $[0, 1]$.

Attack Types Mapping. The NSL-KDD dataset contains 39 types of attack, which belong to 4 categories: Denial of Service (DOS), Probe, Remote to Local (R2L), and User to Root (U2R). We map 39 attack types to 4 categories of attacks and build up an intrusion detection classifier of five-classification (4 attack categories and 1 normal categories).

C-IFGSM. The C-IFGSM method adds *a constraint matrix* C_n to IFGSM method. By setting the constraint matrix C_n, C-IFGSM can restrict the modification of features in original examples. The formula is as follows.

$$X_0^{adv} = X, \ X_{N+1}^{adv} = Clip_{X,\epsilon} X_N^{adv} + \alpha C_n \odot sign\big(\nabla_x J(X_N^{adv}, y_{true})\big) \quad (4)$$

where $C_n \odot sign\big(\nabla_x J(X_N^{adv}, y_{true})\big)$ is the Hadamard product of the constraint matrix and the sign matrix. Hadamard product is a matrix product operation that multiplies the elements in the same position of two matrixes without changing the dimension of the matrix. So the added perturbation depends on the constraint matrix and the sign matrix, which can be expressed as follows:

$$C_n = (c_i)_n, S_n = (s_i)_n \quad (5)$$

By setting the constraint matrix, we can control the noise value added to the original examples. If we do not want to affect the perturbation added to feature x_i, we can set c_i to 1, where the modification will depend on the s_i in sign matrix. If we want to control feature x_j not modified, we can set c_j to 0, where the coefficient α will not work. If we want to control the value of perturbation added to feature x_k, then we can set c_k to a specific value to change the scale of modification. In the following experiment section, we will explore the influence of the selection of modified features on the relationships among features.

Discretization. Because the added perturbation can be a small continues value, so the nominal features and discrete features in the adversarial examples will change to continues feature. For example, the value of nominal feature "protocol_type" may change from "[1, 0, 0, 0]" to "[1.001, 0.001, 0, 0.001]". Obviously, this can not maintain the consistency of muti-type features in original examples and adversarial examples. Therefore, we need to discretize the nominal features and discrete features in the adversarial examples generated by C-IFGSM. The processing of discretization mainly divided into the following two parts.

Discrete Features. For discrete feature x_i whose value can only be 0 or 1, we can discretize it with a boundary of 0.5. The rule is as follows.

$$\text{IF } x_i^{adv} \geq 0.5, \text{ THEN } x_i^{adv} = 1;$$
$$\text{IF } x_i^{adv} < 0.5, \text{ THEN } x_i^{adv} = 0 \tag{6}$$

Nominal Features. For a nominal feature x_j, after one-hot encoding, x_j is convert to $[x_k, x_{k+n}]$, where n is the total number of categories. There is only one feature is assigned 1, and the others are all 0. So we need to discretize the nominal features of adversarial examples as the same, setting the feature with the largest value in $[x_k, x_{k+n}]$ as 1, and setting the rest features as 0. The rule can be expressed as follows.

$$\max(x_i) = 1, i \in [k, k+n]$$
$$x_j = 0, \forall j \in [k, k+n] - i \tag{7}$$

In summary, through the above three processes, we can craft intrusion adversarial examples that satisfy its multiple types of features and the relationships among features. The specific process steps of the algorithm are shown as Algorithm 1.

5   Experiment and Evaluation

We used the method proposed in Sect. 4 to carry out the experiments on NSL-KDD dataset, and evaluated the experiment results in various aspects.

5.1   Experiment Setup

The experiment setup mainly consists of the following steps:

Experiment Environment. The simulation experiment in this article is carried out under the environment of Windows 10 operating system and CPU configuration of Intel(R) Core(TM) i7-4510U @2.00 GHz 2.60 GHz. Python 3.6 was used as the programming language, JetBrains PyCharm as IDE for adversarial examples

---

Algorithm 1. Crafting Adversarial Examples for Intrusion Detection

---

Input: $D_{original}$, C_n, α, T

Output: $D_{adversarial}$

1: $D_{original} = D_{training} + D_{testing}$

2: $DNN \stackrel{D_{training}}{\longrightarrow} F(X)$

3: **for** each example X in $D_{testing}$ **do**

4:     Initialize $X^{adv} = X$

5:     **for** $t = 1, 2, ..., T$ **do**

6:         setting the constraint matrix C_n ;

7:         $X^{adv} = X^{adv} + \alpha C_n \odot sign(\nabla_x J(F(X), y_{true}))$

8:     **end for**

9:     **for** each feature x_i in X^{adv} **do**

10:         **if** x_i^{adv} is a discrete feature **then**

11:

$$f(x_i^{adv}) = \begin{cases} x_i^{adv} = 1 & x_i^{adv} \geq 0.5 \\ x_i^{adv} = 0 & x_i^{adv} < 0.5 \end{cases}$$

12:         **else if** x_i^{adv} belongs to a nominal feature $[x_k^{adv}, x_{k+n}^{adv}]$ **then**

13:

$$f([x_k^{adv}, x_{k+n}^{adv}]) = \begin{cases} \max(x_i) = 1 & i \in [k, k+n] \\ x_j = 0 & \forall j \in [k, k+n] - i \end{cases}$$

14:         **else**

15:             continues

16:         **end if**

17:     **end for**

18: **end for**

19: **return** $D_{advesarial}$

---

generation experiments. The pytorch 0.4.1 machine learning framework was used to build the intrusion detection classification model and generate the intrusion adversarial examples.

NSL-KDD Dataset. Table 3 is the distribution of each category of data in the training and testing sets of NSL-KDD.

Table 3. The data distribution of NSL-KDD dataset

Categories	DOS	Probe	R2L	U2R	Normal	Total
Training set	45,927	11,656	995	52	67,373	126,003
Testing set	7,460	2,421	2,885	67	9,711	22,544
Total	53,387	14,077	3,880	119	77,084	148,547

5.2   Evaluation Measures

According to our definition of intrusion adversarial examples in Sect. 1, the experiments will be evaluated in the following three aspects. They are, whether the generated adversarial examples can make the intrusion detection classifier misclassified, whether they are consistent with the feature types of original examples, and whether the relationship among features in original examples are affected. Below, the evaluation methods are described in detail.

Accuracy of Intrusion Detection. By comparing the classification accuracy of the original testing set and the adversarial example set on the intrusion detection classifiers, the more the accuracy rate decreases, the higher the probability of adversarial example set successfully attacks the model.

Feature Types Matching. By checking the feature types in the adversarial example set, we can know whether they are consistent with the feature types of original examples. There are mainly the following two items: checking whether there is only one bit of the nominal features is 1 and whether the value of discrete features are 1 or 0.

Consistency of Dataset. In order to evaluate whether the intrusion adversarial examples can simulate the real dataset, we compared the adversarial examples with the original dataset in the following two ways.

Euclidean Metric. Euclidean metric is the distance between two matrixes in Euclidean space. The smaller the Euclidean distance is, the more similar the two matrixes are. Therefore, we calculate the Euclidean metric between the two PCC matrixes of the original dataset and the adversarial dataset to compare whether the linear relationship of the two datasets is similar.

Rank of the Matrix. According to the theory of linear algebra, for two matrixes with the same number of rows and columns, they are equivalent if the ranks of them are equal. In this paper, since the original dataset and the adversarial dataset have the same number of examples and features, they can be treated as two homomorphic matrixes. By calculating the rank of the two matrixes, we can determine whether the two matrixes are equivalent, then to determine whether the two datasets are equivalent. In this way, if the two datasets are equivalent, we can assume that the feature relationship of original dataset have not been changed.

5.3   Experiment

According to the algorithm mentioned in Sect. 4, the experiments are also divided into the following two parts.

Deep Learning Based Intrusion Detection. We compared the performance of several widely used neural network models, and selected the model with the highest accuracy as the target of adversarial attack.

It can be seen from Table 4 that DNN performs better than the other two models. Since the models we selected are all basic models, their accuracy is not very high. However, the attack result is not the only criterion for our evaluation, and the adversarial examples are transferable among neural network structures, so we choose the basic DNN model as the target model.

Table 4. The accuracy comparison of intrusion detection classifiers

Algorithms	Test accuracy
Deep Neural Network (DNN)	0.79
Convolutional Neural Network (CNN)	0.77
Multilayer Perceptron (MLP)	0.74

Crafting Intrusion Adversarial Examples. From Sect. 3, we can know that the perturbation added by C-IFGSM algorithm is affected by parameters α, T in the formula, which may affect the relationships among features in adversarial examples. In addition, the modified features restricted by the constraint matrix C_n will also affect the perturbation added to adversarial examples. So we set different values for α, T and constraint parameters for experiments. We evaluated the experiment results and selected the most appropriate parameter setting.

Table 5. The performance comparison of different parameters

Constraint	Parameters	Accuracy	Rank	Euclidean metric
Only traffic features	$\alpha = 1/225, T = 4$	0.75	109	0.34
	$\alpha = 2/225, T = 4$	0.74	109	0.59
	$\alpha = 2/225, T = 8$	0.70	109	1.04
Only numerical features	$\alpha = 1/225, T = 4$	0.59	111	5.15
	$\alpha = 2/225, T = 4$	0.42	112	7.14
	$\alpha = 2/225, T = 8$	0.33	113	9.02
All features	$\alpha = 1/225, T = 4$	0.59	111	5.12
	$\alpha = 2/225, T = 4$	0.41	112	7.20
	$\alpha = 2/225, T = 8$	0.32	113	9.15

We set the constraint matrix to three cases, only modifying the traffic features, only modifying the numerical features and modifying all features. In each case, the parameters are also set to three cases. The larger the values of α and T are, the larger the perturbation is. Because the adversarial examples are generated on the testing set, we compared the results with the original testing set. The rank of the matrix of the original testing set is 112, and its classification accuracy is 0.79.

By analysing Table 5, it can be found that the Accuracy decreases when the noise value (Parameters) increases, but the relationship between features (Euclidean Metric) will change to a greater extent. The Accuracy decreases when the number of modified features increases, but the Rank of the adversarial examples can be equal to the original testing set. In addition, it can be seen that the experiment results are similar in the case of only modifying the numeric features and modifying all the features. It may because the three nominal features in the dataset have a relatively weak influence on the adversarial examples. Considering several indicators in Table 5, we chose "Modifying all features, $\alpha = 2/225, T = 4$" as the best parameter setting.

5.4   Evaluation and Discussion

Compared with Baseline. Since we did not find another special method for crafting intrusion adversarial examples, we compared the performance of C-IFGSM with the IFGSM as a baseline. According to the evaluation methods mentioned above, we compare the adversarial examples generated by the two methods and the original examples to evaluate the effectiveness of C-IFGSM.

Table 6. The performance comparison with the baseline

	C-IFGSM	IFGSM	Original dataset
Classification accuracy	0.41	0.48	0.79
Feature types matching (%)	1	0.69	1
Euclidean metric	7.20	7.82	0
Rank of the matrixes	112	111	112

It can be seen from Table 6 that with the same parameter settings, C-IFGSM is superior to IFGSM in all aspects. In detail, the adversarial examples generated by C-IFGSM makes the intrusion detection classifier get a lower accuracy, a higher feature types matching, a small Euclidean distance, and a same rank of dataset. This also shows that C-IFGSM can better adapt to network traffic datasets and generate high-quality intrusion adversarial examples.

Table 7. The accuracy comparison

Algorithms	Original dataset	Adversarial dataset
Decision tree	0.73	0.25
CNN	0.77	0.68
MLP	0.74	0.73

Verifying the Transferability. In order to verify the transferability of the intrusion adversarial examples, we input the intrusion adversarial examples into other classification models and compared the decreases in the accuracy of the models.

It can be seen from Table 7 that the intrusion adversarial examples generated by C-IFGSM can make an apparent decline on the accuracy of decision tree, and an lesser impact on CNN and MLP models. It indicates that the intrusion adversarial examples generated by C-IFGSM can also attack other models, but the adversarial effect is not very good in other neural network structure models.

6   Conclusion

This paper mainly focuses on the adversarial examples attack in the domain of intrusion detection classifiers. We analyse different data characteristics between images and network traffic data. Based on the analysis of network traffic dataset, we implement a C-IFGSM based adversarial approach to adapt to the multiple types and relationship of features. The experiments prove that the C-IFGSM based method can effectively adapt to network traffic datasets and make high-quality intrusion adversarial examples. In future work, we will further study the constraint-based method to make intrusion adversarial examples that are more similar to the original examples.

Acknowledgement. This research is supported by the National Natural Science Foundation of China (61972025, 61802389, 61672092, U1811264, 61966009), the Fundamental Research Funds for the Central Universities of China (2018JBZ103, No. 2019RC008), Science and Technology on Information Assurance Laboratory, Guangxi Key Laboratory of Trusted Software (KX201902).

References

1. Androutsopoulos, I., Koutsias, J., Chandrinos, K.V., Paliouras, G., Spyropoulos, C.D.: An evaluation of naive Bayesian anti-spam filtering. arXiv preprint cs/0006013 (2000)
2. Buczak, A.L., Guven, E.: A survey of data mining and machine learning methods for cyber security intrusion detection. IEEE Commun. Surv. Tutor. **18**(2), 1153–1176 (2015)
3. Carlini, N., et al.: Hidden voice commands. In: USENIX Security Symposium, pp. 513–530 (2016)
4. Chen, T., Liu, J., Xiang, Y., Niu, W., Tong, E., Han, Z.: Adversarial attack and defense in reinforcement learning-from AI security view. Cybersecurity **2**(1), 11 (2019)
5. Goodfellow, I.J., Shlens, J., Szegedy, C.: Explaining and harnessing adversarial examples. arxiv (2018)
6. Grosse, K., Papernot, N., Manoharan, P., Backes, M., McDaniel, P.: Adversarial examples for malware detection. In: Foley, S.N., Gollmann, D., Snekkenes, E. (eds.) ESORICS 2017. LNCS, vol. 10493, pp. 62–79. Springer, Cham (2017). https://doi.org/10.1007/978-3-319-66399-9_4

7. Kolosnjaji, B., et al.: Adversarial malware binaries: evading deep learning for malware detection in executables. In: 2018 26th European Signal Processing Conference (EUSIPCO), pp. 533–537. IEEE (2018)

8. Kurakin, A., Goodfellow, I., Bengio, S.: Adversarial examples in the physical world. arXiv preprint arXiv:1607.02533 (2016)

9. Li, J., Ji, S., Du, T., Li, B., Wang, T.: Textbugger: Generating adversarial text against real-world applications. arXiv preprint arXiv:1812.05271 (2018)

10. Liang, B., Li, H., Su, M., Bian, P., Li, X., Shi, W.: Deep text classification can be fooled. arXiv preprint arXiv:1704.08006 (2017)

11. Papernot, N., McDaniel, P., Jha, S., Fredrikson, M., Celik, Z.B., Swami, A.: The limitations of deep learning in adversarial settings. In: 2016 IEEE European symposium on security and privacy (EuroS&P), pp. 372–387. IEEE (2016)

12. Rieck, K., Trinius, P., Willems, C., Holz, T.: Automatic analysis of malware behavior using machine learning. J. Comput. Secur. **19**(4), 639–668 (2011)

13. Rigaki, M.: Adversarial deep learning against intrusion detection classifiers (2017)

14. Russakovsky, O., et al.: Imagenet large scale visual recognition challenge. Int. J. Comput. Vis. **115**(3), 211–252 (2015)

15. Samanta, S., Mehta, S.: Towards crafting text adversarial samples. arXiv preprint arXiv:1707.02812 (2017)

16. Sharif, M., Bhagavatula, S., Bauer, L., Reiter, M.K.: Accessorize to a crime: real and stealthy attacks on state-of-the-art face recognition. In: Proceedings of the 2016 ACM SIGSAC Conference on Computer and Communications Security, pp. 1528–1540. ACM (2016)

17. Szegedy, C., et al.: Intriguing properties of neural networks. arXiv preprint arXiv:1312.6199 (2013)

18. Tavallaee, M., Bagheri, E., Lu, W., Ghorbani, A.A.: A detailed analysis of the KDD CUP 99 data set. In: 2009 IEEE Symposium on Computational Intelligence for Security and Defense Applications, pp. 1–6. IEEE (2009)

19. Wang, Z.: Deep learning-based intrusion detection with adversaries. IEEE Access **6**, 38367–38384 (2018)

20. Yang, K., Liu, J., Zhang, C., Fang, Y.: Adversarial examples against the deep learning based network intrusion detection systems. In: 2018 IEEE Military Communications Conference (MILCOM), MILCOM 2018, pp. 559–564. IEEE (2018)

21. Yavanoglu, O., Aydos, M.: A review on cyber security datasets for machine learning algorithms. In: 2017 IEEE International Conference on Big Data (Big Data), pp. 2186–2193. IEEE (2017)

22. Zhang, H., Yu, X., Ren, P., Luo, C., Min, G.: Deep adversarial learning in intrusion detection: A data augmentation enhanced framework. arXiv preprint arXiv:1901.07949 (2019)

23. Zhelo, O., Zhang, J., Tai, L., Liu, M., Burgard, W.: Curiosity-driven exploration for mapless navigation with deep reinforcement learning. arXiv preprint arXiv:1804.00456 (2018)

Deep Reinforcement Learning for Solving AGVs Routing Problem

Chengxuan Lu[1], Jinjun Long[2], Zichao Xing[1], Weimin Wu[1(✉)], Yong Gu[1],
Jiliang Luo[3], and Yisheng Huang[4]

[1] The State Key Laboratory of Industrial Control Technology, Institute of Cyber- Systems
and Control, Zhejiang University, Hangzhou 310027, Zhejiang, China
wmwu@iipc.zju.edu.cn

[2] KENGIC Intelligent Equipment Co., Ltd., Qingdao 266111, Shandong, China

[3] Department of Control Science and Engineering, Huaqiao University, Xiamen 361021,
Fujian, China

[4] Department of Electrical Engineering, Ilan University, Yilan 26047, Taiwan, China

Abstract. The routing of automated guided vehicles (AGVs) is playing an increasingly important role in modern logistics. AGVs routing problem is a complex combinatorial optimization problem. It fails to get the desired results of solving this problem using meta-heuristic algorithms due to its high real-time demand. Large AGVs systems in engineering are usually simplified by adding regulations, which may lead to getting only sub-optimal solutions. In this paper, we present a deep reinforcement learning algorithm to solve the AGVs routing problem. Firstly, the AGVs routing problem is modeled by a Markov decision process (MDP), enabling real-time routing. Secondly, according to the properties of the working scene of AGVs, asynchronous DQN (deep Q-network) is exploited to serve as the base framework of reinforcement learning. More importantly, the map of the working scene is discretized and represented using the embedding technique. Compared with one-hot mode, the input size of the embedding mode is much smaller, greatly improving the training speed. The extracted embeddings are built into conflict vectors, which are finally processed by LSTM (long short-term memory). Experiments show that the proposed algorithm has effectiveness both in real-time responding speed and getting high-quality solutions.

Keywords: AGVs routing problem · Real-time routing · Asynchronous deep Q-network · Embedding

1 Introduction

An AGV is a driverless material handling machine, which can conduct autonomous navigation. Since the introduction of AGVs, the use of them has grown enormously in logistics systems. An AGVs system consists of multiple AGVs running at the same time, which leads to several problems such as scheduling and routing. AGVs scheduling is to arrange a group of AGVs to complete a batch of pick-up and delivery tasks under certain

B. Ben Hedia et al. (Eds.): VECoS 2020, LNCS 12519, pp. 222–236, 2020.
https://doi.org/10.1007/978-3-030-65955-4_16

constraints. Once the scheduling instructions are issued, routing is to find one or more suitable paths for each AGV according to the current traffic situation [1].

AGVs routing algorithms can be roughly divided into exact approaches, heuristics, and meta-heuristics [2]. In the last decades, meta-heuristic is the best developed in the field of approximate optimization [3]. Vos used the MPL (max-plus-linear) algorithm to deal with the scheduling and routing of the AGVs system of a container terminal [4]. Tavakkoli et al. invented a new nonlinear integer model inspired by the cellular manufacturing system [5]. Zanjirani et al. [6] studied the application of the Tabu algorithm and genetic algorithm in the design of an AGVs system. Shirazi et al. [7] considered the load of AGV and proposed a nonlinear multi-objective problem to minimize internal and external circulation flow. Han et al. [8] studied an improved genetic algorithm applied to the AGVs system routing. Zhou et al. [9] accurately established the AGVs system model based on the Petri net adapted to resources, then set up the deadlock control strategy suitable for real-time scenes.

Recent years have witnessed the significant development of deep reinforcement learning (DRL). Deep learning empowers reinforcement learning to solve problems that cannot be solved before, such as learning to play video games directly from pixels [10]. In 2016, DeepMind [12] proposed a hybrid DRL system called AlphaGo, which beat Go world champion. One year later, a higher level of Go-playing software called AlphaGo Zero was presented by DeepMind. In this version, neural networks were fully trained by reinforcement learning [13]. As a result, AlphaGo Zero won 100-0 over AlphaGo after only three days of training. In addition to games, DRL has also achieved excellent results in operational research. Kalakanti et al. [14] proposed a reinforcement learning solver for the vehicle routing problem wherein the optimal route learning problem is cast as a Markov Decision Process. Kintsakis et al. [15] trained a modified sequence-to-sequence neural network architecture via reinforcement learning to perform scheduling decisions as part of a workflow management system.

The study of reinforcement learning in logistics has attracted many researchers [16–18]. Xue et al. [16] addressed a multi-AGV flow-shop scheduling problem with a reinforcement learning method. Kamoshida et al. [17] presented a DRL based AGV route planning method. The algorithm uses raw high-dimensional map information as input instead of hand-engineered low-dimensional state representation, and it enables the acquisition of a successful AGV route planning policy. However, some of the methods [16, 18] may suffer curse-of-dimensionality because of the natural drawback of tabular reinforcement learning. Though the algorithm of Kamoshida et al. [17] used DQN to solve this problem, the use of supervised data will limit the data size and influence the final performance.

To overcome those difficulties, we use embeddings to represent scene state based on asynchronous DQN. This paper is organized as follows. Section 2 introduces the AGVs system and the routing problem that we are going to solve. Section 3 presents asynchronous DQN as our algorithm base framework. Section 4 introduces the embedding technique to represent environment information, and Sect. 5 introduces the neural network architecture of asynchronous DQN. Experiments and analysis are discussed in Sect. 6. Conclusion and future work appear in Sect. 7.

2 AGVs Routing Problem

2.1 AGVs System

In automatic logistics systems, such as factories, warehouses, and container terminals, AGVs are often used to complete the material handling tasks. The control system of AGVs manages multiple AGVs working at the same time and interfaces with other systems. As shown in Fig. 1 is a simple schematic diagram of AGVs system working in a warehouse, in which the AGVs system needs to control AGVs to load, unload, charge, etc.

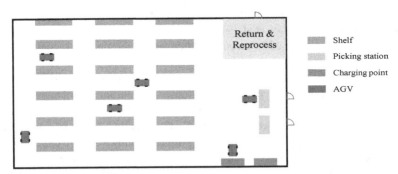

Fig. 1. A simplified AGV working environment. Several AGVs are running in tracks of a warehouse.

This paper takes the system shown in Fig. 1 as its base study object.

2.2 AGVs Routing

After the scheduling instructions are issued, the task of AGVs routing is to find a suitable path from the current location to the destination for each AGV according to the current traffic situation or other information and avoid conflicts and collisions (as shown in Fig. 2) of AGVs moving along the paths. It should access the nodes distributing in each predetermined geographical location. In short, AGV routing is to figure out a series of feasible paths for a group of AGVs.

Fig. 2. Problems in AGVs routing. Left: One AGV is blocked by other AGVs. Right: in a deadlock, both AGVs use the same route in different directions.

There two modes for AGVs routing: static and dynamic. The input data in a static situation is constant when vehicles are running. Hence, static methods cannot adapt to the changes in an AGVs system. While for dynamic routing, planning and implementation evolve simultaneously. In this paper, we use DRL because of its fast response, which is suitable for dynamic routing mode.

3   DRL Framework

In this section, we first use MDPs to transform AGVs routing problem into a reinforcement learning problem. In addition, Asynchronous DQN is exploited as the base reinforcement learning framework. Part 3.4 shows how to force agents to finish tasks quickly by the reward setting. Lastly, we use parameter sharing to solve the multi-agent problem.

3.1   Reinforcement Learning

Reinforcement learning is to learn how to map from state to optimal action to maximize the weighted reward of each step. Agents will never be told what action to take, but they will try to find the action that can produce the maximum reward.

Markov decision processes (MDPs) are an ideal model of reinforcement learning that can make accurate theoretical statements. MDPs describes learning from interaction to achieve goals. In the process of interaction, agents select actions according to states, and the environment responds to actions then display new states to agents. The environment will also give rewards, which are values agents need to maximize. Agents and the environment interact in a series of discrete-time steps:

$$S_0, A_0, R_1, S_1, A_1, R_2, S_2, A_2, R_3, \ldots \tag{1}$$

The goal of the agent is to maximize the *expected return*. In this paper, we use the expected *discounted return*:

$$G_t \doteq R_{t+1} + \gamma R_{t+2} + \gamma^2 R_{t+3} + \ldots \tag{2}$$

$q_\pi(s, a)$ is the *action-value function* mapping (s, a) to value under policy π:

$$q_\pi(s, a) \doteq \mathbb{E}_\pi [G_t | S_t = s, A_t = a] \tag{3}$$

3.2   Modeling Routing Problem Based on MDPs

As shown in Eq. (1), an MDP is a discrete process, but AGVs routing problem is generally considered as a beforehand planning problem. Therefore, we decompose one route of an AGV into a series of decisions of time sequence.

Figure 3 shows an MDP for an AGV routing. We set the action space as:

$$\mathcal{A}\{up, down, left, right, stop\}; \tag{4}$$

In some states, some actions in action space are equivalent to *stop*; for example, if an agent (AGV) chooses action *up* at point 1, it will still stay at point 1. With the

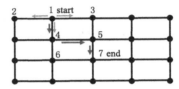

Fig. 3. AGVs routing based on an MDP. An AGV routing problem is first discretized into several steps and then formed as an MDP. There are five actions in every step.

above definition, an MDP can be described as follows. Assume the start position of an agent was point 1, and its destination was point 7. When at point 1, the agent chose *down* and performed this action, and the agent reached point 4. Then the agent chose *right* and reached point 5. Similarly, the agent finally reached point 7 by those kinds of step-by-step decision-making mode.

3.3 Use Asynchronous DQN as the Base Framework

DQN is suitable for the AGVs routing problem whose actions are discrete. As Fig. 3 presented, there are five optional actions the agent can choose. These can be formed into five Q-values of DQN. In addition, DQN uses two methods to solve the instability of Q-learning [10]. The first one is experience replay, that is storing data in memory and sample from memory randomly. Using experience replay can average the distribution of many previous states, smooth the learning process, and avoid parameter oscillation or divergence. The second one uses a target network whose parameter is substituted by the parameter of the evaluating network periodically. This method can further improve stability.

The training processes in DRL are always time-consuming. As shown in Table 1, we combine the asynchronous method [11] and DQN to improve training speed. The asynchronous method is a lightweight framework of DRL, which can make full use of multi-core CPU. First, multi actor-learners can be put in a single machine, which reduces the communication costs between actor-learners. Second, multi actor-learners running in parallel is beneficial for exploration. In addition, data produced by different actors will improve diversity, and hence improve stability.

3.4 Reward Setting

In reinforcement learning, the purpose of the agent is the maximization of the cumulative sum of rewards. The rewards were set +1 for the win and −1 for the loss in AlphaGo [12]. In this paper, we set the tasks of an AGV episodic. An AGV moving from the start to the end is regarded as completing a task. To force an agent to finish a task as quickly as possible, we give +1 reward when the AGV reaches its destination; otherwise, give a small negative reward (such as −0.01) as a punishment.

3.5 Multi-agent Reinforcement Learning

Efficiently completing tasks is the only goal of the AGVs control system. Therefore, considering from the point of view of the designer, the AGVs routing problem is a fully

Table 1. Asynchronous DQN

Each actor-learner thread of asynchronous DQN
Assume global shared parameters θ, θ^-, global shared counter $T = 0$, updating period for θ^- is C
Assume thread training period is U, thread counter $t = 0$
Initialize thread replay buffer D to capacity N
Initialize $\theta^- \leftarrow \theta$
Initialize thread-specific $\theta' \leftarrow \theta$
Initialize s
Repeat
Choose a according to $Q(s; \theta')$
Perform a, receive r, s'
Store (s, a, r, s') in D
$t \leftarrow t + 1, T \leftarrow T + 1, s \leftarrow s'$
If $t \bmod U == 0$
$\theta' \leftarrow \theta$
Sample a batch of (s_j, a_j, r_j, s_{j+1})
$y_j \leftarrow \begin{cases} r_j & \text{if episode terminates at step } j + 1 \\ r_j + \gamma\max_a Q(s_{j+1}, a; \theta^-) & \text{otherwise} \end{cases}$
Perform asynchronous update of θ according to $d\theta'$ of $(y_j - Q(s_j, a_j; \theta'))^2$
If $T \bmod C == 0$ perform $\theta^- \leftarrow \theta$
Util $T > T_{max}$

cooperative multi-agent problem. For cooperative multi-agent DRL, parameter sharing can make the training of the same type of agents more efficiently [19]. This method enables the policy to be trained using the experiences of all agents simultaneously. Moreover, it still allows different behaviors between agents. Because corresponding observations or states of agents are different, and the observations contain their indices. In the parameter sharing method, control is decentralized, but learning is not.

4 Feature Processing

In Sect. 3, asynchronous DQN is exploited as the base DRL framework, but the feature processing and neural network of DQN have not been introduced. In this section, we present a sound method of feature processing in the AGVs routing problem. Feature processing plays an important role in machine learning. The characteristics of data have a direct impact on the performance of the model. We discretize the map into grids and use the embedding technique to represent every grid. The input data size of the embedding method is much less than the one-hot method. In Sect. 5, the neural network architecture is built based on embedding representation.

4.1 Discretization of Continuous Features

The working scenes of AGVs are generally a two-dimension plane, so the Cartesian coordinate can accurately mark the position of points. In theory, point coordinates can represent the exact positions of AGVs. However, the motion of AGVs is usually restricted by tracks. For example, coordinates A:(1.1, 3.2) and B:(1.05, 3.2). There is only 0.05 difference on the x-axis numerically between point A and point B, but the two coordinates may represent two completely different kinds of areas. Point A may be on a track, but B may on an obstacle. This property makes it difficult for a neural network to map the right location from corresponding coordinates. Deep neural networks are capable of nonlinear approximation. Still, the track-to-obstacle can be regarded as a kind of jump characteristic that brings a great challenge to the neural network training. In our experiments, the model with coordinate inputs does not converge even in small environments with obstacles. In this paper, we propose discretization to process features.

In some machine learning practices, the discretization of continuous variables is often used to facilitate problem handling. Discretization reduces data size, and hence accelerate calculations and training [20]. As presented in Fig. 4, we discretize the environment map into grids.

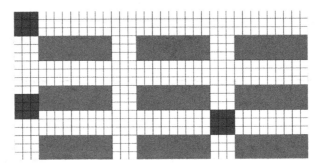

Fig. 4. Discretize the map into grids. Red grids denote AGVs; gray grids denote obstacles; white grids denote tracks. (Color figure online)

The map is cut into a grid-net formed by multiple regular arrangements of grids with the same size. The gray parts indicate obstacles where AGVs cannot reach. The others represent tracks whose widths are three times of the length of a grid. Tracks signed with red indicates that AGVs occupy these areas. Each AGV occupies 3×3 grids. There is no overlap between grids; each grid is independent and has a unique number (ID). One grid can only be occupied by one AGV at the same time. The information in the environment (positions of AGVs and their destinations) can be represented easily and accurately with the above discretization.

4.2 One-Hot Code

The discretization of the map is discussed in Sect. 4.1, but the representation of environment information has not been totally solved.

One-hot code is a sparse representation of data. In this coding mode, data is formed into a binary string, where only one bit is 1 and the others are 0. Four kinds of one-hot codes can represent one map grid.

The first one indicates the occupation situations of AGVs in this gird. If the upper limit of the number of AGVs is k, then the length of the first code is set to $k + 1$, and the extra one is used to represent the case of no AGV. Similarly, the following part is formed by $k + 1$ one-bit-long one-hot codes (because different AGVs can have the same destination), indicate destinations of AGVs. Geographic characteristics are represented by the third part, whose length is 1. The third part represents the geographic characteristics. Besides, because parameter sharing is used to solve the problem of multi-agent DRL, input data should specify the currently controlled AGV. Therefore, in the fourth part, we use a one-bit code to inform this information.

Finally, as presented in Fig. 5, those four codes are concatenated together to form a grid code, whose length is $2k + 4$. If the map is cut into $n \times n$ grids, the entire map code will be formed into a tensor of size $(n, n, 2k + 4)$, which is like an image. Therefore, we used a convolutional neural network to process these inputs in experiments.

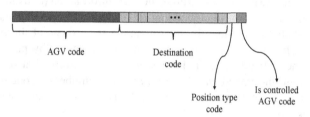

AGV code   Destination
                  code

                            Position type   Is controlled
                            code   AGV code

Fig. 5. The one-hot code for one gird. We use four kinds of codes to represent the information of one grid. And codes of all the grids form into the input data.

4.3 Embedding Code

In a language model, a word can be represented by a one-hot vector. But one of the shortcomings of the one-hot code is that it isolates words from one word to another, resulting in a reduction in the generalization performance of the algorithm. In fact, a word can be represented by an embedding vector. Embedding is one of the most important ideas in the field of natural language processing. Word embedding was originally created to overcome the curse-of-dimensionality in neural language models [21].

For word embedding, every word is mapped into a vector. Similarly, a map grid can also be represented by an embedding vector.

As shown in Fig. 6, an AGV occupies nine grids, and these nine grids are occupied only by this AGV. Hence the set of these nine gird numbers $(1, 2, ..., 9)$ can uniquely mark the position of this AGV. The relationship between grids has an analogy to the relationship between words. For example, 'apple' is similar to 'banana', but not close with 'book'. Analogously, the relationship between adjacent grids should be more intimate than the two that are far from each other.

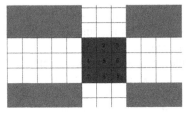

Fig. 6. Occupation grids of an AGV. An AGV occupies nine grids (1, 2, ..., 9) of the map.

In fact, for the routing problem, the location of AGVs and their destinations are all the dynamic information of the environment. Because only the positions of AGVs and their destinations can change, other objects (such as obstacles, tracks) are fixed. Therefore, for the routing problem, we can use the current position and destination of an AGV as the only information of this AGV, then use such information of all AGVs as the input data of the algorithm. Since grid embeddings can represent the position of an AGV, a destination can also be represented in this way.

Figure 7 presents the process of forming an AGV information (e) using embeddings. In this figure, one index is a grid embedding ID related to the corresponding position. Firstly, an AGV-indices is formed by concatenating the indices of the AGV position and destination. Then the AGV information vector is obtained by propagating those indices through the embedding layer. The set of all AGV information vectors is global information. Assume the embedding length is dim, AGV number is k, one AGV occupies $m \times m$ grids and its destination occupies $m \times m$ grids, then the input size of embedding mode is $k \times dim \times m^2 \times 2$.

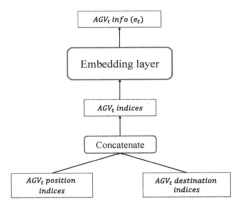

Fig. 7. Form an AGV information using embeddings. The t_{th} AGV-indices is formed by concatenating the indices of the AGV position and destination. The AGV info (denoted as e_t) is gathered from the grid embedding layer according to the AGV indices.

4.4 Comparison of Input Size of One-Hot and Embedding

Assume that the map is divided into $n \times n$ grids, one AGV occupies $m \times m$ grids, the maximum number of AGVs is k, and the dimension of grid embedding is dim. According to Sect. 4.2, the input size of the one-hot code is $n^2(2k + 4)$; the input size of embedding is $k \times dim \times m^2 \times 2$. The gap between the two methods will be obvious after bringing in some values. For example, $n = 100, m = 3, k = 10, dim = 32$, then the one-hot input size is 240,000, and embedding input size is 5,760. The embedding input is about 41.7 times smaller than the one-hot input. Especially when the tracks are sparse, the gap will be even greater. Section 6.1 also shows that embedding mode is also better than the one-hot mode in training speed. Therefore, our proposed model uses embedding to form the environment information.

5 Neural Network Architecture

Section 4 shows the advantages of embedding mode in feature processing. As illustrated in Fig. 8, the neural network architecture of our asynchronous DQN obtains Q-values by processing the input data from embedding mode. There are roughly two parts in the neural network architecture. The lower part is the formation of conflict vectors, which is described in Sect. 5.1. The upper part is the processing of the conflict vectors, which is described in Sect. 5.2.

5.1 Formation of Conflict Vectors

Parameter sharing is used to solve the multi-agent problem in this paper, and each observation in parameter sharing is for one specific agent. That is, this agent gets the global information, and then the agent makes its own decision. The input data should be formed from the perspective of the ego of this agent. At first, the currently controlled AGV should be discriminated from the others. As shown in Fig. 8, an e denotes an AGV information, whose formation is illustrated in Sect. 4.3. e_{this} represents the information of the currently controlled AGV; for other AGVs, the t_{th} AGV information is noted as e_{other_t}. v_{this} and v_{other} are obtained by propagating e_{this} and e_{other} through the first fully connected layer. After that, repeat v_{this} $k - 1$ times and concatenate with v_{other_t} separately to form the initial conflict vectors. Finally, the conflict vectors are obtained by propagating the initial conflict vectors through the second fully connected layer.

The reason for naming the conflict vector is that there is a competitive relationship between the current controlled AGV and other AGVs.

5.2 Conflict Vectors Processing

The easiest way to process conflict vectors (sequences) is by using fully connected layers. However, a fully connected layer is not suitable for dealing with sequence data. The parameter volume of fully connected layers would be huge, and it is proportional to the number of conflict vectors. Recurrent neural networks are a class of neural networks for processing sequence data. They take advantage of parameter sharing, which enables

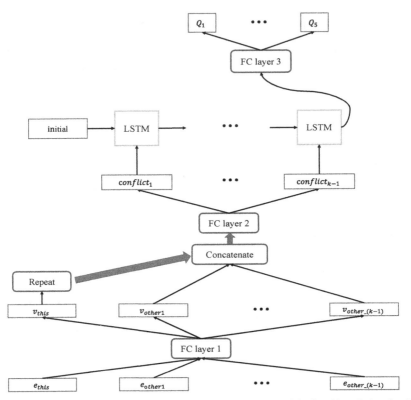

Fig. 8. The neural network architecture of asynchronous DQN. FC is the abbreviation for 'fully connected'. An e denotes an AGV information. 'this' is linked with the currently controlled AGV, and 'other' is linked with AGVs except for the currently controlled AGV. Repeated v_{this} and v_{other} form the initial conflict vectors. Finally, LSTM encodes those conflict vectors and generates Q-values.

the model to adapt to data of different lengths. Simple recurrent models are suffered from long-term dependencies [22]. In this paper, LSTM (long short-term memory), which is good at long-term dependencies, is used to process conflict vectors.

The upper part of Fig. 8 shows the structure of conflict vectors processing. LSTM processes multiple vectors of the same type. This model not only utilizes the advantage of parameter sharing of recurrent neural networks but can filter and memorize important information through various gates of LSTM. Finally, the final step output of LSTM is formed into five Q-values of DQN by a fully connected layer.

6   Experiments

We conducted several experiments with an AGV system simulator to determine the validity of our proposed algorithm. We use the *average number of tasks completed by one AGV every 1000 steps* to measure performance of algorithms, because it indicates the working efficiency of AGVs. The hardware configuration for the experiments was set up double Intel Xeon Platinum 8280L, one NVIDIA GeForce RTX 2080Ti.

6.1 Comparison of One-Hot and Embedding

The network for one-hot mode was a convolutional neural network with two convolutional blocks and two fully connected layers. The map was discretized into 8×4 grids, and the track area accounted for 43% of the total map area. There were three AGVs in this scene. The embedding length was set to 64.

Figure 9 shows the comparison of training time between in embedding mode and one-hot mode. Embedding mode is about seven times faster than one-hot.

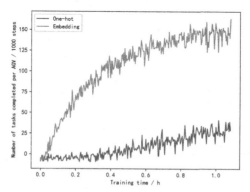

Fig. 9. Embedding compared with one-hot in training speed (track proportion: 62%).

Figure 10 shows the situation when tracks are sparse. Embedding mode is more than 8.5 times faster than one-hot. In this condition, the advantages of embedding are more evident because the number of embeddings is only related to the size of the tracks. By contrast, one-hot input is constrained by a 3-dimension structure. When a little track is added around the scene, the one-hot mode has to stretch the length and width to cover the entire scene, while only several grids need to be appended for embedding mode.

6.2 Results

The training process of the scene with 22 AGVs is shown in Fig. 11. There were 28×14 grids in this scene, and the embedding length was set to 64. We used Adam optimizer with the initial learning rate of 0.00001. During the initial stage, the training curve was relatively smooth because the initial epsilon (ϵ) annealed from 0.9 to 0.1 as the training progresses. Big epsilon will encourage exploration at the beginning of the training. As the number of steps in completing one task (episode) is relatively small, we set the discount rate $\gamma = 0.9$. It took about 50 h for the asynchronous DQN to achieve its best performance. We collected data of 176200 tasks to train our model.

Compared with other reinforcement learning algorithms [16–18] for AGVs scheduling or routing, our method is able to handle scenes with more AGVs.

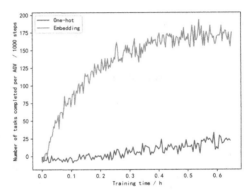

Fig. 10. Embedding compared with one-hot in training speed (track proportion: 43%).

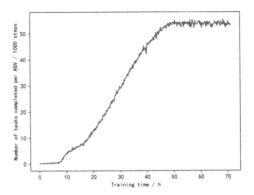

Fig. 11. Training process of a scene with 22 AGVs.

As shown in Table 2, our method also outperforms the regulation method in the same AGVs system simulator. The regulation method uses mechanisms such as adding one-way road, anti-deadlock, unlock arbitration to simplify the routing problem. The regulation method is one of the most commonly used algorithms in engineering applications. The performance of the random policy is about 0.355. Compared with 54.2 of asynchronous DQN, there is about a 153 times gap. Asynchronous DQN is also about 1.18 times better than the regulation method in performance.

Table 2. Performance comparison between random, regulation, and asynchronous DQN

	Average performance
Random	0.355
Regulation	46.1
Asynchronous DQN	54.2

Moreover, DRL has a high responsive ability compared with searching algorithms (such as exact approaches, heuristics, and meta-heuristics). To complete a decision, a trained DRL model only need to be fed with the state data to perform a forward propagation, which takes only several milliseconds.

7 Conclusion

In this paper, we propose a more efficient and responsive algorithm for solving the AGVs routing problem. The problem is first transformed into a discrete time-step decision-making process, and modeled by an MDP. Next, the asynchronous technique is used to improve CPU utilization. After the map discretization, the embedding technique for grid representation is used to improve the training speed of the algorithm. Finally, LSTM is exploited to process features. The experiment results show that our model has advantages over conventional methods both in responding speed and getting more optimal solutions. In the future, we aim to use more advanced DRL algorithms to improve algorithm performance and training speed.

Acknowledgement. This work is supported in part by National Key R&D program of China under Grant 2018YFB1305900, the Key Research and Development Program of Guangdong Province of China under Grant 2019B010120001, and the Project of State Key Laboratory of Industrial Control Technology, Zhejiang University, China under Grant ICT1921.

References

1. Qiu, L., Hsu, W.J., Huang, S.Y., et al.: Scheduling and routing algorithms for AGVs: a survey. Int. J. Prod. Res. **40**(3), 745–760 (2002)
2. Fazlollahtabar, H., Saidi-Mehrabad, M.: Methodologies to optimize automated guided vehicle scheduling and routing problems: a review study. J. Intell. Robot. Syst.: Theory Appl. **77**(3–4), 525–545 (2015)
3. Osman, I.H., Laporte, G.: Metaheuristics: a bibliography. Ann. Oper. Res. **63**(5), 513–623 (1996)
4. Vos, A.: Scheduling Automated Guided Vehicles in Container Terminals Using Max-plus-linear Systems (2015)
5. Tavakkoli-Moghaddam, R., Aryanezhad, M.B., Kazemipoor, H., Salehipour, A.: Partitioning machines in tandem AGV systems based on "balanced flow strategy" by simulated annealing. Int. J. Adv. Manuf. Technol. **38**(3), 355–366 (2008)
6. Farahani, R.Z., Laporte, G., Miandoabchi, E., Bina, S.: Designing efficient methods for the tandem AGV network design problem using tabu search and genetic algorithm. Int. J. Adv. Manuf. Technol. **36**(9), 996–1009 (2008)
7. Shirazi, B., Fazlollahtabar, H., Mahdavi, I.: A six sigma based multi-objective optimization for machine grouping control in flexible cellular manufacturing systems with guide-path flexibility. Adv. Eng. Softw. **41**(6), 865–873 (2010)
8. Han, Z., et al.: Multi-AGV path planning with double-path constraints by using an improved genetic algorithm. Plos One **12**(7), e0181747 (2017)
9. Zhou, M., Wu, N.: System Modeling and Control with Resource-Oriented Petri Nets. CRC Press, Inc (2009)

10. Mnih, V., et al.: Human-level control through deep reinforcement learning. Nature **518**(7540), 529–533 (2015)
11. Mnih, V., et al.: Asynchronous methods for deep reinforcement learning. In: 33rd International Conference on Machine Learning (2016)
12. Silver, D., et al.: Mastering the game of Go with deep neural networks and tree search. Nature **529**(7587), 484–489 (2016)
13. Silver, D., Schrittwieser, J., Simonyan, K., et al.: Mastering the game of Go without human knowledge. Nature **550**(7676), 354–359 (2017)
14. Kalakanti, A. K., Verma, S., Paul, T., et al.: RL SolVeR pro: reinforcement learning for solving vehicle routing problem. In: 2019 1st International Conference on Artificial Intelligence and Data Sciences (AiDAS) (2019)
15. Kintsakis, A. M., Psomopoulos, F. E., Mitkas, P, A.: Reinforcement Learning based scheduling in a workflow management system. Eng. Appl. Artif. Intell. 81, 94–106 (2019)
16. Xue, T., Zeng, P., Yu, H.: A reinforcement learning method for multi-AGV scheduling in manufacturing. In: 2018 IEEE International Conference on Industrial Technology (ICIT). IEEE (2018)
17. Kamoshida, R., Kazama, Y.: Acquisition of automated guided vehicle route planning policy using deep reinforcement learning. In: 2017 6th IEEE International Conference on Advanced Logistics and Transport (ICALT), pp. 1–6. Bali(2017)
18. Zhao, M., Li, X., Gao, L., et al.: An improved Q-learning based rescheduling method for flexible job-shops with machine failures. In: 2019 IEEE 15th International Conference on Automation Science and Engineering (CASE). IEEE (2019)
19. Gupta, J.K., Egorov, M., Kochenderfer, M.: Cooperative multi-agent control using deep reinforcement learning. In: Sukthankar, Gita, Rodriguez-Aguilar, Juan A. (eds.) AAMAS 2017. LNCS (LNAI), vol. 10642, pp. 66–83. Springer, Cham (2017). https://doi.org/10.1007/978-3-319-71682-4_5
20. Mittal, A., Cheong, L. F.: Employing Discrete Bayes Error rate for discretization and feature selection. In: IEEE International Conference on Data Mining. IEEE (2002)
21. Bengio, Y., et al.: A neural probabilistic language model. J. Mach. Learn. Res. **3**(6), 1137–1155 (2003)
22. Bengio, Y., Simard, P.: Frasconi: learning long-term dependencies with gradient descent is difficult. IEEE Trans. Neural Netw. **5**(2), 157–166 (1994)

Research on Vehicle Detection Based on Visual Convolution Network Optimization

Liu Nanyan and You Jingyang[✉]

Xi'an University of Science and Technology, Xi'an, China
youjyang@qq.com

Abstract. Aiming at the problem that the vehicle detection algorithm based on convolutional neural network is too deep in the network layer, resulting in low training efficiency, this paper proposes a visualization method to adjust the structure of convolutional neural network, so as to improve training efficiency and detection effect. Firstly, the existing convolutional neural network model for image classification is visualized using the intermediate layer visualization method of convolutional neural network. Then, the layers of the convolutional neural network model are analyzed to select the layer with the best visualization effect for network reconstruction, so as to obtain a relatively simplified network model. The experimental results show that the similar multi-target detection method proposed in this paper has obvious improvement in training efficiency and accuracy.

Keywords: Convolutional neural network · Visualization · CAM · Intermediate layer activation diagram

1 Introduction

Vehicle detection has always played an important role in parking lot management and intersection monitoring and is an indispensable part in intelligent transportation. In recent years, a variety of vehicle detection technologies have been developed, including ring coil detection, ultrasonic detection, geomagnetic detection, video detection and so on. Compared with other detection technologies, image detection technology is more mature and superior. Image detection has the advantages of low development cost, high efficiency and high detection accuracy.

Vehicle classification and traffic flow statistics of highway video are problems of moving object detection, identification and tracking, which can be realized by traditional image method and modern deep network. Traditional image methods have relatively high real-time performance due to the small amount of computation. Modern deep network has great advantages in the accuracy of background segmentation and target classification.

The traditional method for moving object detection is to do difference between successive frames of the image, remove the background or model the foreground background of each pixel. Classical methods, such as frame difference method [1], optical flow method [2], background subtraction method [3], and Gaussian background modeling [4], have high real-time performance, but there are still a variety of noise interference,

B. Ben Hedia et al. (Eds.): VECoS 2020, LNCS 12519, pp. 237–243, 2020.
https://doi.org/10.1007/978-3-030-65955-4_17

such as background interference, light intensity interference, etc. These environmental factors are difficult to bypass and greatly reduce the robustness of the algorithm. However, vehicle detection algorithms based on convolutional neural network, such as the improved YOLO, SSD and VGG-16, have the problems of too many network layers and high training cost. In this paper, the intermediate layer visualization method of convolutional neural network is proposed to optimize the network in the process of image classification so as to reduce the training cost and improve the training efficiency.

2   Optimize the Convolutional Neural Network by Visualization

2.1   Visual Analysis Method

Compared with other image classification of convolutional neural network, VGG-16 has a distinct structure, which is convenient for visual analysis of the effect of feature classification of each layer of targets. At the same time, the convolution kernel with the size of 3 × 3 is adopted. Under the same perception field, the smaller convolution kernel increases the network depth and reduces the number of parameters.

In this paper, the vehicle classification model is trained by VGG-16 network. The convolutional layer of the classification model is visualized through the activation diagram of the middle layer and the thermal diagram of the class activation. The network structure and depth that should be adopted are analyzed, so as to optimize the research object in this paper.

The intermediate layer activation diagram can show the feature diagram of each channel output by the convolutional layer in the network. The activation graph of the middle layer is the visualization of the feature graph. It can obtain the process in which the input image is decomposed into filters, so as to monitor the learning process of the network and play a guiding role in parameter regulation. CAM (class activation Map) [5] is to calculate the classification importance of each position of the input image, and mark the importance of each area to classification with different colors. The class activation thermal diagram can visualize the characteristics through which the learned model identifies the target.

Total of 13 convolutional layers in the VGG-16 model are shown in Fig. 1 [6]. Accuracy and loss rate tend to remain unchanged after 100 trainings. Therefore, this paper adopts the model after 300 trainings to visualize the classification results of some vehicle images. The convolutional layer of VGG-16 network can be divided into 5 blocks according to the location of maxpooling layer. Therefore, the activation diagram of the middle layer of the convolutional layer at the second, fourth, seventh and thirteenth layers of VGG-16 is selected in this paper as the activation diagram of the middle layer at this stage (there are too many feature diagrams in the fourth, seventh and thirteenth layers, and only some feature diagrams are shown). A vehicle image in the original drawing is intercepted as a visualization use case, as shown in Fig. 2. It can be clearly seen that the feature map has the best effect in the convolutional layer of the fourth and seventh layers. However, at the 10th convolutional layer, the convolution effect decreases, and features such as headlights in the original image begin to be lost gradually. However, at the 13th convolutional layer, no effective feature map is obtained.

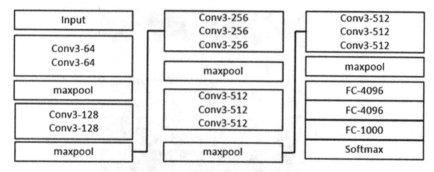

Fig. 1. VGG-16 network model.

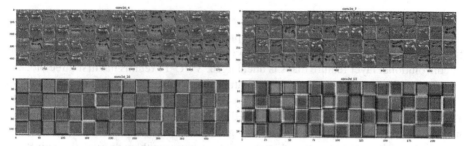

Fig. 2. VGG-16 intermediate layer activation diagram for an object

The class visualization of the convolutional layer 4, 7, 10 and 13 of the model is presented by the CAM. In 5 different vehicle features are obtained, as shown in Fig. 3. It can be seen from the thermal diagram that the features of the convolutional layer in the fourth layer get the best effect, and the CAM concentrates the main features of the classification on the front cover and roof. In the CAM of the seventh convolutional layer, it can be seen that, but the comparison between the characteristic regions of the fourth layer and other regions decreases. On the tenth layer, the outline of vehicle features can be barely seen, while on the thirteenth convolutional layer, there is no effective feature, resulting in the same invalid thermal diagram output by different vehicle images. Therefore, the conclusion is the same as that obtained by analyzing the activation diagram of the intermediate layer, and relatively good results can be obtained when the target is convolved at the fourth layer in the VGG-16 network.

2.2 Optimize the Network with Visualization

According to the conclusion of the above feature map visualization, the VGG-16 network is too deep for the simple target detection in this paper, and the features cannot be extracted well. Based on this conclusion, this paper adopts the method of reducing the number of network layers and the number of convolutional layer feature maps to design the network. This paper chose the best network structure by visually comparing the three optimized training results. These three network models are shown in Fig. 4,

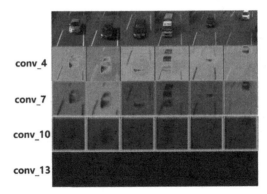

conv_4

conv_7

conv_10

conv_13

Fig. 3. VGG-16's CAMs for objects

respectively: VGG-16 retains the first two pieces and increases the number of convolutional layer feature maps; The first two pieces of VGG-16 are retained, and the number of convolutional layer feature maps is unchanged. VGG-16 retains the first three blocks, and the number of convolutional layer feature maps remains unchanged. These three network names are tentatively named network 1, network 2, and Network 3. Network 1 and network 3 are respectively compared with network 2 in terms of changing the number of convolution kernels and reducing the number of convolutional layers. This paper will continue to compare and analyze the gap between them through the activation diagram of the middle layer and CAM.

Input
Conv3-128 conv2d_1 Conv3-128 conv2d_2
maxpool
Conv3-256 conv2d_3 Conv3-256 conv2d_4
maxpool
FC-1000
Softmax

Input
Conv3-64 conv2d_1 Conv3-64 conv2d_2
maxpool
Conv3-128 conv2d_3 Conv3-128 conv2d_4
maxpool
FC-1000
Softmax

Input
Conv3-64 conv2d_1 Conv3-64 conv2d_2
maxpool
Conv3-128 conv2d_3 Conv3-128 conv2d_4
maxpool
Conv3-256 conv2d_5 Conv3-256 conv2d_6 Conv3-256 conv2d_7
maxpool
FC-1000
Softmax

Fig. 4. The three network models

In the last convolutional layer of the three networks, intermediate layer activation diagram of the same vehicle image is shown in Fig. 5. Through comparison, it can be seen that the gap between network 1 and Network 2 is not too large, and other areas such as headlights of the front of the car are featured. However, the characteristic area of network 3 is gradually diluted, and the contrast between the characteristic area and other noise areas is no longer obvious.

Using the same 5 vehicle images, the classification basis of the model is visualized through the class activation thermal diagram. This time, only the class activation thermal

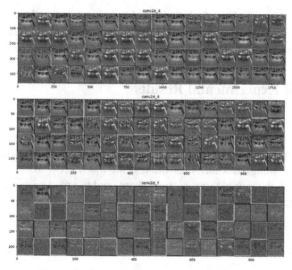

Fig. 5. intermediate layer activation diagrams

diagram of the last convolutional layer of the three networks is used as the model performance reference. The class activation thermal diagram shown by the three network models for the targets under different interference is shown in Fig. 6

Fig. 6. Thermal maps of different network models

It can be seen from the CAM that the characteristics of network 1 and network 2 for target classification are roughly the same, both are to identify the head and headlights. By comparison, it can be seen that the features extracted by network 2 to judge the target are more sufficient. However, network 3 almost takes the full map as the focus to identify vehicles, and the characteristics of vehicles are relatively more important than other areas. In the actual use, stains on the road surface may be identified as vehicles. Meanwhile, in the last image of empty road, network 3 is still identified as a vehicle, which is obviously wrong. Therefore, through the above visual analysis, it can be known that the classification model of network Model 2 is more accurate. Therefore, network model 2 will be adopted as the network model for practical use in this paper.

3   Analysis of Experimental Results

The Dataset adopted in this paper was the bitVehicle Dataset [7]. 7580 images were used as training set and 1371 images as test set. The image resolution of the training set is 1600 * 1200, and after preprocessing, the image becomes a three-channel image of 224 * 224. The image resolution of the test set was 1920 * 1080. After the results were preprocessed, the model test was carried out, and the road without the presence of vehicles was spliced from the consistent images of part of the road surface, and the road surface was cut as part of the data set, a total of 432 pieces [8].

In order to verify the effectiveness of the optimization method proposed in this paper, the evaluation of this problem is divided into accuracy Acc, Precision and Recall. Its calculation formula is as follows:

$$Acc = \frac{N_{TP} + N_{TN}}{N_{TP} + N_{FP} + N_{TN} + N_{FN}} \tag{1}$$

$$Precision = \frac{N_{TP}}{N_{TP} + N_{FP}} \tag{2}$$

$$Recall = \frac{N_{TP}}{N_{TP} + N_{FN}} \tag{3}$$

The number of N_{TP} is correctly classified to positive samples, N_{FP} is incorrectly classified to positive samples, N_{TN} is correctly classified to negative samples, and N_{FN} is incorrectly classified to negative samples (Table 1).

Table 1. Different networks' performance

Networks	Acc(%)	Precision (%)	Recall(%)
VGG-16	87.96	95.70	89.25
Network 1	88.13	97.16	88.39
Network 2	**89.02**	**98.03**	**88.71**
Network 3	87.85	97.88	87.59

Through the table, the recognition accuracy rate of network 2 adopted in this paper can reach 89.02% in actual use, Accelerate by the Geforce RTX 2060 gpu, training process once is used for an average of 55 s. Iteration after 250 time, loss value and accurate basic can achieve stable convergence, a total of 16513 s, compared VGG- 16 average round need 129 s, a total of 38749 s, the training time reduced by 57.38%.

The experimental results show that, in view of the problem of vehicle detection, this paper optimizes the network structure by using the visual optimization network method. At the same time, compared with the classification network with deeper convolutional layers, the operation time is reduced, the operation cost is saved, and the accuracy is slightly improved.

4 Conclusion

Aiming at the problem that the vehicle classification network layer is too deep in vehicle detection, this paper simplifies the VGG network model and eliminates unnecessary convolutional layer through the intermediate layer visualization technology of convolutional neural network, so as to improve efficiency and reduce training cost. The experimental results show that the visualized optimization method of convolutional neural network is reasonable and effective to deal with the problem of Vehicle detection in this paper, and a good detection effect is obtained on the BIT-Vehicle data set. At the same time, the method of visual adjustment of convolutional neural network proposed in this paper has some shortcomings. For example, after training, the convolutional neural network is manually adjusted through visualization, and in future studies, the subset selection method and other algorithms are used to automatically optimize the network.

References

1. Tesauro, G.: Temporal difference learning and TD-Gammon. Commun. ACM **38**(3) (1995)
2. Lucas, B.D., Kanade, T.: An iterative image registration technique with an application to stereo vision. In: Proceedings of the 7th International Joint Conference on Artificial Intelligence, vol. 2, pp. 674–679 (1981)
3. Max, X., Grimson, W.E.L.: Edge-based rich representation for vehicle classification. In: Tenth IEEE International Conference on Computer Vision (ICCV'05), vol. 2, pp. 1185–1192 (2005)
4. Stauffer, C., Grimson, W.E.L.: Adaptive background mixture models for real-time tracking. Computer Vision and Pattern Recognition. https://doi.org/10.1109/cvpr.1999.784637
5. Selvaraju, R.R., Cogswell, M., Das, A., Vedantam, R., Parikh, D., Batra, D.: Grad-CAM: visual explanations from deep networks via gradient-based localization. Springer **128** (2) (2020)
6. Simonyan, K., Zisserman, A.: Very deep convolutional networks for large-scale image recognition. Comput. Sci. (2014)
7. Engineering: New Engineering Data Have Been Reported by Researchers at University of Malaysia Pahang (Vehicle Type Classification Using an Enhanced Sparse-filtered Convolutional Neural Network With Layer-skipping Strategy). Comput. Netw. Commun. (2020)
8. Ferreira, A.J., Figueiredo, M.A.T.: Boosting Algorithms: A Review of Methods, Theory, and Applications[M]/ /Ensemble Machine Learning, pp. 35–85. Springer, New York (2012)

A Self-adaptive Multi-hierarchical Modular Neural Network for Complex Problems

Zhang Zhao-zhao, Wang Qiu-wan[✉], and Zhu Ying-qin

Xi'an University of Science and Technology, Xi'an 710600, China
18208207036@stu.xust.edu.cn, 1531662162@qq.com

Abstract. Due to the fact that the number of function models and the structure of the sub-model of the modular neural network are difficult to determine when applied to complex problems. This paper presents a self-adaptive multi-hierarchical modular neural network structure design method. In this method, a fast find of density peaks cluster algorithm is adopted to determine the number of the function modules, and a conditional fuzzy clustering algorithm is used to further divide the training samples of each function module into several groups to determine the number of sub-modules in each function module. For each sub-module, an incremental design of radical basis function (RBF) network network algorithm based on train error peak is applied to construct the structure of sub-modules which can self-adaptively build the structure of the sub-modules based on the training samples that allocated to the sub-modules. In sub-modules integration, a sub-module integrate approach based on relative distance measure is applied which can select different sub-modules from different function modules to collaboratively learning the training samples. Experiment results demonstrate that the self-adaptive multi-hierarchical modular neural network can not only solve the complex problems that the fully coupled RBF difficult to deal with, but also can improve the learning accuracy and generalization performance of the network.

Keywords: Multi-hierarchical · Modular neural network · Self-adaptive

1 Introduction

It is well known that feed-forward neural networks, such as radical basis function (RBF) network, is considered as universal approximator and applied to solve various problems in many fields, such as pattern recognition, nonlinear compensations, and robotic manipulators [1–4]. However, RBF is a fully coupled network in which each input sample activates all hidden nodes [5]. This characteristic of RBF not only leads to a large calculation cost but also makes knowledge accumulation extremely difficult, since the

Support by the National Natural Science Foundation of China (No.61440059), the Natural Science Foundation of Liaoning Province (No.201602363), China Scholarship Council (No.201508210045) and the Basic Research Plan of Nature Science in Shaanxi Province of China (No.2020JM-522).

© Springer Nature Switzerland AG 2020
B. Ben Hedia et al. (Eds.): VECoS 2020, LNCS 12519, pp. 244–256, 2020.
https://doi.org/10.1007/978-3-030-65955-4_18

RBF tend to forget previously learned mappings rather quickly when exposed to new mappings [6].

Modular design is an effective method for alleviating this forgetfulness and decreasing the computation costs common for fully coupled neural network [7]. The concept of modularity in artificial neural networks was inspired by biological and psychological research proving that modularity is the key to the efficiency and intelligence of the human brain [8]. The conventional modular neural networks (MNNs) adopt the principle of "divide and conquer", which means that each MNN system consists of several specialized modules, each of the modules performs a subtask of the original complex task, each of the modules has simpler structure, and each evaluates the same or distinct input samples without communicating with other modules [9]. After each module obtains its independent results, an integration unit combines the results in a predefined method to generate the overall output of the complex system [10]. There are many advantages to adopting a modular approach to design neural network structures, such as efficiency, simplicity of structure, ease of evaluation, fault tolerance, better extendibility, design and implementation flexibility, and robustness [11]. Because the MNN has so many advantages, so it is widely used in various fields. Monlina [12] proposed a MNN architecture for learning of grasping tasks, Nikola [13] presented a MNN to price the S&P-500 European call options, however, the MNN architecture design methods described above require highly domain knowledge. Tseng [14] applied the MNN to pattern profiling problems, and the number of the sub-modules in MNN is determined by random classification of the training samples space, however, for many practical problems, especially for classification problems, this method can not improve the learning performance of the MNN necessarily. Bo [15] presented a multi-module neural network based on the visual system work principle, this MNN have an extremely learning ability for complex problems, however, the number of the sub-modules in this MNN is determined by adopting K-Means cluster algorithm which is very sensitive to the initial cluster center, and the number of the clusters must be determined in advance. Moreover, the structure of the sub-modules in those MNN described above must determined in advance, therefore, we have to spend lot of time and many experiments to determine an appropriate sub-module structure when those MNNs were applied to practical applications.

Since the brain networks demonstrate the property of hierarchical modularity, within each module there will be a set of function modules, and within each function module a set of sub-modules [16, 17]. In this paper we present a brain-like self-adaptive multi-hierarchical modular neural network (BMNN), there are some new characteristics in the BMNN compared with conventional MNN. (1) BMNN owns a brain-like multi-hierarchical structure, which will improve its information processing capability. (2) the number of the function modules in BMNN is automatically determined by the training sample space. (3) the structure of the sub-modules is self-adaptively determined by the training samples. (4) unlike the conventional MNN learning method, for each learning sample, one or more function module will participation in the learning process, this learning method in BMNN will improve the learning accuracy, especially for the information of the classification boundary, and it also can improve the generation ability of BMNN.

The remainder of the paper is organized as follows: Sect. 2 describes the structure of the BMNN. Section 3 discusses the BMNN's structure design method. Section 4 introduces the applications of the BMNN. Section 5 summarizes and concludes.

2 The Structure of the BMNN

Without loss of generality, BMNN is a multi-input single-output system. Each layer of the BMNN performs specific operations in tandem to realize a learning task. The structure of the BMNN proposed by us is shown in Fig. 1, BMNN has an obviously hierarchical structure, the whole module is divided into F function modules, and each function module includes several sub-modules. The final output of BMNN can be acquired by integration of subsequent levels of neurons. It should be noted, unlike conventional MNN [18], in BMNN learning process, for a given training sample, especially in the classification boundary, there will be a number of sub-modules from different function modules coordination to complete learning task.

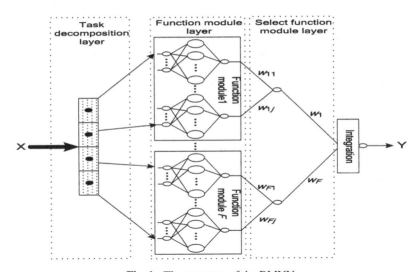

Fig. 1. The structure of the BMNN.

In BMNN, each sub-module is a RBF network, its task is to learn the training sample assigned from task distribution layer, and its structure is self-adaptively determined by those training samples. Without loss of generation, suppose the structure of the H_i sub-module in F function module is $I - M - 1$ (I input units, M hidden units, 1 output unit), the training sample assigned to H_i sub-module is $x_k = (x_{k,1}, x_{k,2}, \ldots, x_{k,I})$, then the output of the H_i sub-module is follow

$$O_{h_i,k} = \sum_{h=1}^{M} w_h \phi_h(x_k) + w_0 \tag{1}$$

$$\phi_h(x_k) = exp(-\frac{\|x_k - c_h\|^2}{\sigma_h}) \tag{2}$$

where w_h presents the weight on the connection between the RBF unit h and network output; w_0 is the bias weight of output unit; c_h and σ_h are the center and width of RBF unit, respectively.

3   Construction of the BMNN Structure

The structure construction method of the described BMNN includes: how to divide the function modules and the sub-modules, how to select sub-module from each function module for each training sample, how to integration the learning result, and how to self-adaptively construction the structure of the sub-modules based on the training sample that assigned task distribution layer.

3.1   Division of the Function Modules

Clustering algorithm is often applied to determine the number of sub-modules in the absence of learning objects background knowledge [19]. Alex presents a cluster algorithm by fast search and find of density peaks [20], the core idea of this algorithm is that cluster centers are characterized by a higher density than their neighbors and by a relatively large distance from points with higher densities. This cluster idea is used in this paper to identify the cluster centers of the training samples to determine how many function modules that the BMNN should have. Two parameters ρ_i and δ_i are used in this algorithm, where ρ_i is the local density of data point i and δ_i is distance of the data point i from points of higher density. Suppose the training sample set of the BMNN is $S = \{(x_k, y_k), x_k \in R^l, y_k \in R^1, k = 1, \ldots, N\}$, for each training sample x_i, the ρ_i and δ_i are defined as

$$\rho_i = \sum_{\substack{j=1 \\ j \neq i}}^{N} \exp -\left(\frac{d_{ij}}{d_c}\right)^2 \tag{3}$$

$$\delta_i = \min_{j:\rho_j > \rho_i} d_{ij} \tag{4}$$

where $d_{ij} = dist(x_i, x_j)$ represents the Euclidean distance of data point i and j ; and d_c is a cutoff distance.

Suppose the cluster centers is $\{c_1, c_2, \ldots, c_F\}$ according to above cluster algorithm, then there will be F cluster centers, establishing the training samples S into F fuzzy set as the follow

$$f_{ik} = \exp\left(\frac{-\|x_k - c_i\|^2}{0.02}\right) \tag{5}$$

where f_{ik} is the fuzzy membership degree of the training sample x_k to i^{th} fuzzy set, and $i = 1, \ldots, F$. As shown in (5), because each of the cluster of c_i have established their fuzzy set for all training samples, so the BMNN can be divided into F function modules based on the spatial distribution of the training samples, and each function module has its own training sample set.

3.2 Division of the Sub-modules in Each Function Module

The division of the function modules initially realized division of the training sample, but for complex issues owns large training samples, the absolute number of the training samples assigned to each function module is still very large. In order to improve the learning performance of the BMNN, we need to divide the training samples of each function module training samples.

For each function module training sample fuzzy set, conditional fuzzy clustering the training sample x_k in fuzzy set of each function module as the follow

$$u_{ijk} = \frac{f_{ik}}{\sum_{m=1}^{H_i} \left(\frac{\|x_k - v_{ij}\|}{x_k - v_{im}} \right)^2} \tag{6}$$

$$v_{ij} = \frac{\sum_{k=1}^{N} \left(u_{ijk} \right)^2 x_k}{\sum_{k=1}^{N} \left(u_{ijk} \right)^2} \tag{7}$$

where v_{ij} is the ith $(i = 1, \ldots, F)$ function module training samples fuzzy set corresponding to the jth $(j = 1, \ldots, H_i)$ cluster centers of the input fuzzy sets S; H_i is the cluster center number of the ith function module training sample set; $U = [u_{ijk}] \in R^{F \times H_i \times N}$ is the clustering segmentation matrix of the ith function module training sample set, and u_{ijk} represents the fuzzy membership of x_k to j^{th} sub-module of the ith function module, which satisfy

$$\sum_{j=1}^{H_i} u_{ijk} = f_{ik}, u_{ijk} \in [0, 1] \tag{8}$$

The method aforementioned can divide the whole training sample set S into H_T training sample subsets, which satisfy

$$H_T = \sum_{i=1}^{F} H_i \tag{9}$$

Based on the results of the conditional fuzzy clustering, we can establish several sub-modules for each function module, and each sub-module owns its own training sample set. That is to say, there will be F function modules, and each function module have H_i sub-modules. In this paper, FM_j $(i = 1, \ldots, F)$ represents the ith function module, and SM_{ij} represents the jth $(j = 1, \ldots, H_i)$ sub-module of ith function module.

3.3 Select Sub-modules from Function Modules

The essence of the selection sub-modules from functional module is to determine which sub-modules will be selected to learning the training sample x_k in the learning process of the BMNN. According to the aforementioned training samples divided method, there exists some affiliations between each training sample x_k and each sub-module SM_{ij}, therefore, the selection of sub-module from function module is to determine the likelihood of x_k belongs to sub-module SM_{ij}, if the distance of x_k is close to the center of v_{ij}, the conditional fuzzy cluster center of SM_{ij}, then the likelihood of x_k belongs to SM_{ij} is larger. In this paper, the relative distance is adopted to measure the likelihood of x_k belongs to SM_{ij}.

$$J_i = \sum_j^{H_i} w_{ij}d_{ij}; \sum_j^{H_i} w_{ij} = 1; w_{ij} \in [0, 1] \tag{10}$$

$$d_{ij} = \|x_k - v_{ij}\| / da_{ij}; da_{ij} = \frac{1}{N_{ij}} \sum_{m-1}^{N_{ij}} \|x_m - v_{ij}\| \tag{11}$$

where da_{ij} is the average distance of the training samples in SM_{ij}; d_{ij} is the relative distance measure of x_k to v_{ij} (the center of SM_{ij}); N_{ij} is the training sample number of SM_{ij}; and w_{ij} is the membership of x_k for SM_{ij}.

Minimizing the performance index function J_I with Lagrangian multiplier method can solve w_{ij}:

$$w_{ij} = \begin{cases} 1, & when\ d_{ij} = 0. \\ \dfrac{\left(\frac{1}{d_{ij}}\right)}{\sum\limits_{j=1}^{H_i}\left(\frac{1}{d_{ij}}\right)}, & otherwise. \end{cases} \tag{12}$$

Obviously, if d_{ij} is larger then w_{ij} is smaller, and the likelihood of x_k belongs to SM_{ij} is smaller, on the contrary, if d_{ij} is smaller then w_{ij} is larger, and the likelihood of x_k belongs to SM_{ij} is larger. We can select a most suitable sub-module to learn the training sample x_k from several sub-modules in each function module based on the principle of maximum membership, and the selected sub-module is SM_{is}, where $w_{is} = 1$, and $w_{ij,i \neq s} = 0$. Thus, the output of the function module is the output of the SM_{is}. However, this is only a primary selection, not all selected sub-modules SM_{is} from each function module are suitable for take part in the learning process. Therefore, the sub-modules that are selected but not suitable for processing x_k must be filtered. The method of filtering selected sub-modules is the same way as the aforementioned sub-module selection method. Establish the performance index function for the selected sub-modules:

$$J = \sum_i^F w_i d_i, \sum_i^F w_i = 1, w_i \in [0, 1] \tag{13}$$

$$d_i = \|x_k - c_i\| / da_i, da_i = \frac{1}{N_i} \sum_{j=1}^{N_i} \|x_k - c_i\| \tag{14}$$

Where d_i represents the relative distance measure of x_k to c_i (the center of FM_i); da_i represents the average distance of training samples in FM_i. N_i is the number of the training sample in FM_i.

Minimizing the performance index function J, use Lagrange multiplier method can figure out w_i as

$$w_i = \begin{cases} 1, & when\ d_i = 0. \\ \dfrac{\left(\frac{1}{d_i}\right)}{\sum\limits_{i=1}^{F}\left(\frac{1}{d_i}\right)}, & otherwise. \end{cases} \tag{15}$$

Obviously, if the relative distance of d_i is larger then the w_i is smaller, this means that the sub-modules in this function module are not suitable to learn x_k. Because each function module has only one sub-module will be selected to learning the training sample x_k, therefore, the w_i is actually the membership degree of x_k to FM_i. Take into account the overlapping of the training samples among the function modules, a threshold K can be set, the sub-modules which satisfies the condition $w_i \geq K$ will be selected to learning x_k. Through the above-described selection method, for a given training sample x_k, the varying amounts of sub-modules will be involved in learning process, especially for the training samples in classification boundaries, there will be several sub-modules from different function modules to learn it.

3.4 Integration the Output of Sub-modules

Suppose the training sample is x_k, let $w = (w_1, \ldots, w_F)$, if $w_i < K$ then let $w_i = 0$, normalize w, then the output of BMNN is

$$Y = \sum_{i=1}^{C} w_i o_i \tag{16}$$

where O_i is the output of the SM_{is} as shown in (1), w_i is the weight value of SM_{is}, which is the ith component of the normalized w. The value of the unselected sub-modules $w_i = 0$, so it make no contribution to the output of the BMNN.

3.5 Self-adaptively Construction the Structure of Sub-modules

In BMNN, each sub-modules is a RBF network, how to construct an appropriate structure of RBF network according to the learning task is a difficult problem, Wilamowski improved the Levenberg-Marquardt(LM) learning algorithm, and Yu proposed an ErrCor algorithm which is an incremental design of RBF networks. The basic idea of the ErrCor algorithm is to use RBF units with kernel function (1) to create a peak/valley shape to compensate for the largest error in the error surface at the beginning of each iteration, and it is able to design the most compact RBF structure.

The LM algorithm parameters update rule is given by

$$\triangle_{k+1} = \triangle_k - (\mathbf{Q}_k + \mu_k \mathbf{I})^{-1} \mathbf{g}_k \tag{17}$$

where vector Δ consists of parameters, including centers c, widths σ, and weights w; Q is the quasi Hessian matrix; I is identify matrix; μ is the combination coefficient; and g is the gradient vector.

Quasi-Hessian matrix Q can be calculated as the sum of subquasi-Hessian matrix q_p

$$Q = \sum_{p=1}^{P} q_p; \ q_p = j_p^T j_p \tag{18}$$

and gradient vector g can be calculated as the sum of subgradient vector η_p

$$g = \sum_{p=1}^{P} \eta_p; \eta_p = j_p^T e_p \tag{19}$$

and the training error e_p can be calculated as

$$e_p = y_p - o_p \tag{20}$$

where y_p is desired output; and o_p is actual output.

The elements of Jacobian row j_p can be calculated by

$$j_{p,n} = \frac{\partial e_p}{\partial \Delta_n} \tag{21}$$

For a given training sample X_p, considering the RBF network parameters w_h, $c_{h,i}$ and σ_h, the elements of the Jacobian row can be organized as

$$\begin{aligned}
j_{p,n} = [& \frac{\partial e_p}{\partial w_0}, \frac{\partial e_p}{\partial w_1} L \frac{\partial e_p}{\partial w_h} L \frac{\partial e_p}{\partial w_h}, \frac{\partial e_p}{\partial c_{1,1}} L \frac{\partial e_p}{\partial c_{1,i}} L \\
& \frac{\partial e_p}{\partial c_{1,I}} L \frac{\partial e_p}{\partial c_{h,1}} L \frac{\partial e_p}{\partial c_{h,i}} L \frac{\partial e_p}{\partial c_{h,I}} L \frac{\partial e_p}{\partial c_{H,1}} L \\
& \frac{\partial e_p}{\partial c_{H,i}} L \frac{\partial e_p}{\partial c_{H,i}}, \frac{\partial e_p}{\partial \sigma_1} L \frac{\partial e_p}{\partial \sigma_h} L \frac{\partial e_p}{\partial \sigma_H}]
\end{aligned} \tag{22}$$

Integrating (1), (2) and (20), with differential chain rule, the Jacobian row elements for X_p in (22) can be rewritten as

$$\frac{\partial e_p}{\partial w_h} = -\varphi_h(X_p), \frac{\partial e_p}{\partial w_0} = -1 \tag{23}$$

$$\frac{\partial e_p}{\partial c_{h,i}} = -\frac{2 w_h \varphi_h(X_p)(x_{p,i} - c_{h,i})}{\sigma_h} \tag{24}$$

$$\frac{\partial e_p}{\partial \sigma_h} = -\frac{w_h \varphi_h(X_p) \|X_p - c_h\|^2}{\sigma_h^2} \tag{25}$$

With (23)–(25), all the elements of Jacobian row J_p for the given training sample X_p can be calculated. After applying all the patterns, quasi-Hiessian matrix Q and gradient vector g are obtained by (18) and (19), so as to apply the update rule (17) for parameter adjustment. It can self-adaptively construct the structure of the sub-modules according to the training samples from task decomposition layer by applying the aforementioned ErrCor algorithm.

4 Experiments and Analysis

In order to demonstrate the performance of the BMNN, in this section, the BMNN is applied to many well-known nonlinear bench tests. The testing environment of the proposed BMNN consists of the Windows 7 64-bit operating system, an Intel Core i7-2600 CPU @ 3.4-GHz processor, and 8-GB RAM.

4.1 Two-Spiral Problem

The two-spiral problem is primarily used as a benchmark for pattern classification. It can also be used as an approximation problem where patterns on one spiral should produce $+1$ outputs, while patterns on the other spiral should produce -1 outputs. This problem is widely used as a challenging benchmark to evaluate the efficiency of learning algorithms and their network architectures. The method of generating training samples is as follow

$$\begin{cases} \theta = i \times \dfrac{\pi}{16} \\ r = 6.5 \times (104 - i)/104 \\ x = r \times \sin(\theta) \\ y = r \times \cos(\theta) \end{cases} \tag{26}$$

where $i = 0, \ldots, 192$, so (x, y) and $(-x, -y)$ can constitute two kinds of category, a total of 388 patterns; test samples is $x = -6.6 : 0.1 : 6.5$ and $y = -6.5 : 0.1 : 6.5$, a total of 17161 test patterns. In this experiment, the training sample is divided into 3 function modules by task decomposition layer, set the parameters $K = 0.1$, and the number of sub-module in each function is 2.

Figures 2, 3, 4, 5, 6 and 7 shows the learning performance of the 6 sub-modules in 3 function modules. A few facts can be observed from those figures. First, each sub-module in BMNN can accurately learning their own task that assigned from task decomposition layer, Second, the training sample division method proposed in this paper can allocate different train patterns to different sub-modules based on their space distribution. It is interesting to see from those figures that the learning object of each sub-module has no twining phenomenon, that is to say the learning object of each sub-module is simple, this means that the training sample division method proposed in this paper can really divide the complex task into several simple task. Third, each sub-module can self-adaptively construct its own structure according to the training samples allocated from task decomposition layer. We can also see from those figures that the training samples between different sub-modules is partly overlap, this is because that the fuzzy strategy is used in training sample division method, this method enables the training samples in classification boundaries to be cooperating learned by different sub-modules from different function module, this learning method not only can improve the learning accuracy but also cam improve the generalization ability of the BMNN.

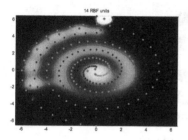

Fig. 2. Learning result of sub-module 1

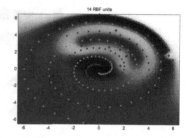

Fig. 3. Learning result of sub-module 2

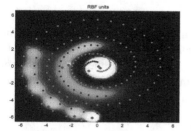

Fig. 4. Learning result of sub-module 3

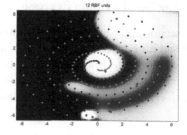

Fig. 5. Learning result of sub-module 4

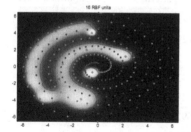

Fig. 6. Learning result of sub-module 5

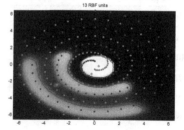

Fig. 7. Learning result of sub-module 6

After the learning process, the hidden unites of each sub-module are 14, 14, 16, 12, 16, 13, respectively. For the same problem, the RBF-MLP networks required at least 74 RBF units to solve the two-spiral problem. It was reported by Neruda that the two-spiral problem was solved using 70 hidden RBF units. Using the orthonormalization procedure, the two-spiral problem can be solved with at least 64 RBF hidden units. Therefore, compared with the learning method aforementioned, the BMNN has powerful learning ability. Figure 8 shows the test effect of BMNN, and the correct classification rate of 17161 test patterns touched 99.83%, this means that the proposed BMNN has a strong generalization performance.

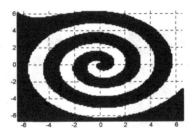

Fig. 8. Test result of BMNN

4.2 The Real Life Data Regression

This section compares BMNN with well-known algorithms on traditional benchmarks form various repositories. These are real life problems with many dimensions and with number of patterns from hundreds to thousands. Table 1 shows the specifications of the benchmark data sets. In our experiments, all of the inputs have been normalized into the range [−1,1] while the outputs have been normalized into [0,1].

Table 1. Specification of real life data sets

Real life problem	Train patterns	Test patterns	Input dimensions
Abalone	2000	2177	8
Delta ailerons	3000	4129	5
Delta elevators	4000	5517	6
Computer activity	4000	4192	8
Census	10000	12784	8
Bank domains	4500	3692	8
California housing	8000	1246	8

The comparison of root mean square error (RMSE) on several algorithms is shown in Table 2. It can be noticed from Table 2 that the test RMSE of proposed BMNN on all of the data set are smaller than the other algorithms. A comparison of training times for different algorithms on all of the data sets can be observed in Table 3. Again, the proposed BMNN has a faster training time than other algorithm. There are two reasons why proposed BMNN outperforms other algorithms. First, each sub-module in BMNN learning a relatively simple task, and the training samples of each sub-module is less than the monolithic module, so the learning speed is faster than other RBF networks. Second, the BMNN applied a fuzzy strategy to division the training samples, and the training samples in classification boundary well be collaboratively learned by several sub-modules from different function modules, therefore the generalization performance of the BMNN is outperforms other commonly used algorithms.

Table 2. Test RMSE comparison of several algorithms

Real life problem	RAN	ErrCor	BMNN
Abalone	0.0978	0.0765	0.0501
Delta ailerons	0.0552	0.0431	0.0280
Delta elevators	0.0733	0.0573	0.0375
Computer activity	0.0649	0.0507	0.0328
Census	0.0905	0.0707	0.0461
Bank domains	0.0579	0.0452	0.0294
California housing	0.1434	0.1012	0.0655

Table 3. Training time comparison of several algorithms

Real life problem	RAN(s)	ErrCor(s)	BMNN(s)
Abalone	105.17	4.808	2.070
Delta ailerons	114.12	5.219	3.760
Delta elevators	131.46	5.997	3.902
Computer activity	120.94	5.519	3.153
Census	241.89	11.06	7.480
Bank domains	147.55	6.750	4.330
California housing	212.44	9.710	7.422

5 Conclusion

For the fully coupled neural network RBF is difficult to deal with complex issues, and the problems that it is difficult to determine the number of function modules and sub-modules when constructing the modular neural network structure. This paper presents a self-adaptive multi-hierarchical modular neural from the angle of the brain-link information processing. The proposed BMNN solves the problem of how to determine the number of number of sub-modules and the structure of the sub-modules in the absence of learning objects background knowledge, moreover, the proposed BMNN requires few artificial parameters compared with conventional MNN structure design methods, therefore, the proposed method really achieves the black box effective of neural network in some degree. Experiments demonstrate that the BMNN can not only solve the complex problems that the fully coupled RBF difficult to deal with, but also can improve the learning accuracy and generalization performance of the network.

References

1. Xu, L., Qian, F., Li, Y., Li, Q., Yang, Y., Xu, J.: Resource allocation based on quantum particle swarm optimization and RBF neural network for overlay cognitive. Neurocomputing **173**(3), 1250–1256 (2016). Author, F., Author, S.: Title of a proceedings paper. In: Editor, F., Editor, S., (eds.) Conference 2016, LNCS, vol. 9999, pp. 1–13. Springer, Heidelberg (2016)

2. Quan, L., Fei, X., Qi-Ming, F., et al.: Collaborative Q-V value function approximation model based on adaptive normalized radial base function network. Chin. J. Comput. **38**(7), 1386–1396 (in Chinese)

3. Xiong, T., Bao, Y., Hu, Z., Chiong, R.: Forecasting interval time series using a fully complex-valued RBF neural network with DPSO and PSO algorithms. Inform. Sci **305** (2015)

4. Li-Cheng, J., Shu-Yun, Y., Fang, L., et al.: Seventy years beyond neural networks: retrospect and prospect. Chin. J. Comput. **39**(8), 1697–1716 (2016) (in Chinese)

5. Yu, H., Reiner, P.D., Xie, T., Bartczak, T., Wilamowski, B.M.: An incremental design of radial basis function networks. IEEE Trans. Neural Netw. Learn. Syst. **25**(10) (2014)

6. Liang, L., Guo, W., Zhang, Y., Zhang, W., Li, L., Xing, X.: Radial Basis Function Neural Network for prediction of medium-frequency sound absorption coefficient of composite structure open-cell aluminum foam. Appl. Acoust. **170**, 107505 (2020)

7. Jacobs, R.A, Michael, I, Jordan, A.: Modular connectionist architecture for learning piecewise control strategies. In: Proceedings of the American Control Conference, pp. 343–224. Boston, USA (1991)

8. Li, W., Li, M., Zhang, J., Qiao, J.: Design of a self-organizing reciprocal modular neural network for nonlinear system modeling. Neurocomputing **411** (2020)

9. Qiao, J.F., Zhang, Z.Z., Bo, Y.C.: An online self-adaptive modular neural network for time-varying systems. Neurocomputing **125**(11), 7–16 (2014)

10. Smetana, S., Seebold, C., Heinz, V.: Neural network, blockchain, and modular complex system: the evolution of cyber-physical systems for material flow analysis and life cycle assessment. Resour. Conservat. Recycl. **133** (2018)

11. Happel, B.L.M., Murre, J.M.J.: Design and evolution of modular neural network architectures. Neural Netw. **7**(6–7) (1994)

12. Molina-Vilaplana, J., Feliu-Batlle, J., López-Coronado, J.: A modular neural network architecture for step-wise learning of grasping tasks. Neural Netw. **20**(5) (2007)

13. Nikola, G., Ramazan, G., Dragan, K.: Option pricing with modular neural networks. IEEE Trans. Neural Netw. **20**(4) (2009)

14. Tseng, H.C., Almogahed, B.: Modular neural networks with applications to pattern profiling problems. Neurocomputing **72**, 2093–2100 (2009)

15. Ying-Chun, B., Jun-Fei, Q., Gang, Y.: A multi-modules cooperative neural networks. CAAI Trans. Intell. Syst. **3**, 225–230 (2011) (in Chinese)

16. Melin, P., Miramontes, I., Prado-Arechiga, G.: A hybrid model based on modular neural networks and fuzzy systems for classification of blood pressure and hypertension risk diagnosis. Expert Syst. Appl. **107**, 146–164 (2018)

17. Goltsev, A., Gritsenko, V.: Modular neural networks with radial neural columnar architecture. Biol. Inspired Cognitive Arch. **13**, 63–74 (2015)

18. Stoffel, M., Gulakala, R., Bamer, F., Markert, B.: Artificial neural networks in structural dynamics: a new modular radial basis function approach vs. convolutional and feedforward topologies, Comput. Meth. Appl. Mech. Eng. **364**, 112989 (2020)

19. Srivastava, V., Tripathi, B.K., Pathak, V.K.: Evolutionary fuzzy clustering and functional modular neural network-based human recognition. Neural Comput. Appl. **22**(1) (2013)

20. Rodriguez, A., Liao, A.: Clustering by fast search and find of density peaks. Science **344**, 1492–1496 (2014)

Multi-scale Risk Assessment Model of Network Security Based on LSTM

Yinfei Lv, Huorong Ren[(✉)], Xuefeng Gao, Tong Sun, Haopeng Zhang, and Xinyu Guo

School of Electro-Mechanical Engineering, Xidian University, Xi'an, China
`yinfeilv@qq.com, hrren@xidian.edu.cn`

Abstract. As the problem of network security becomes more and more serious, how to accurately perceive the current network security situation and discover the attack behavior in time has become the focus of research in the field of network security. This paper proposes a multi-scale risk assessment model for network security based on Long Short Term Memory neural network (LSTM). The model utilizes the wavelet transform to decompose network traffic time series into sub-sequences of various scales. The LSTM network is used to predict the sub-sequence of wavelet decomposition. By comparing the difference between the actual sub-sequence and the predicted sub-sequence, the model determines whether there is "anomaly" in the network traffic time series. The anomaly detection results of each sub-sequence are summarized into a network security risk value to assess the risk of network security. The introduction of the multi-scale technology improves the detection accuracy of network traffic time series anomaly detection and effectively enhances the reliability of the risk value.

Keywords: Time series · Risk assessment · LSTM · Wavelet decomposition

1 Introduction

The rapid development of the Internet has changed the way of human production and life, and promoted the social transformation and development. But at the same time, the scale and complexity of the network are increasing, and the means of network attack emerge one after another. Common attack behaviors are "Denial of Service" and "Distributed Denial of Service". The purpose of a Denial of Service (DoS) attack is to make a computer or network unable to provide a normal Service. Distributed Denial of Service (DDoS) means that multiple attackers in different locations attack one or more targets at the same time, or that one attacker controls multiple machines in different locations and uses those machines to attack the victim simultaneously. Network attack behavior causes great harm to the society and people, how to effectively identify the attack behavior in the massive network traffic data and evaluate the current network security situation in a timely manner is very important.

Network traffic time series is a kind of data with nonlinear, non-stationary and long-term memory characteristics. As network traffic time series are noisy, non-parametric dynamic system, the analysis of the time series is a challenging task.

© Springer Nature Switzerland AG 2020
B. Ben Hedia et al. (Eds.): VECoS 2020, LNCS 12519, pp. 257–267, 2020.
https://doi.org/10.1007/978-3-030-65955-4_19

The time series anomaly detection technology extract useful information efficiently from massive data. There are three types of time series anomaly detection: clustering-based anomaly detection, distance-based anomaly detection and predicted-value-based anomaly detection. With the development of artificial intelligence, the time series anomaly detection technology based on predicted value is constantly improved. The anomaly detection technology based on the LSTM network is a typical example. Developed from recursive neural network, it solves the problem of limited data in the process of time series. However, there is still much room for improvement in anomaly detection model based on LSTM at a single-scale. For example, we can improve the detection rate of "anomalies" and reduce the rate of false detection.

There are multi-scale effects in nature [1, 2], and time series is no exception. In other words, different observation results can be obtained by observing time series at different scales. The development of the wavelet theory provides a theoretical basis for practitioners to extract the multi-scale characteristics of time series. In this paper, the db4 wavelet is used to decompose network traffic time series to obtain sub-sequences at different scales, and then each sub-sequence is reconstructed.

In this work, the LSTM network is selected to predict the reconstructed sub-sequences. Compared with traditional artificial neural network time series prediction and back propagation (BP) neural network time series prediction [3], the LSTM network effectively avoids problems such as "limited long-term memory data", "gradient disappearance", and "gradient explosion". This choice effectively improves the prediction accuracy of time series and the performance of time series anomaly detection, and increases the reliability of the risk assessment model of network security.

The main contribution of this paper is to introduce multi-scale analysis on the basis of the risk assessment model of network security based on the LSTM network to realize multi-scale anomaly detection of network traffic time series, thus improving the reliability of the model. Experimental results show that the proposed method can effectively improve the reliability of the evaluation model.

The organization of the paper is as follows. First, in Sect. 2, we introduce the related technologies and concepts of time series anomaly detection as well as the LSTM network and the risk assessment model based on the LSTM network of network security. This section helps the readers understand how to make a risk assessment based on the LSTM network. In Sect. 3 we explain why and how to introduce multi-scale technology, recall the basic theory of wavelet analysis and multi-scale risk assessment model. Section 4 is the experimental chapter. We first use the LSTM network to conduct a single-scale risk assessment of network traffic time series and analyze its assessment effect, namely, the reliability of the model. Then the network traffic time series is decomposed and reconstructed by wavelet, and then the risk is evaluated. Finally, the reliabilities of single-scale and multi-scale evaluation model are compared. In Sect. 5 we review the relevant work, and summarize our contributions.

2 A Single-Scale Risk Assessment Model for Network Security Based on LSTM

Time series is a period of data with obvious time characteristics. Production practice, scientific research, stock trading and other human activities produce a large number of time series. Only a few of these data are valuable, and one of high research values is abnormal data.

2.1 Time Series Anomaly Detection

Traditional anomaly detection methods are based on statistics to judge anomaly, but with the explosive growth of data, a large number of time series with complex internal rules have emerged. The traditional statistical model has been unable to meet the requirements. At present, there are three types of anomaly detection technologies: anomaly detection based on distance, anomaly detection based on clustering, and anomaly detection based on predicted value. The model proposed in this paper is an anomaly detection technique based on predicted value.

The core idea of time series anomaly detection technology based on distance is to calculate the distance between data points, and the distance of abnormal data is often far from the normal data. These distances can be either Euclidean or Markov. Korr et al. [4] first proposed a distance-based anomaly detection method. Later, Ramaswamy et al. [5] proposed a K-nearest neighbor distance to identify anomaly by describing the sparsity of data points.

Distance-based detection technology takes the distance of the whole time series as the judgment basis of anomaly detection, while clustering-based detection technology focuses on the distribution of local data. In the latter, the anomaly is determined by the density relation between the data point to be detected and the data in the neighborhood of the data point. Agyemang et al. [6] first proposed the concept of "local anomaly".

The idea of time series anomaly detection based on predicted value is that a period of time series conforms to a change model. We can study the normal data, build the model and fit it. If there is a big difference between the predicted data and the actual data, the data is judged to be abnormal. Among many prediction models, the ARMA [7] model and ARCH model [8, 9] have better prediction performance. With the rapid development of artificial intelligence, more and more detection methods have been proposed, such as those based on support vector machine [10] and BP neural network [11]. The LSTM network proposed in [12] is developed from recursive neural network, which overcomes the difficulty of recursion neural network to predict long-term time, and solves the problems of "gradient disappearance" and "gradient explosion". The LSTM network is used to effectively predict the network traffic in [13].

2.2 Time Series Anomaly Detection Model Based on LSTM

Compared with other neural networks, the LSTM networks avoid the problem of "limited long-term memory data" and make the prediction results of time series more practical, thus improving the detection rate of time series anomalies and reducing the rate of false detection.

The process of anomaly detection using the LSTM prediction model is as follows. Firstly, the historical network traffic time series is input into the time series anomaly detection model based on LSTM to obtain the prediction series $y(t)$. The difference series $e(t)$ between the test time series $x(t)$ and the forecast time series $y(t)$ are calculated. Then the standard deviation $s(t)$ and the mean $m(t)$ are calculated for $e(t)$ in a sliding window of a certain length. When the absolute value of $e(t)$ exceeds the absolute value of $m(t)$ by N times the standard deviation of $s(t)$, the current data is considered "anomaly". N is defined as the "anomaly" sensitivity coefficient.

Figure 1 shows the diagram for time series anomaly detection.

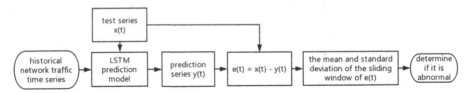

Fig. 1. Diagram of time series anomaly detection model based on LSTM

2.3 A Single-Scale Risk Assessment Model

The single scale risk assessment model is based on the time series anomaly detection model. The number of "anomaly" detected in a certain length of time series is defined as the risk value. The more the number of "anomaly" detected, the greater the probability of the existence of risks in this network, and vice versa. The risk assessment process for network security is shown in Fig. 2.

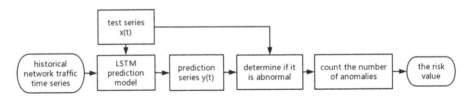

Fig. 2. Diagram of a single-scale network security risk assessment model

The effect of time series anomaly detection is directly related to the reliability of network security risk assessment model. If the anomaly detection model fails to effectively detect all the "anomalies", the risk assessment model will result in low assessment values. On the contrary, if the anomaly detection model mistakenly considers normal data as "abnormal", it will lead to higher risk values from the risk assessment model. In either case, the reliability of the risk assessment model is reduced.

3 Multi-scale Risk Assessment Model of Network Security Based on LSTM

In nature and human society, many things observed at different scales lead to different results, this phenomenon is called multi-scale effect. Multi-scale analysis is also known as multi-resolution analysis. To conduct multi-scale analysis on time series is to decompose the time series at different resolutions to obtain the high frequency part and the low frequency part. The high frequency part reflects the change in the details of time series, while the low frequency part reflects the general change trend of time series.

Different observation results can be obtained by observing time series at different scales, and a comprehensive analysis of these results can improve the effect of time series anomaly detection. At present, multi-scale analysis techniques mainly include two categories: empirical mode decomposition [14] and wavelet analysis [15].

Empirical mode decomposition is considered to be based on the Fourier transform of linear and steady-state spectrum analysis of a major breakthrough, the method is based on their own time scale characteristics to signal Decomposition, without the need to pre-set any basis function. In theory, this method can be applied to the decomposition of any type of signal, so it has very obvious advantages in processing non-stationary and nonlinear data. It is suitable for the analysis of nonlinear and non-stationary signal sequences and has a high signal-to-noise ratio.

3.1 Wavelet Analysis

The wavelet analysis is developed on the basis of the short-time Fourier analysis, which not only makes up for the shortcoming that Fourier analysis will lose the characteristic information of the signal changing with time, but also has the advantages of observing the signal in different scales. Mallat [16] and Daubechies [17] are the founders of the wavelet theory. At present, the wavelet analysis has become an important tool for time series analysis. In addition to its excellent performance in removing outliers and suppressing baseline drift, the wavelet tool can also observe arbitrary details of the signal. Therefore, wavelet analysis, also known as the mathematical microscope, provides convenience for us to discover weak anomalies [18].

3.2 Multi-scale Risk Assessment Model

Due to the existence of multi-scale effect, "anomaly" may show different characteristics at different scales. By introducing the wavelet analysis technology into the time series anomaly detection model based on LSTM, we can not only suppress the influence of noise on the detection results, but also detect the insignificant "anomaly". The multi-scale technology improves the effect of anomaly detection in time series, and then increases the reliability of risk assessment model of network security. The multi-scale risk assessment model mainly includes three parts:

Wavelet Decomposition: The evaluation model first decomposes a network traffic time series into a number of sub-sequences through the wavelet tool, and then reconstructs these sub-sequences.

Sub-sequence Anomaly Detection: Next, the reconstructed sub-sequences are put into the time series anomaly detection model based on LSTM to detect the abnormal data of each sub-sequence. Since the sub-sequences after wavelet decomposition and reconstruction have different sensitivities to "anomaly", the sensitivity of each sub-sequence to "anomaly" can be taken as the weight value to determine whether each sub-sequence is abnormal or not. Therefore, we can set different anomaly detection thresholds to improve the performance of the model.

Evaluate the Multi-scale Risk Assessment Value: If a time window of the network traffic time series is detected to be "anomaly" by more than three sub-sequences, the test time series is considered to be "anomaly" in this window. We define "the number of anomalies" as the risk value of a network security. The higher the risk value of a network security, the higher the probability that it will be attacked.

The multi-scale assessment process of the risk value of network security based on LSTM is as follows (Fig. 3).

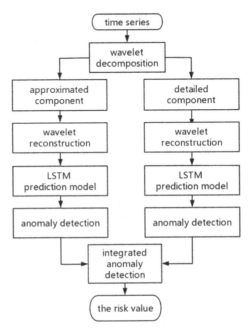

Fig. 3. Diagram of a multi-scale risk assessment model for network security

4   Experiment

The Center of Excellence for Information Security (ISCX) at the University of New Brunswick (UNB) in Canada released an intrusion detection dataset called CICIDS2017 in 2017 [19]. Network traffic data is recorded over time and contains time information, so it can be sorted according to packet timestamp into a continuous time series. The CICIDS2017 dataset covers common network attacks such as DoS Hulk and DDoS. The dataset started at 09:00 on Monday, July 3rd and continuously ran for 5 days, ending at 17:00 on Friday July 7th. Attacks were subsequently executed during this period. Monday is the normal day and just includes the benign traffic. The DoS Hulk and DDoS are executed in Wednesday and Friday respectively. To verify the generalization performance of the model, we added the Dos Hulk on Wednesday and DDoS on Friday to the benign traffic data on Monday, and then evaluated the risk of this composite data.

We select "Bwd Packet Length Std" in the dataset as the feature sequence to evaluate the current network security status.

Python3.7, Sklearn, Keras, and other libraries were used in this experiment. Sklearn and Keras are libraries used to build LSTM-based risk assessment model.

4.1 Data Synthesis

Figure 4 shows the data of this experiment.

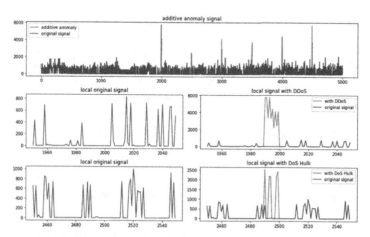

Fig. 4. Illustrations of experimental data (Color figure online)

The data shown above is derived from CICIDS2017, with abnormal fluctuations inserted around the 2000th, 2500th, 3000th, 3500th, 4000th and 4500th points. The blue line in the first sub-graph represents the benign traffic data, and the red part is the added anomalies. The type and location of the anomaly are shown in the next four sub-graphs. According to the assumptions in this article, the risk value for this period network traffic data is 6.

4.2 Single Scale Assessment of Network Security Risk

Figure 5 shows the results of network traffic time series anomaly detection. The purple and green parts in the first sub-graph represent the added anomalies, and the red parts in the second sub-graph represent the anomalies detected by the model.

As can be seen from the Fig. 5 and Fig. 6, in a single-scale, although the time series anomaly detection model based on LSTM detects all "abnormal" data, a large amount of "normal" data is considered "abnormal" data. The default risk value of this experiment is 6, while the risk value of network security obtained by this model is well over 6. Obviously, this risk assessment model is not reliable.

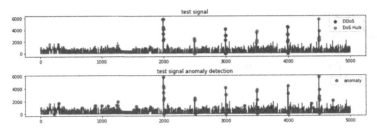

Fig. 5. Single-scale anomaly detection results (Color figure online)

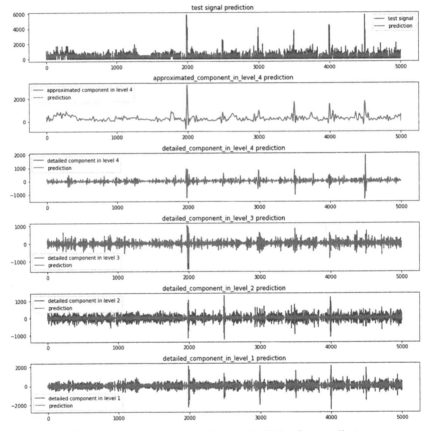

Fig. 6. Sub-sequence prediction results (Color figure online)

4.3 Multi-Scale Risk Assessment of Network Security

Figure 6 shows the prediction results of the LSTM network prediction model on the sub-sequences obtained by wavelet decomposition and reconstruction. The wavelet function of Daubechies is constructed by Inrid Daubechies, a world-famous wavelet analysis scholar. It is abbreviated as dbN, and N is the order of wavelet. Db4 wavelet is used in this experiment.

In the Fig. 6, the blue lines represent the test sequence and the orange lines represent the prediction sequence. Seen from the prediction results, the more detailed parts are more difficult to predict, but the overall prediction effect is good enough to meet the requirements of anomaly detection.

Figure 7 shows the detection results of the time series multi-scale anomaly detection model based on LSTM.

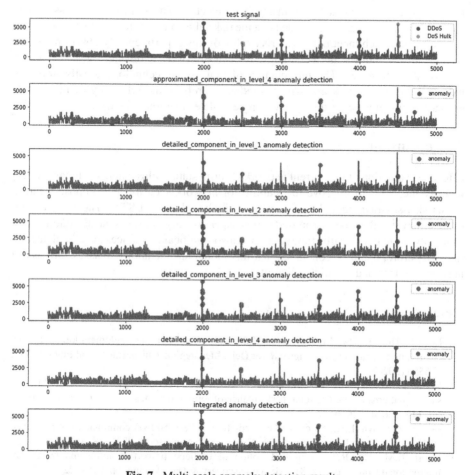

Fig. 7. Multi-scale anomaly detection results

The first sub-graph shows that the added "anomalies" are located at the 2000th, 2500th, 3000th, 3500th, 4000th and 4500th points, where the "anomaly" at the 2500th point is less obvious.

The second sub-graph is an anomaly detection on the scale of the approximate part. The model detects all "anomalies" at this scale, but erratically detects "normal" data as "abnormal" data. That's because on this scale, there's not much difference between "abnormal" and "normal". In the third sub-graph, anomaly detection at the first level of detail misses the "anomalies" at the 3000th point and the 4000th point. The fourth sub-graph is the anomaly detection at the second level of detail. Although all "anomalies" are detected at this scale, there are cases of false detection. In the fifth sub-graph, all "anomalies" are detected at the third level of detail, but there are also cases of false detection. In the sixth sub-graph, not only are not all "anomalies" detected in the fourth scale of the detail section, but there are also cases of false detection.

The seventh sub-graph is the comprehensive detection effect. It can be seen that all the "anomalies" are detected without any false detection. The multi-scale risk assessment model of network security concluded that the risk value of the current security state of the network is 6, which meet the preset risk value of this experiment. By observing a time series with the multi-scale technology, we can discover "anomalies" at some scales, but not at other scales. The introduction of multi-scale technology greatly improves the reliability of the risk assessment model. Therefore, the reliability of multi-scale evaluation model is better than that of single scale evaluation model.

5 Conclusion

Effective evaluation of the current network security status helps managers to take timely security measures to improve the stability of network security. We propose a multi-scale risk assessment model of network security based on the LSTM. The experimental results show that the model proposed in this paper not only improves the detection rate of "anomaly" of network traffic time series, but also significantly reduces the false detection rate. Therefore, it can be concluded that this model has the ability to assess the risk of network security and has high reliability.

References

1. Li, W.F., Huang, C.M., Li, W.J., et al.: Multi-scale effects of urban agglomeration on thermal environment: a case of the Yangtze River Delta Megaregion, China. Ence Total Environ. **713**, 136556 (2020)
2. König, F., Rosenkranz, A., Grützmacher, P.G., et al.: Effect of single-scale and multi-scale surface patterns on the frictional performance of journal bearings a numerical study. Tribol. Int. **143**, 106041 (2019)
3. Zhang, Y.D., Wu, L.N.: Stock market prediction of S&P 500 via combination of improved BCO approach and BP neural network. Expert Syst. Appl. **36**(5), 8849–8854 (2009)
4. Knorr, E.M., Ng, R.T.: A unified notion of outliers: properties and computation. In: International Conference on Knowledge Discovery & Data Mining, pp. 219–222 (1997)
5. Ramaswamy, S., Rastogi, R., Shim, K.: Efficient algorithms for mining outliers from large datasets. In: ACM SIGMOD International Conference on Management of Data, pp. 427–438. ACM (2000)
6. Agyemang, M., Ezeife, C.I.: Lsc-Mine: algorithm for mining local outliers. In: International Conference on Information Resource Management Association, pp. 5–8 (2004)
7. Singh, R.K., Rani, M., Bhagavathula, A.S., et al.: The prediction of COVID-19 pandemic for top-15 affected countries using advance ARIMA model. JMIR Public Health Surveill. **6**(2), e19115 (2020)
8. Corba, B.S., Egrioglu, E., Dalar, A.Z.: AR-ARCH type artificial neural network for forecasting. Neural Process. Lett. **51**(1), 819–836 (2020)

9. Chung, S., Hwang, S.Y.: A profile Godambe information of power transformations for ARCH time series. Commun. Stat. Theory Methods **46**, 6899–6908 (2016)
10. Li, J., Dai, Q., Ye, R.: A novel double incremental learning algorithm for time series prediction. Neural Comput. Appl. **31**, 6055–6077 (2018)
11. Li, Y.C., Huang, J., Chen, H.J.: Time series prediction of wireless network traffic flow based on wavelet analysis and BP neural network. J. Phys. Conf. Ser. **1533**(3), 032098 (2020)
12. Hochreiter, S., Schmidhuber, J.: Long-short-term memory. Neural Comput. **9**(8), 1735–1780 (1997)
13. Lu, H., Yang, F.: Research on network traffic prediction based on long-term short-term memory neural network. In: 2018 IEEE 4th International Conference on Computer and Communications (ICCC), pp. 1109–1113 (2018)
14. Huang, N.E., Shen, Z., Long, S.R., et al.: The empirical mode decomposition and the Hilbert spectrum for nonlinear and non-stationary time series analysis. Proc. Math. Phys. Eng. ences **1998**(454), 903–995 (1971)
15. Zhang, J., Li, H., Shi, X., et al.: Wavelet-nonlinear cointegration prediction of irrigation water in the irrigation district. Water Resour. Manage. **33**(8), 2941–2954 (2019)
16. Maallat, S.G.: A theory for multi-resolution signal decomposition: the wavelet representation. IEEE Trans. Pattern Anal. Mach. Intell. **11**(7), 674–693 (1989)
17. Daubechies, I., Christopher, H.: Ten Lectures on Wavelets. Society For Industrial, Philadelphia (1992)
18. Yang, J., Wang, J.H., Zhang, X.M., et al.: Analysis of voltage sag source location based on wavelet multiresolution method. Dianli Xitong Baohu yu Kongzhi/Power Syst. Prot. Control **38**(22), 90–95 (2010)
19. Sharafaldin, I., Lashkari, A.H., Ghorbani, A.A.: Toward generating a new intrusion detection dataset and intrusion traffic characterization. In: International Conference on Information Systems Security & Privacy (2018)

Weighted Lightweight Image Retrieval Method Based on Linear Regression

Lina Zhang[⊠], Xiangqin Zheng, Xuan Dang, and Jiehui Zhang

College of Computer Science and Technology, Xi'an University of Science
and Technology, Xi'an 710600, China
jszhangln@qq.com

Abstract. With the rapid development of Internet technology, the number of images has shown explosive growth. Content-based image retrieval is an important research topic in the field of computer vision, and it aims to efficiently retrieve target images in massive image databases. Due to the different image characteristics in various fields, the previous image retrieval based on single content feature (color, shape, texture, etc.) can no longer meet the application requirements of image retrieval in related fields. In order to efficiently retrieve specific target images in related fields, based on the typical image content feature namely the color moments, image hash feature including perceptual hash, average hash, and difference hash was respectively fused with the color moments to retrieve images in this paper. Aiming to improve the retrieval efficiency, a weighted lightweight image retrieval method based on linear regression was proposed. Linear regression analysis was performed on image perceptual hash, average hash and color moments. The similarity obtained by the faster hash feature was used to replace the color moments. Finally, the hash feature was merged with the new color moments to retrieve the image. Experimental results show that compared with the direct fusion of image hash feature and color moments, the weighted lightweight image retrieval method based on linear regression proposed in this paper can improve the retrieval efficiency while maintaining the retrieval accuracy.

Keywords: Image retrieval · Image hashing · Color moments · Linear regression

1 Introduction

Digital images are the most common information carrier in multimedia data and are widely used in various fields based on communication networks. Image retrieval [1] that aims to quickly and efficiently find the target images from the massive image databases has become one of the research hotspots in recent

Supported by Natural Science Basic Research Program of Shaanxi(No.2017JQ6026) and Yulin Science and Technology Plan Production-University-Research Program (No.2014CXY-08-01).

© Springer Nature Switzerland AG 2020
B. Ben Hedia et al. (Eds.): VECoS 2020, LNCS 12519, pp. 268–280, 2020.
https://doi.org/10.1007/978-3-030-65955-4_20

years. The early image retrieval is mainly based on the text. Image's query is performed by asking questions in the form of keywords. Since distinct users have different text descriptions of the same image, it results in insufficient retrieval results. Actually, text-based image retrieval has major drawbacks. Content-based image retrieval is one of the current mainstream research methods[2–5]. It uses the image content information (such as color, shape, texture features, etc.) to obtain feature vectors to calculate the similarity between images.

At present, how to improve retrieval efficiency and reduce the computational complexity in the retrieval process is a major challenge faced by image retrieval technology. In response to these problems, researchers have proposed various image retrieval methods [6–9]. The work in [10] proposes a content-based image retrieval using the color, texture, and edge features of the image. This method improves the retrieval accuracy to a certain extent. However, the effect is not significant. The study in [11] develops an image retrieval method with improved features, where they improve the comprehensive features of color and texture. The image similarity is calculated by obtaining the average feature vector of the color region and the low-frequency texture feature vector. The retrieval accuracy obtained by this method has also been improved. However, the improvement is less obvious. Besides, image hashing technology is one of the research hotspots in recent years [12,13], and image retrieval based on image hashing has furthermore attracted much attention [14,15]. The approach in [14] proposes a latent semantic minimum hash for image retrieval. This method attempts to obtain a more compact hash code that can accurately describe image features, namely the generated hash code has more semantic properties. However, it causes a high computational complexity. At present, content-based image retrieval [16] has become a research trend [17,18]. Compared with text-based retrieval, content-based image retrieval greatly reduces the workload in the retrieval process. Moreover, it avoids the subjectivity of manual description.

It can be observed from the above that how to reduce retrieval time based on improving retrieval accuracy is one of the problems that need to be solved at present. This paper proposes a weighted lightweight image retrieval method based on linear regression. Since the color moments consume too much time in the retrieval process compared with the hash algorithm, we use weighting to fuse the image hash and color moments. The perceptual hash and average hash with higher retrieval accuracy and less time consumption are selected to replace the color moments, and the new color moments and hash are merged to retrieve the image. The experimental results show that compared with the direct fusion of image hashing algorithm and color moments, the weighted lightweight image retrieval method based on linear regression can further improve retrieval efficiency while maintaining the retrieval accuracy.

2   Preliminaries

2.1   Image Hash Algorithm

The perceptual hash algorithm [8,9,19] is robust. It uses the discrete cosine transform(DCT) to obtain the low-frequency component of the image for processing and calculation. The specific steps are shown as below.

(1) Reduce the image: reduce the size of the image to 32×32 to facilitate the DCT calculation later;
(2) Convert to grayscale image: convert the zoomed image into a 256-level grayscale image;
(3) Calculate the DCT: calculate the DCT coefficients matrix;
(4) Reduce DCT: the size of the DCT coefficients matrix is determined by step (1), which is 32×32. We select the 8×8 area in the upper left corner of the low-frequency part of the image for subsequent calculations;
(5) Calculate the average value: calculate the average value of all pixels after the DCT is reduced;
(6) Further reduce DCT: record as 1 if it is greater than the average value, otherwise record as 0;
(7) Obtain the information fingerprint: combine 64 information bits and arrange them into a matrix according to the same rule.
(8) Compare the fingerprints: calculate the Hamming distance between the information fingerprint of the two images obtained in the previous step. The larger the calculated value, the less similar the two image are. On the contrary, it means that the two images are more similar. When the result is 0, it means that the two images are exactly the same.

The average hash algorithm [20] mainly uses the low-frequency information of the image to identify the similarity, which is based on comparing each pixel of the gray image with the average value. The steps are similar to perceptual hashing, mainly including the following aspects: (1) Scale the image to a ratio of 8×8. (2) Convert to a grayscale image. (3) Calculate the average value of all pixels of the grayscale image and comparing each pixel value with the average value (If it is greater than the average value, replace it with 1, otherwise with 0).(4) Obtain the information fingerprint and calculate Hamming distance to judge the similarity of two images.

Compared with the average hash algorithm, the hash value generated by the differential hash algorithm [20] will not change with the image scaling and aspect ratio. Enhancing or reducing the brightness or contrast, or modifying the color will not have a significant impact on the hash value. In addition, gamma correction and complex adjustments to the color configuration will not affect the results. The main steps include: (1) Reduce the image to a 9×8 scale. (2) Convert it into a grayscale image. (3) Calculate the difference value. (4) Obtain the fingerprints and calculate the Hamming distance to judge the similarity of two images.

2.2   Color Model Conversion

HSV color space [21] is a uniform color space, and RGB space to HSV space is a non-linear transformation. In order to accord with the human visual system perception characteristics, RGB color space is converted to HSV color space. The values of these three components R, G, B in the normalized RGB model belong to $[0, 1]$, and in the corresponding HSV model they can be expressed as Eqs. (1), (2) and (3).

$$V = \frac{R + G + B}{3} \tag{1}$$

$$S = 1 - \frac{3[min(R, G, B)]}{R + G + B} \tag{2}$$

$$H = \frac{cos^{-1}\left\{ \frac{[(R-B)+(R-B)/2]}{(R-G)^2+(R-G)\sqrt{(R-G)}} \right\}}{360} \tag{3}$$

Suppose HSV space colors $c1 = (h1, s1, v1)$ and $c2 = (h2, s2, v2)$, the similarity S between $c1$ and $c2$ can be defined as Eq. (4):

$$S = 1 - \frac{1}{\sqrt{5}}\sqrt{(s1cos(h1) - s2cos(h2))^2 + (s1sin(h1) - s2sin(h2))^2 + (v2 - v1)^2} \tag{4}$$

where $S \in [0, 1]$, the closer S is to 1, the more similar $c1$ and $c2$ are.

2.3   Similarity Calculation

In this paper, Hamming distance is used to calculate the similarity between the images retrieved with three hash algorithms, and Manhattan distance is used to calculate the similarity between the images retrieved with color moments. Their definitions are shown in Eqs. (5), (6) and (7).

(1) Hamming Distance

Hamming distance represents the number of different bits in the same position of two strings with the same length. The definition is shown in Eq. (5):

$$d(x, y) = \sum(x[i] \oplus y[i]) \tag{5}$$

where $i = 0, 1, 2, \cdots, n - 1$. x, y are both n-bit binary codes.

(2) Manhattan Distance

Manhattan Distance is also called chessboard distance. It means the sum of the projection distances of the line segment formed by two points on the axis on the fixed rectangular coordinate system of Euclidean space. The definition can be expressed as Eqs. (6) and (7).

First, the Manhattan distance between the two points $A(x_1, y_1)$ and $B(x_2, y_2)$ on a two-dimensional plane is:

$$d_{12} = |x_1 - x_2| + |y_1 - y_2| \tag{6}$$

Then, the Manhattan distance between the two dimensional vectors $A(x_{11}, x_{12}, \cdots, x_{1n})$ and $B(x_{21}, x_{22}, \cdots, x_{2n})$ is:

$$d_{12} = \sum_{k=1}^{n} |x_{1k} - x_{2k}| \tag{7}$$

where $k = 1, 2, \cdots, n$.

3   The Proposed Algorithm

Image hashing algorithm is an irreversible calculation of digital digests of original image data. It has the characteristics of one-way and fragility, which can guarantee the uniqueness and immutability of original data. The perceptual hashing technology is to extract the robust fingerprint of the information and compare the fingerprint with the fingerprint extracted in advance. When the content of the information has not changed (such as applying moderate repairs, beautification, format adjustments, etc. to the information), its fingerprint value remains unchanged or does not change much. When the information content is tampered with, the fingerprint value will change significantly. In some practical applications, we need to judge whether the content of the image is similar, but also whether the color of the image is similar. In this case, there exists some certain limitations for perceptual hashing technology. In contrast, color moments can represent the color characteristics of an image, and the retrieval accuracy of images with significant color characteristics is better. However, when we use single color feature to retrieve images, the retrieval efficiency is quite low. In this paper, we merge it with the image hash algorithm to retrieve image.

A weighted lightweight image retrieval method based on linear regression is proposed in this paper. First, the image hash and color moments are merged separately for image retrieval. The Hamming distance a, p, d obtained from the retrieval of average hash, perceptual hash and difference hash and the Manhattan distance c obtained from the retrieval of color moments are weighted. Then, the three weighted Hamming distances are respectively fused with the weighted Manhattan distance. The results of the fusion are represented by $z1, z2, z3$ and used as new measures of similarity.

$$
\begin{aligned}
z1 &= \alpha_1 p + \beta_1 c \\
z2 &= \alpha_2 a + \beta_2 c \\
z3 &= \alpha_3 d + \beta_3 c
\end{aligned}
\tag{8}
$$

In Eq. (8), $\alpha_i, \beta_i, i = 1, 2, 3$ respectively represent the weight, and $\alpha_i + \beta_i = 1$. The value of α_i are $0.2, 0.3, 0.4, \cdots, 0.8$ in the experiment. The three variables

$z1, z2, z3$ represent the similarity obtained when the three hash algorithms are separately fused with the color moments to retrieve the image.

The fusion of hash algorithms and color moments improves retrieval accuracy to a certain extent. However, the retrieval speed is slow. In practice, it can be considered to perform linear regression analysis on the distance data obtained by color moments, perceptual hash and average hash. Then, we use the perceptual hash and average hash to replace the color features. When we use the experimental results to perform a similar retrieval process, the retrieval time is shortened to a certain extent. It can be expressed as Eq. (9):

$$c' = \alpha'_0 + \beta'_0 p + \beta'_1 a$$
$$s = \alpha' p + \beta' c' \tag{9}$$

then, bring c' into s and merge similar items to obtain:

$$s = \beta' \alpha'_0 + (\alpha' + \beta' \beta'_0) p + \beta' \beta'_1 a \tag{10}$$

where c' represents the new color moments represented by perceptual hash and average hash; α'_0, β'_0 and β'_1 represent the regression coefficients; α' and β' are the new weights; s represents the result of fusing the perceptual hash with the new color moments and is used as a new similarity calculation method.

The experimental results show that compared with the retrieval that directly combines the hash and color moments, the retrieval efficiency obtained by regression analysis has been greatly improved.

3.1 Similarity Acquisition

First of all, image retrieval is performed by using perceptual hashing algorithm, average hashing algorithm, difference hashing algorithm and a single color moment method. In the retrieval process, the distances between the target image and the images in the image database are calculated. Thus, multiple sets of distance data or similarities can be obtained. As shown in Table 1, the smaller the distance is, the more similar the two images are.

Table 1. Distances (similarities) obtained by three hash algorithms and color feature

AH	PH	DH	CM
a_1	p_1	d_1	c_1
a_2	p_2	a_2	c_2
a_3	p_3	d_3	c_3
$\vdots$	$\vdots$	$\vdots$	$\vdots$
a_n	p_n	d_n	c_n

In Table 1, AH, PH, DH, and CH refer to average hash, perceptual hash and difference hash, respectively. $a_1, \cdots, a_n, p_1, \cdots, p_n, d_1, \cdots, d_n, c_1, \cdots, c_n$ respectively represent the distance or the similarity obtained by different algorithms in the retrieval process. In this paper, we use a weighted method to fuse three hash algorithms with color moments to retrieve similar images. Then, linear regression analysis is performed on color moments, perceptual hash and average hash with better retrieval results. Finally, we fuse the hash algorithm with the color moments obtained by regression analysis to retrieve image.

3.2 Multivariable Linear Regression Model

A multivariable linear regression model can be expressed as Eq. (11):

$$Y = \gamma_0 + \gamma_1 X_1 + \gamma_2 X_2 + \cdots + \gamma_p X_p + \varepsilon \tag{11}$$

where Y and $X_i, i = 1, 2, \cdots, p$ represent the actual variables; γ_i and ε represent the regression coefficients and error terms, respectively. In this paper, Y represents the dependent variable CM and $X_i, i = 1, 2$ represents any two of the three independent variables AH, PH and DH. For example, we choose AH and PH for experiments.

We use least squares approximation to fit the linear regression model in this paper. Then, the least squares estimate of the parameters is shown in Eq. (12)

$$\hat{\gamma} = (X^T X)^{-1} X^T Y \tag{12}$$

In this paper, we use the least squares method for linear regression analysis to obtain the formula $s = k_0 + k_1 a + k_2 p$ by using the obtained multiple sets of distances or similarities, where a and p respectively represent the Hamming distance obtained by average hash and perceptual hash algorithm; k_0, k_1 and k_2 are the regression coefficients obtained by linear regression; s is the new color feature expression. Finally, the image is retrieved by fusing the hash algorithm and new color feature in a weighted manner. The experimental results show that this method improves the retrieval efficiency to a certain extent.

4 Experimental Results and Analysis

First, we briefly introduce the experimental environment: CPU is Intel-Core i5-4210U; memory is 4G; the operating system is Windows10, and programming environment is MATLAB R 2014a.

The image categories of the image library used in the experiment are selected based on the public data sets such as CIFAR-100 data set[22] and CalTech data set[23]. It contains nearly 500 images, which involves eight categories of images including flowers, dinosaurs, horses, airplanes, buses, wahaha, motorcycles and Mars.

In this paper, we mainly analyze the performance of the proposed algorithm from two aspects: retrieval time and retrieval accuracy. Among them, the retrieval accuracy can be calculated from Eq. (13):

$$W = \frac{m}{N} \times 100\% \tag{13}$$

where m represents the number of images retrieved in a specific category, N represents the total number of images of this type, and W is the retrieval accuracy.

4.1   Results of Separate Retrieval for Each Algorithm

The retrieval time and accuracy of the three hash algorithms and single color feature are shown in Table 2 and Fig. 1, respectively.

It can be observed from Table 2 and Fig. 1 that the retrieval speed of the three hash algorithms is relatively fast, and the retrieval speed based on a single color feature is very slow. Compared with the hash algorithms, the retrieval based on color moments has higher accuracy when retrieving eight types of images. However, the retrieval time is longer. Furthermore, in some retrieval methods that incorporate color features, retrieval efficiency will be reduced. In order to further improve the retrieval efficiency, we merge the three hash algorithms and color moment separately.

4.2   Hash and Color Fusion

Tables 3, 4 and 5 show the accuracy and time consumption of the retrieval after the three hash algorithms including perceptual hash, average hash and difference hash are fused with color moments, respectively, where the values of PH, AH, DH and CM represent their respective weights. Note that different types of images have different characteristics. Therefore, the table shows the distribution of weights when the retrieval accuracy is the highest when retrieving different

Table 2. Time consumption of three hash algorithms and color feature retrieval (in seconds)

Image category	PH	AH	DH	CM
Airplanes	4.8906	4.8125	4.6406	60.4375
Wahaha	4.8906	4.9531	4.7344	59.9219
Flowers	4.875	5.5156	4.625	60.5313
Dinosaurs	4.7031	4.6094	4.5469	60.2656
Buses	4.8125	4.5781	4.5625	60.0938
Horses	4.7188	4.3281	4.6875	59.8906
Motorcycles	4.9844	4.6563	4.625	60.0938
Mars	4.6719	4.6563	4.6094	59.6094

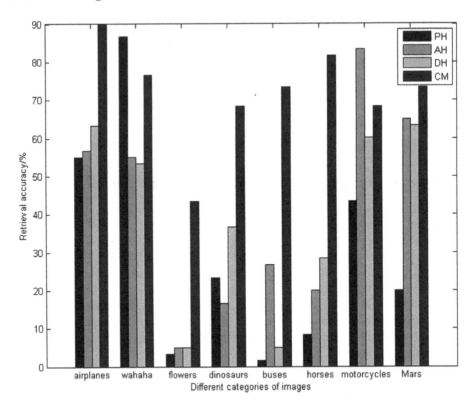

Fig. 1. The accuracy of different algorithms to retrieve different types of images

Table 3. Search results after the fusion of perceptual hash and color moments

Image category	PH	CM	Accuracy/%	Time/s
airplanes	0.3	0.7	95	62.5
Wahaha	0.5	0.5	83.33	62.1406
Dinosaurs	0.2	0.8	66.67	62.4844
Buses	0.3	0.7	81.67	63.875
Horses	0.4	0.6	81.67	62.375
Motorcycles	0.2	0.8	68.33	62.2344
Mars	0.3	0.7	71.67	62.5313

types of images. Figure 2 (a) is the result of retrieving "aircraft" based on color moments, (b) is the retrieval result after perceptual hash and color moments are fused. It can be seen that the retrieval accuracy after fusion is improved compared with the retrieval based on single color moments. However, the retrieval efficiency is still low. In this paper, linear regression is used to replace color moments with hash algorithms to improve retrieval efficiency.

Table 4. Search results after the fusion of average hash and color moments

Image category	AH	CM	Accuracy/%	Time/s
Airplanes	0.2	0.8	91.67	62.4219
Wahaha	0.2	0.8	73.33	62.2301
Dinosaurs	0.2	0.8	66.67	62.2188
Buses	0.3	0.7	78.33	62.6563
Horses	0.2	0.8	78.33	61.25
Motorcycles	0.7	0.3	88.33	61.5313
Mars	0.5	0.5	75	61.7031

Table 5. Search results after the fusion of difference hash and color moments

Image category	DH	CM	Accuracy/%	Time/s
Airplanes	0.2	0.8	93.33	65.3906
Wahaha	0.2	0.8	70	63.9531
Dinosaurs	0.3	0.7	83.33	63.0313
Buses	0.3	0.7	75	62.2656
Horses	0.3	0.7	83.33	62.2813
Motorcycles	0.6	0.4	80	62.7813
Mars	0.5	0.5	75	65.7344

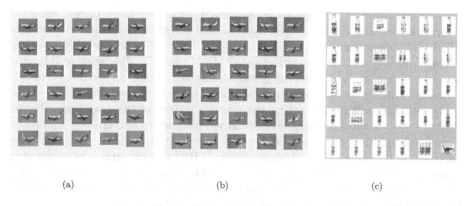

    (a)   (b)   (c)

Fig. 2. Part of the results of retrieving the "airplanes" and "wahaha". (a) is part of the result of retrieving "airplanes" based on color moments, (b) is part of the retrieval result after the fusion of perceptual hash and color moments, and (c) is part of the search results after the fusion of perceptual hash and color moments obtained by linear regression (Color figure online)

4.3   Fusion of Hash and Color Moments Obtained by Linear Regression

In this paper, color moments are used to represent color feature vector. Since the first, second, and third-order color moment of HSV need to be calculated, the retrieval time based on the color moments is relatively large. A linear regression method is used to express the color moments with hash algorithms which have faster retrieval speed to further improve the retrieval efficiency.

We have done a lot of experiments and they have proven that the retrieval efficiency is improved to a certain extent when perceptual hashing and average hashing are used to replace color moments and then the new color moments are combined with perceptual hashing.

As shown in Table 6 and Fig. 2 (c), we use "wahaha" as the target image to retrieve images, and perform linear regression analysis on the distance data obtained by perceptual hash, average hash and color moments. Note that the values of PH and s represent the weights. Finally, the regressed results are then fused with the perceptual hash to retrieve "wahaha". The retrieval results are shown in Table 6 and Fig. 2 (c), where s represents the regressed result that is the new color feature.

Table 6. Search results after the fusion of perceptual hash and color moments obtained by linear regression

Image category	PH	s	Accuracy/%	Time/s
Wahaha	0.6	0.4	83.33	7.7188

The algorithm in this paper is compared with four more classic retrieval algorithms, namely the image retrieval method based on the improved features (IF) in [11], the color histogram features (TCHF) in [24], the gray mean feature (TGMF) in [24] and the algorithm (LBP+DCT) in [25]. The work in [11] proposes an image retrieval method based on improved color and texture feature, where accuracy rate and retrieval rate have been improved accordingly. The study in [24] dwells on extracting the average feature of the gray image block. However, the method requires excessively simple feature and ignores important related information such as color, which reduces the retrieval accuracy. The approach in [25] extracts local binary mode (LBP) features and applies discrete cosine transform (DCT) to the image, then, the feature vector of the image texture is obtained. However, DCT cannot completely eliminate the influence of high-frequency noise, which still affects the retrieval accuracy.

We primarily evaluate the performance of the proposed algorithm from two aspects: retrieval time and retrieval accuracy. Three aspects including the feature dimensions, the average retrieval time and the highest retrieval accuracy are compared between the related works. The results are shown in Table 7.

Table 7. Comparison with related works

Method	Feature dimensions	Average retrieval time/s	The highest retrieval accuracy/%
IF	315	0.0861	80
TCHF	75	0.0263	<60
TGMF	85	0.0510	<80
LBP+DCT	469	0.1329	<60
The proposed method	64	7.0	>90

It can be concluded from Table 7 that the retrieval method proposed in this paper is superior to the above several retrieval methods in the feature dimensions and the highest retrieval accuracy. However, the average retrieval time is slightly longer. Generally speaking, the method in this paper is a lightweight retrieval method. When the number of images in the image database increases, the error will increase rapidly. When the number of images in the image database is relatively small, the proposed method has certain practicability.

5   Conclusion

In this paper, firstly, image hash including perceptual hash, average hash and difference hash are fused with the color moments to retrieve images, respectively. To further improve retrieval efficiency, a weighted lightweight image retrieval method based on linear regression is proposed. The image's perceptual hash, average hash and color moments are subjected to linear regression analysis. Then, the similarities obtained by faster hash algorithms are used to replace the color moments. Finally, the hash algorithm is merged with the new color moments to retrieve image. Experimental results show that the weighted lightweight image retrieval method based on linear regression has improved retrieval efficiency. Since we use linear regression to represent the color moments in this paper, when the number of images in the image database is large, such as every additional 100 images, the error will rapidly expand. Our future work is to consider how to use machine learning to replace linear regression and find the relationship between efficient hashing algorithms and image content features.

References

1. Rui, Y., Huang, T.S., Chang, S.-F.: Image retrieval: current techniques, promising directions, and open issues. J. Vis. Commun. Image Represent. **10**, 39–62 (1999)
2. Piras, L., Giacinto, G.: Information fusion in content based image retrieval: a comprehensive overview. Inform. Fusion **37**, 50–60 (2017)
3. Raghuwanshi, G., Tyagi, V.: A novel technique for content based image retrieval based on region-weight assignment. Multimed. Tools Appl. **78**(2), 1889–1911 (2018). https://doi.org/10.1007/s11042-018-6333-6
4. Cui, S., Xiong, S., Liu, C., Chen, M.: A survey of content-based medical image retrieval methods. J. Chongqing Univ. Technol. (Nat. Sci.) **12**, 113–121 (2018)

5. Zhou, W., Li, H., Tian, Q.: Recent advance in content-based image retrieval: a literature survey (2017). arXiv preprint arXiv:1706.06064

6. Peng, J., Su, Y., Xue, X.: SAR image feature retrieval method based on deep learning and synchronic matrix. Comput. Sci. **46**(S1), 196–199+204 (2019)

7. Zhang, C., Yang, X., Qiquan, X., Chen, S.: Hash fast retrieval and image matching based on sift feature. Modern Electron. Techn. **42**(12), 127–131 (2019)

8. Shuang, Z.J.C., Lili, H.: Clothing image retrieval method based on deep learning. Comput. Syst. Appl. **28**(03), 229–234 (2019)

9. He, X., Tang, Y., Wang, L., Chen, P., Yuan, G.: Estimating graphlets via two common substructures aware sampling in social networksmultitask hierarchical image retrieval technology based on faster rcnnh. Comput. Sci. **46**(03), 303–313 (2019)

10. Pavithra, L.K., Sharmila, T.S.: An efficient framework for image retrieval using color, texture and edge features. Comput. Elect. Eng. **70**, 580–593 (2018)

11. Qiao, H., Deng, Z., Xue, J., Song, Q.: Research of image retrieval method based on improved feature. J. Northwest. Polytech. Univ. **36**(4), 742–747 (2018)

12. Liu, Z.: Research on key techniques of image perceptual hashing. J. Harbin Inst. Technol. (2013)

13. Niu, X., Jiao, Y.: An overview of perceptual hashing. Acta Electron. Sinica **7**, 1405–1411 (2008)

14. Xiaoqiang, L., Zheng, X., Li, X.: Latent semantic minimal hashing for image retrieval. IEEE Trans. Image Process. **26**(1), 355–368 (2016)

15. Zhu, L., Shen, J., Xie, L., Cheng, Z.: Unsupervised visual hashing with semantic assistant for content-based image retrieval. IEEE Trans. Knowl. Data Eng. **29**(2), 472–486 (2016)

16. Alzu'bi, A., Amira, A., Ramzan, N.: Content-based image retrieval with compact deep convolutional features. Neurocomputing **249**, 95–105 (2017)

17. Ahmed, K.T., Ummesafi, S., Iqbal, A.: Content based image retrieval using image features information fusion. Information Fusion **51**, 76–99 (2019)

18. Varish, N., Kumar, S., Pal, A.K.: A novel similarity measure for content based image retrieval in discrete cosine transform domain. Fundam. Inform. **156**(2), 209–235 (2017)

19. Zhang, W., Kong, X., You, X.: Secure and robust image perceptual hashing. J. Southeast Univ. (Nat. Sci. Ed.), (S1), 188–192 (2007)

20. Fei, M., Ju, Z., Zhen, X., Li, J.: Real-time visual tracking based on improved perceptual hashing. Multimedia Tools Appl. **76**(3), 4617–4634 (2016). https://doi.org/10.1007/s11042-016-3723-5

21. Wang, H., Yin, W., Wang, L., Jianghao, H., Qiao, W.: Fast edge extraction algorithm based on HSV color space. J. Shanghai Jiaotong Univ. **53**(07), 765–772 (2019)

22. Krizhevsky, A., Nair, V., Hinton, G.: Cifar-10 and cifar-100 datasets, 6:1 (2009). https://www.cs.toronto.edu/kriz/cifar.html

23. Griffin, G., Holub, A., Perona, P.: Caltech-256 object category dataset (2007)

24. Chathurika, K.B.A.B., Jayasinghe, P.K.S.C.: A revised averaging algorithm for an effective feature extraction in component-based image retrieval system. In: 2015 IEEE International Advance Computing Conference (IACC), pp. 1153–1157 (2015)

25. Kobayashi, K., Chen, Q.: Image retrieval using features in spatial and frequency domains based on block-division. In: 2015 International Conference on Computational Science and Computational Intelligence (CSCI), pp. 448–453 (2015)

Author Index

Printed in the United States
By Bookmasters